Problems and Methods in the Study of Politics

The academic study of politics seems endlessly beset by debates about method. We hear of qualitative scholars versus quantitative scholars; of institutionalists versus students of political culture; of rational choice theorists versus everyone else. At the core of all these debates is a single unifying concern: should political scientists view themselves primarily as scientists, developing ever more sophisticated tools and studying only those phenomena to which such tools may fruitfully be applied? Or should we instead try to illuminate the large, complicated, untidy problems thrown up in the world, even if the chance of offering definitive explanations is low? Is there necessarily a tension between these two endeavors? Are some domains of political inquiry more amenable to the building up of reliable, scientific knowledge than others, and if so, how should we deploy our efforts? In this collection of essays, some of the world's most prominent students of politics offer original discussions of these pressing questions, eschewing narrow methodological diatribes to explore what political science is and how political scientists should aspire to do their work.

Problems and Methods in the Study of Politics is essential reading for anyone who cares about advancing rigorous understanding of the most important political problems and choices of our time. It is the first in a series of projected volumes growing out of the Yale University Political Science Initiative on "Rethinking Political Order." Future volumes will include *Crafting and Operating Institutions*; *Identities, Affiliations, and Allegiances*; *Order, Conflict, and Violence*; *Representation and Popular Rule*; and *Distributive Politics*.

IAN SHAPIRO is William R. Kenan, Jr. Professor of Political Science at Yale University, where he also serves as Henry R. Luce Director of the Yale Center for International and Area Studies. His many books include *The State of Democratic Theory*, *The Moral Foundations of Politics*, *Democratic Justice*, and (with Donald Green), *Pathologies of Rational Choice Theory*.

ROGERS M. SMITH is the Christopher H. Browne distinguished Professor and Chair of Political Science at the University of Pennsylvania. His previous books include *Civic Ideals: Conflicting Visions of Citizenship in US History*, *Liberalism and American Constitutional Law*, and *Stories of Peoplehood: The Politics and Morals of Political Membership*.

TAREK E. MASOUD is a doctoral student in Political Science at Yale University.

Problems and Methods in the Study of Politics

Edited by
Ian Shapiro, Rogers M. Smith, and Tarek E. Masoud

CAMBRIDGE
UNIVERSITY PRESS

CAMBRIDGE UNIVERSITY PRESS
Cambridge, New York, Melbourne, Madrid, Cape Town, Singapore,
São Paulo, Delhi, Dubai, Tokyo

Cambridge University Press
The Edinburgh Building, Cambridge CB2 8RU, UK

Published in the United States of America by Cambridge University Press, New York

www.cambridge.org
Information on this title: www.cambridge.org/9780521539432

First published 2004

A catalogue record for this publication is available from the British Library

ISBN 978-0-521-83174-1 Hardback
ISBN 978-0-521-53943-2 Paperback

Transferred to digital printing 2009

In memory of Gabriel A. Almond

Contents

Part II. Redeeming rational choice theory?

Part III. Possibilities for pluralism and convergence

Contributors

TRUMAN F. BEWLEY, Department of Economics, Yale University

BRUCE BUENO DE MESQUITA, Hoover Institution, Stanford University

WILLIAM E. CONNOLLY, Department of Political Science, Johns Hopkins University

GARY COX, Department of Political Science, University of California, San Diego

ROBERT A. DAHL, Department of Political Science, Yale University

ELISABETH ELLIS, Department of Political Science, Texas A&M University

JOHN FEREJOHN, Department of Political Science, Stanford University

ALAN S. GERBER, Department of Political Science, Yale University

DONALD P. GREEN, Department of Political Science, Yale University

EDWARD H. KAPLAN, School of Management, Yale University

MARGARET LEVI, Department of Political Science, University of Washington

TAREK E. MASOUD, Department of Political Science, Yale University

JOHN MEARSHEIMER, Department of Political Science, University of Chicago

ANNE NORTON, Department of Political Science, University of Pennsylvania

FRANCES FOX PIVEN, Department of Political Science, City University of New York

ADOLPH REED, JR., Department of Political Science, New School University

SUSANNE HOEBER RUDOLPH, Department of Political Science, University of Chicago

ALAN RYAN, New College, Oxford University

IAN SHAPIRO, Department of Political Science, Yale University

RUDRA SIL, Department of Political Science, University of Pennsylvania

ROGERS M. SMITH, Department of Political Science, University of Pennsylvania

LISA WEDEEN, Department of Political Science, University of Chicago

Acknowledgments

This volume grew out of a conference on "Problems and Methods in the Study of Politics" held at Yale in December 2002. Thanks are due to Yale University, and particularly the Institution for Social and Policy Studies, for financial support. We thank the editors of *Political Theory* for permission to reprint material from Ian Shapiro's essay, "Problems, methods, and theories in the study of politics, or: What's wrong with Political Science and what to do about it" (30 (4) August 2002). Thanks are due also to Sandra Nuhn and Marilyn Cassella for sterling organizational assistance. Finally, we are pleased to record our gratitude to John Haslam of Cambridge University Press for his interest in the project from its inception, and for helping to ensure the timely appearance of the volume.

1 Introduction: problems and methods in the study of politics

Ian Shapiro, Rogers M. Smith, and Tarek E. Masoud

Political science, particularly in the United States, is often said to be a fractured discipline – perpetually split among warring camps (or "separate tables" to use the late Gabriel Almond's evocative phrase (1988)). Partisans articulate their positions with passion and intensity, yet the nature of what divides them is hard to pin down. At times we hear of a stand-off between "qualitative" scholars, who make use of archival research, ethnography, textual criticism, and discourse analysis; and "quantitative" scholars, who deploy mathematics, game theory, and statistics. Scholars in the former tradition supposedly disdain the new, hyper-numerate, approaches to political science as opaque and overly abstract, while scholars of the latter stripe deride the "old" ways of studying politics as impressionistic and lacking in rigor. At other times the schism is portrayed as being about the proper aspirations of the discipline – between those who believe that a scientific explanation of political life is possible, that we can derive something akin to physical laws of human behavior, and those who believe that it is not. For partisans of the latter view, the stochastic nature of politics and the unpredictability of human action mean that the best we can do is explain specific events – with as much humility and attention to context as possible. At still other times the rivals are portrayed as "rational choice theorists," whose work is animated by the assumption that individuals are rational maximizers of self-interest (often economic, sometimes not), and those who allow for a richer range of human motivations.

There is some truth to all these dichotomous accounts of disagreement, but they are subsumed under a larger, more fundamental question about the proper place of problems and methods in the study of politics. This volume asks: Which should political scientists choose first, a problem or a method? Here too, there are fierce proponents of both approaches. Those who advocate "problem-driven" work claim that it is most important to start with a substantive question thrown up in the political world and then seek out appropriate methods to answer it. These scholars contend that only a problem-driven political science is likely to contribute much of practical or intellectual importance to the broader communities in which we work. Critics charge that the practitioners of this approach still have little if anything to offer that is more rigorous than the best writings of journalists and historians. The first imperative today, they contend, must be to make political science more of a science. Consequently, they argue that, for now, political scientists must focus on developing more rigorous methods, restricting their terrain of study to topics to which those methods can fruitfully be applied.

The easy and perhaps correct response is to say that political scientists should both develop more scientific methods and choose important questions; but decisions about obtaining and committing professional resources create pressures for individuals, departments, institutions, and the profession as a whole to establish priorities. Besides, it is not clear that either is desirable. A problem-driven approach might lead to a kind of disciplinary chaos, with scholars drawing on methods as diverse as textual criticism and formal modeling to offer a jumble of theories between which no adjudication or comparison is possible. On the other hand, a method-driven political science would preclude some "unscientific" types of substantive questions altogether, crippling the discipline by constraining its scope. There is no simple way to resolve this debate, but the essays in this volume, drawn from a conference held at Yale University in December 2002, attempt to grapple with these questions and suggest corrigible answers.

The book is divided into three parts. The first explores the problem–method divide, and includes essays that describe the rift, reconceptualize it, and illustrate pathologies of both problem- and method-driven work. Part II explores rational choice theory – which has been subject to the most opprobrium for supposed inattentiveness to substantive problems – and attempts to defend it against attacks and reformulate it in response to those attacks. The chapters in Part III explore the possibilities for methodological pluralism, asking whether such a pluralism is possible or even advisable, what such a pluralism might look like, and the changes that pluralism would necessitate in the way scholars approach their own work.

I. Description, explanation, and agency

One response to the problem–method dichotomy as it is framed above is to argue that all empirical research is theory laden, that there is no such thing as a purely problem-driven perspective, since our choices about what to study are invariably influenced by our prior theoretical and methodological commitments. Shapiro takes up this question in chapter 2. He concedes that theory can sway one's choice of subject, but argues that there is a difference between researchers who choose subjects because they are of intrinsic interest and those who approach them as opportunities for the vindication of their pet theories. He argues that scholars who seek to illuminate the political world should direct themselves to questions thrown up in that world, come to grips with previous attempts to answer the question, and then, if previous answers are inadequate, offer new ones using whatever methodological tools are most appropriate. Method, he tells us, should be subordinate to problem, not the other way around.

But the fact that all observation is theory laden means that even when a problem has been selected on its own merits, there will likely be disagreement among scholars over how to characterize that problem. Since every piece of social reality is given to multiple descriptions, Shapiro points out, scholars must dedicate as much energy to getting the right descriptive cut at a phenomenon as they do to explaining it. Adjudicating among different descriptions is difficult, and Shapiro registers skepticism that a single standard is possible or even desirable. He explores the suggestion that a description's ability to allow scholars to generate non-trivial predictions might offer a useful standard by which to judge descriptive efforts, but he notes that this is an almost insurmountable standard for most current political research. He concludes that, in the end, description might be a virtue in its own right, especially when it involves "problematizing redescription" which encourages scholars to debunk fallacies, and force critical reappraisals of received wisdom.

In chapter 3, Rogers M. Smith suggests that political science – as it is practiced in the United States – suffers not only from an insufficient focus on real-world problems, but also from undue continued attention to outdated ones. According to Smith, the agenda of American political science remains stuck in the events and concerns of the last century, particularly the Cold War. He points out that this is reflected nowhere as much as in the way the discipline is organized: the fields of American politics, comparative politics, and international relations speak to an era in which the nation-state was the main unit of analysis, the United States was the exceptional nation, and the politics of other nations were of interest only in comparison to it or as friends and enemies on the international stage.

But, Smith tells us, the political developments of the later years of the last century, and the opening years of this one, have ushered in a new set of concerns that should become more central to the study of politics. The rise of new nation-states, the dismantling of old ones, the displacement of populations, the successes of minority groups in pressing for political recognition, and the appearance of transnational groups and movements lend new urgency to questions of "political membership, status, and identity." Political scientists have not entirely ignored such questions, especially in the realm of nationalism, but Smith argues that "there have been few thoroughgoing efforts to unify theories of national identity formation with accounts of the creation of groups, social movements, and gender, racial, cultural, and religious identities." But if political scientists are to explore the possibility of a more "unified" theory of identity formation, Smith points out, they will not be able to rely primarily on the standard tools of rational choice analysis and large-N statistical models. Instead, they will have to employ methods that are "richly interpretive" – including "textual analyses, ethnographic field work, biographical studies, in-depth interviews, individual and comparative case studies, participant observation research" – which will allow researchers to understand why certain identities appeal over others, and how those identities "are likely to shape conduct."

Anne Norton takes issue with Shapiro and Smith's exhortations to problem-drivenness in chapter 4, noting that scholarly attention to real-world problems has not always (or even often) been salutary. Though she agrees with much of the criticism of method-driven scholarship – pointing out that blind commitment to a particular method "leads to proselytizing, imperialism, and coercive assimilation" – she notes that scholars who view themselves as solving social, political, and economic problems often end up in the service of the state and corporate interests. Through an examination of three episodes – colonialism, reconstruction after the Civil War, and the Cold War – she shows how scholars who engaged the great issues of the day easily fell into the role of providing cover and legitimation to the dominance practices of the powerful. Norton also detects an arrogance at the heart of the problem-solving enterprise – such scholars, she says, are animated by a belief that "their acts will follow their will, that they are able to determine the consequences of their actions, and control the uses of their work." In reality, Norton tells us, such faith is woefully misplaced, as attempts to solve problems invariably create new and unforeseen ones.

Having argued that both problem- and method-driven scholarship facilitate domination – the former of powerful interests, the latter of one methodological camp over another – Norton attempts to redirect the attention of political scientists to a more benign enterprise. Instead of problem solving, she beckons scholars to engage in what Habermas called "world disclosure." In this she echoes Shapiro's call for scholars to dedicate their intellectual energies to the description of political phenomena – both as a first step to explaining them and as a means of questioning accepted narratives and myths. According to Norton, "world disclosure" is the only way that we can address questions that admit of no answers, to address what she points to as the enduring puzzles of our enterprise: "What is being? What is thinking? What is justice?"

Norton's skepticism of problem-driven social science is amplified by Frances Fox Piven in chapter 5. Piven dissects the uses of what she calls "policy science" – that is, social scientific research conducted in the service of government policy. The standard critique of the marriage of social science and policy is that the linear cause-and-effect relationships posited by social scientists are seldom accurate reflections of reality, and that attempts to build policies on such fictive foundations are thus doomed to failure. Piven registers her sympathy with this critique, but advances an additional one. In reality, she tells us, social science's usefulness to policymakers is not as a guide to the crafting of policy, but as a means of providing scientific justification to partisan initiatives. As she puts it, policy science "comes to be used as a political strategy, providing arguments or stories to elicit support from among those in a wider public whose participation can influence the outcome in a political contest." Piven explores the use of policy science in the debates around welfare reform policy during the Clinton Administration, demonstrating how social scientists were recruited into the nakedly political effort to cut aid to the poor, how they rushed out flawed research in order to legitimize government policy, and how marginalization was the fate of those who did not toe the line.

Adolph Reed Jr. continues this exploration of the nexus of power and social science in chapter 6, but he does so by engaging in the kind of problematizing redescription that Shapiro advocates. He shows how the study of black politics has, from its inception as an academic field in the early twentieth century, unreflectively perpetuated a myth fostered by black elites – that of blacks as an undifferentiated group with unified interests best served by an enlightened leadership. In the political arena, blacks who dissented from the reigning orthodoxy and attempted to advance claims based on interests other than black identity were tarred as misguided or deviant, and black scholars typically followed suit, ignoring the diversity of the black community and refusing to question the "accountability and legitimacy" of the community's leadership. In a sense, the dominant myths of black politics provided a kind of theoretical framework that drove much of black scholarship, precluding potentially fruitful lines of inquiry that were fundamentally at odds with it. Reed argues for a new, more democratic approach to the study of black politics, one that focuses not on the promotion of an artificial consensus based on a fictitious racial solidarity, but on encouraging norms of participation and open debate, and on pressing for the accountability of black leaders.

In chapter 7, John Ferejohn voices his mistrust of the supposed dichotomy between problem- and method-driven work. Though he agrees with Shapiro that every social phenomenon admits of multiple descriptions, and hence multiple explanations, he rejects the idea that one can adjudicate among them, doubting that there can be said to exist a single, correct description or explanation of anything. Method-driven and theory-driven work, he tells us, may yield "uneven or partial explanations," but social phenomena are multifaceted, and the accumulation of one-sided explanations increases our understanding of the totality of the object under study.

Having rejected the problem–method dichotomy as unhelpful, Ferejohn proceeds to identify and explore another disciplinary divide – that between positivist social scientists and those who emerge from a more humanistic tradition, including political philosophers and ethnographers. He reconceptualizes the schism as being between those who seek "external" and "internal" explanations. Traditional social scientists, he tells us, are externalists; they look for causal explanations of human action. Internalists, on the other hand, "explain action by showing it as justified or best from an agent's perspective" – they identify an actor's reasons for behaving in a certain way. These two endeavors, he says, are often seen to be in conflict; after all, if a causal account of action can be advanced, there is no need for a justificatory, internal account, since the actor had no choice but to act in a given way. The debate is really one about agency – the human capacity to act intentionally is the keystone of internalist explanation, while externalists disregard it as unnecessary to explain action.

According to Ferejohn, rational choice theory provides a model for how the two types of explanation can be bridged. He departs from the standard interpretation of rational choice theory, which holds that individuals are simply hardwired to behave rationally, and instead views rationality as a norm to which actors may or may not adhere, depending on the attractiveness of the norm and their capacities for acting in accordance with it. Rational choice explanations, Ferejohn tells us, are thus both internal and external; they are internal in that those who act rationally do so intentionally, in accordance with a norm of rationality; and they are external in that actors' capacities to act in accordance with the norm, and the attractiveness of the norm, are shaped by causal processes. This reconceptualization of rational choice explanations, Ferejohn says, should serve to demonstrate that practices as different as descriptive ethnography and theoretical economics "are, at bottom, normative enterprises aimed at showing how the agents – real or imagined – can act and have reasons to act in ways that comply with norms they accept."

II. Redeeming rational choice theory?

As Ferejohn's essay shows, the place of rational choice theory in the study of politics is a vexing one for scholars. Indeed, many critiques of method- and theory-driven work were initially advanced as critiques of rational choice theory. The most notable such critique, Green and Shapiro's (1994) assessment of rational choice theory's contribution to our body of political knowledge, and their conclusion that the contribution has thus far been paltry, spawned a bevy of rejoinders. In chapter 8, Gary Cox moves beyond reflexive defenses of rational choice theory to explore what, if anything, makes rational choice scholarship different from other forms of social research. He tells us most scholars – not to mention novelists, journalists, and historians – employ some variant of the rational actor assumption, inasmuch as they rely on a model of human psychology in which actors behave in ways consistent with their preferences and beliefs about the world. Under this broad classification, only radical behavioralists (who eschew any assumptions about unobservable mental states), and those who argue that human psychology is too rich to be captured by reductionist assumptions, are outside the rational choice paradigm. The most vocal critics of rational choice theory, Cox tells us, are really critics of a particular brand of the theory, now outmoded, that assumes that actors have unlimited computational abilities and perfect information. Game theoretic models make assumptions about what actors know and how efficiently they act on it, but, Cox tells us, such assumptions should be substantively grounded and adjudged on a case-by-case basis. He argues that wholesale denunciations of game theory are "unlikely to produce methodological or substantive advance, relative to specific diagnoses of what is wrong with the current models and what might be done to improve or replace them."

Alan Ryan concurs with Cox's expansive definition of rational choice explanation in chapter 9, arguing that all attempts at explanation start from a premise of rationality, or as he puts it, rationalizability – we advance accounts that rationalize an agent's actions on the basis of his goals and beliefs. Whether action is said to be taken out of a commitment to some norm, or out of naked self-interest, or in order to express one's allegiances, the story being told is a rationalizing one. But such stories only get us so far, Ryan tells us. Scholars inevitably must direct themselves to asking why an agent was motivated by one reason over another (and whether his professed motivations were in fact his real ones). In this sense, rational actor explanation is only "prima facie" explanation – in order to advance full accounts of human actions, it is necessary to understand the genesis of the beliefs and preferences that are said to cause those actions.

Rational choice explanations are often criticized as being inattentive to the history and context surrounding political phenomena. In chapter 10, Margaret Levi describes an analytic approach that marries rational choice assumptions with historical analysis and in-depth case studies. These "analytic narratives," as they are called by Levi and cocontributors to a 1998 volume, offer explanations that are both sensitive to history and context and, because of their underlying theoretical framework, generalizable (Levi *et al.* 1998).

According to Levi, the analytic narrative approach is explicitly problem driven, in that it begins with historical puzzles that cry out for explanation. Cases are not selected with a view toward fitting them to off-the-shelf models; instead, models are constructed in close proximity to the details of the actual case. The method does, however, limit the range of phenomena that can be studied – the emphasis on narratives of specific events confines scholars to those events whose causal narratives are likely to be tractable, precluding attempts to answer grand questions of broad historical sweep, or in which contingency and unintended consequences play too prominent a role. But, as Levi points out, "by focusing on the institutional details that affect strategic interactions, choices, and outcomes, it can be useful in suggesting likely outcomes under given initial conditions." Thus, analytic narratives can be especially useful for problem-driven scholars who wish to offer advice on the likely outcomes of proposed, real-world policies. The superior usefulness of rational choice theory – and game theoretic models – for policymakers is the focus of Bruce Bueno de Mesquita's contribution.

In chapter 11, Bueno de Mesquita claims that in the last few decades policymakers have employed rational choice models successfully to predict outcomes in a range of political situations. He argues that these successes – and the fact that several firms make a brisk business out of making political predictions using such models – effectively refute claims of rational choice's limited real-world utility. But the predictive power of game-theoretic, rational choice models is not their only virtue, he tells us. The mathematical reasoning employed in such models allows scholars to achieve more logical rigor and argumentative coherence than with what he calls "ordinary language" approaches. The complexity of the political world, often invoked to suggest the impotence of reductionist mathematical models, is actually what makes such models more useful for political analysis than ordinary language, because of the reduced risk of logical error.

Yet Bueno de Mesquita does not prescribe these methods for all lines of inquiry. He argues that the choice of method will invariably rely on the substantive concerns of the researcher; large-N statistical research for those who wish to establish general relationships between political phenomena, historical case studies for those who want to explain specific events, and textual criticism and discourse analysis for those who wish to advance normative arguments. Additionally, he argues that different methods should be used in tandem – for example, while game theory can explain actions given an actor's beliefs and desires, other approaches to the study of politics (he singles out constructivism) can help scholars understand from whence such beliefs and desires emerge.

III. Possibilities for pluralism and convergence

Even if we accept the formulation that scholars should seek to answer important questions, and that methods should be selected with a view toward what will best help us answer the chosen questions, there is unlikely to be agreement on which problem–method combinations will be most apt. It is conceivable, even likely, that scholars driven by the same substantive interest will approach their questions from methodologically distinct positions, and that each will have strong reasons for choosing one method and not the other. In chapter 12, Alan S. Gerber, Donald P. Green, and Edward H. Kaplan advance a claim for their own methodological predisposition – field experimentation. They argue that the lack of experimental controls and randomization procedures in non-experimental, observational research means that scholars have little means of assessing the bias inherent in such studies. Accordingly, they tell us, whatever we "learn" from such research is likely to be illusory. The authors admit, however, that experimental research is often costly or (especially in such areas as international relations) impossible, and so concede that observational research will remain the only option for some fields. Still, in the end such fields may be doomed to inexactitude, since at some point a ceiling is reached where refinements to estimates of causal parameters cannot be made without experimental research.

It is hard not to conclude from Gerber, Green, and Kaplan's essay that methodological pluralism is a sad fact of life, to be tolerated only because some lines of inquiry are inhospitable to field experimentation. In chapter 13, Lisa Wedeen articulates a more sanguine view of the usefulness of different methodologies. The same substantive problem can be explored with different methods, and "different epistemological commitments sometimes yield divergent questions, sources of data, and explanations of political life." She examines three different approaches to the study of democracy: large-N statistical studies that seek to establish relationships between democracy and economic factors; a Wittgensteinian approach that seeks to understand the meaning of such terms as democracy, representation, and accountability; and an interpretive approach that seeks to uncover how democracy is lived and practiced.

Taking Przeworski *et al.'s* (2001) study of democracy and economic growth as her starting point, she illustrates how large-N studies require scholars to make simplifications and use proxies that may not adequately capture their variables of interest. Thus, democracy is reduced by Przeworski *et al.* to the holding of competitive elections, whereas in reality elections may capture only a small part of what democracy is about. Wedeen examines the "qat chew" in Yemen – gatherings of men in which a mild narcotic is consumed – and how such gatherings entail a form of political participation that cannot be captured by any study that equates democracy with elections. Wedeen does not try to pronounce one method superior to others, but rather endorses a methodological pluralism, echoing and illustrating Ferejohn's conviction that methodologically diverse accounts of a single phenomenon can build our understanding of it.

In chapter 14, Rudra Sil explores attempts to resolve the tension between calls for methodological pluralism and the idea of social science as a "unified" enterprise. Those who try to unify social science, like Gerber *et al.*, push for a single methodology, or like King, Keohane, and Verba (1994) offer general methodological principles that purport to hold for all types of research. At the other extreme are those who resign themselves (or actively encourage) a methodological anarchism "consisting of incommensurable research products that can never be evaluated relative to each other." Sil notes that the former forces upon all the ontological and epistemological commitments of a few, while the latter bodes ill for disciplinary progress. Instead, Sil promotes what he calls "constrained pluralism" in which scholars embedded in "research communities" with distinct substantive and methodological orientations participate in a "complex division of labor," each providing different insights into different types of problems. He suggests that breakthroughs in our understanding of real-world phenomena occur when eclectic scholars draw upon the insights, practices, and empirics of different research communities to suggest new understandings and lines of inquiry.

William E. Connolly's contribution in chapter 15 is also concerned with the possibility of methodological pluralism. He argues that pluralism is only possible if scholars reconceptualize their methodological commitments, understanding them as "existential faiths" rooted in unverifiable beliefs about the world and the way it works. Methodological debates, in Connolly's view, are akin to religious ones, and like religious ones, conversions are possible, but faith is often durable even in the face of disconfirming evidence. Connolly advances his own orientation to the study of politics – he calls it "immanent naturalism" – which eschews the notion that the world is a well-ordered (or divinely designed) place governed by laws that can be determined by social scientists, and instead is animated by a belief that the world is in a state of constant and unpredictable flux.

Whatever one's beliefs about the world, Connolly argues, they predispose him or her to study certain sets of problems. The immanent naturalist, for example, looks to the emergence of new and unforeseen social and political movements. Connolly suggests that the best – indeed only – way to navigate between differing methodological commitments is to adopt an ethos of "generosity and forbearance" – to advocate as forcefully as possible for one's method, but to recognize that it is itself rooted in a contestable faith.

Elisabeth Ellis' essay in chapter 16 can be read as an endorsement of Connolly's yearning for scholarly modesty. Turning our attention to the discipline of political theory, she argues that the insistence upon "incontrovertible standards of right" is as problematic as the insistence upon a single method in the social sciences. Both forms of absolutism, she argues, should be replaced by what she calls a "provisional outlook." In the social sciences, such an outlook involves being "fallibilist, sensitive to a wide variety of explanators, and open to the proposition that significant factors may not be measurable in any ordinary sense." In political theory, provisionalism entails the recognition of value pluralism and the fact that makeshift compromises of higher principles will often be necessary to secure good results. As Ellis puts it, "provisionalism allows the theorist to focus on the midrange problem of maintaining the possibility of progress, rather than on determining particular policy or even regime type outcomes and expecting the practical problems to be resolved later."

Chapter 17 is an account of a roundtable discussion between Robert A. Dahl, Truman F. Bewley, Susanne Hoeber Rudolph, and John Mearsheimer that took place at the conclusion of the Yale conference from which this volume arises. Participants in the roundtable were asked to reflect on what they had learned from the essays described above, and to advance their own ways of thinking about the tradeoffs between problems and methods. Dahl notes that political science, which he defines as the "systematic study of relations of power and influence among human beings," is an endeavor of enormous complexity, since the relationships between human beings and groups of human beings are nearly infinite, and contingency and agency play a large role in political events. The field's unmatched complexity, he argues, should render political scientists doubtful about the possibilities of emulating fields like physics, which grapple with less complex subjects than the myriad interactions between purposive humans. Reductionism of the sort that endogenizes complex outcomes to a single factor at some distant point in the causal chain is unlikely to illuminate political questions and the explanations tendered are often absurd on their face – he cites the example of the natural scientist who ascribes the entire body politic to Darwinian competition.

Instead of searching fruitlessly through all of this complexity for a general theory of politics, Dahl suggests that the field should dedicate itself to achieving "good human ends" – a reformulation, of sorts, of the problem-driven perspective articulated by Shapiro and Smith. Good ends, Dahl argues, are achieved by a political science that is pluralistic – that retains places of equal worth for normative inquiry, and also to small, often-derided incremental increases in our knowledge about the political world that can be put to concrete use in enhancing human welfare. He concludes on a hopeful note, pointing out that political science is "richer . . . broader . . . more systematic . . . [and] more solidly grounded in human experience" than it was during the middle of the last century. "Our field is dauntingly complex," he says, "yet we're far more aware of the complexities than we were 50 years ago."

Truman F. Bewley, an economist who has worked in the more formal, mathematical traditions that some political scientists seek to emulate, and who has engaged more recently in what could be called qualitative or ethnographic research in his quest to understand why business leaders do not cut wages during recessions, wonders why political scientists pay so little attention to what his research has pointed to as important motivators of human behavior: demands for reciprocity and fairness, feelings of group solidarity, and the sense of justice. He points out that the rational maximization of self-interest is unlikely to motivate in all circumstances, and suggests that political scientists can only know whether a rationality assumption is apt by getting close to their data and interrogating political actors as to their motives. He rejects the notion, common in economics and in some sectors of political science, that one should not do so "because they'll deceive you, or they won't know what their own motives are, they don't understand them, they'll exaggerate their own roles and glorify themselves." He suggests that one should be critical when interpreting respondents' claims, but that when one engages with a large number of subjects in a variety of circumstances, a logically coherent overall story can emerge.

Statistical analyses, Bewley notes, can be useful – but for certain problems, such as the question of downward wage rigidity, they are inadequate. Not only are statistical relationships unhelpful in determining causation, but statistical proxies often do not capture changes in their underlying variables, and so run the risk of mischaracterizing the phenomenon under study. For example, statistical data showed that wages did not fall during recessions, but they could not capture whether people were increasingly reassigned to other, lower-paying jobs, offered reduced hours, or different shifts – all of which would entail real reductions in an individual worker's pay without altering the wage scale for particular jobs.

Finally, Bewley responds to Bueno de Mesquita's assertion that arguments should be framed mathematically in order to avoid the propensity for logical flaws that besets "ordinary language" analysis. He argues that there is ultimately no distinction between verbal and mathematical reasoning, and that at its most fundamental level mathematics must make use of proofs and definitions that are all framed verbally. Far from being immune to logical flaws, mathematicians "make mistakes all the time," and these mistakes are typically brought out in debate between mathematicians, "just as they are in any verbal discussion."

Susanne Hoeber Rudolph wonders whether a convergence between diverse "modes of inquiry, literatures, and epistemologies" in political science is possible or even desirable. She envisions political science as divided into two epistemic communities: those who work in a scientific mode, seekers after such things as "certainty," "parsimony," the accumulation of knowledge, and causal relationships; and those whose work might be classified as interpretive and qualitative, and who respect contingency and agency, and employ thick description to discern meaning in political life. Rudolph points out that convergence between these two communities seems a difficult proposition, and bridge-building efforts are often thinly veiled hegemonic projects aimed at forcing one epistemic community to accept the orthodoxies of the other. Still, the fact that each epistemic community has begun to adopt and employ the vocabulary of the other suggests that something is taking place. Rudolph tells us that this linguistic exchange is unlikely to signal convergence, but rather a "mutual epistemic legibility," which is, perhaps, all we can hope for.

John Mearsheimer takes issue with Rudoloph's bipolar conception of the discipline, arguing that he finds himself on both sides of the divide she identifies. Though he employs historical case studies – and as such is labeled a "qualitative" scholar – he also employs the rationality assumption, values parsimonious theories, and pursues causal relationships, all hallmarks of the "scientific" approach in Rudolph's classification. Mearsheimer suggests that Rudolph's attention to the methodological or epistemological cleavages dividing political scientists is symptomatic of the broader failure of the chapters in this volume to address the larger, more fundamental question of whether political scientists should aim to solve problems or to simply hone their methodological implements. Instead, he tells us, the contributions merely debate the attributes of different methods. The question, Mearsheimer argues, comes down to one of audience; political scientists, he declares, speak mainly to themselves, abdicating any responsibility to grapple with the real world issues that non-specialists care about. He urges political scientists to privilege problems over methods, to address broad audiences rather than each other.

Mearsheimer then addresses a possible rejoinder – that scholars who attempt to address the great issues of the day may find themselves in the role of journalists. He points to the examples of Albert Wohlstetter and Thomas Schelling, both of whom made enormous theoretical contributions to the study of nuclear deterrence and international relations, but did so in close proximity to "the big problems of the day in the real world."

Mearsheimer's comments bring us full circle. One might concur with his critique of the utility of debating the merits of different methods in the absence of a well-specified problem, but Mearsheimer's call for problem-drivenness is not as straightforward as it first appears. After all, even when a scholar is driven to solve a "real world problem," her characterization of that problem will still invariably be colored by her methodological and theoretical priors, returning us to the domain of argument among partisans of various methods. Additionally, there is unlikely to be agreement on what constitutes an important, consequential problem, given the diversity of ideological commitments and tastes among scholars. And finally, the issue Norton raises, of whether we should see the contributions of scholars like Schelling and Wohlstetter to policies affecting "big problems" in the "real world" as salutary or pernicious, obviously is also likely to receive sharply different answers.

The reality that any given real world problem can be characterized in different ways, depending on the assumptions that the researcher brings to the inquiry, and that different characterizations can generate very different responses, have important implications for the possibility and desirability of methodological pluralism. We may wish to adjudicate among different descriptive cuts at a particular problem, but it is unlikely that we can formulate methodologically or politically neutral standards by which such adjudication should take place. In the end we are left with the argument advanced by Ferejohn in chapter 7: that methodological pluralism may be desirable because of the impossibility of giving complete descriptions of any phenomenon. Each attempt at explaining a phenomenon calls forth a different, methodologically and politically inflected description of that phenomenon, and in the absence of compelling reasons to privilege one over others, the best counsel one might offer to political scientists is to attend to as many of them as possible.

REFERENCES

Almond, Gabriel A. 1988. "Separate Tables: Schools and Sects in Political Science." *PS: Political Science* 21(4): 828–42.

Bates, Robert, Avner Greif, Margaret Levi, Jean-Laurent Rosenthal, and Barry Weingast. 1998. *Analytic Narratives*. Princeton: Princeton University Press.

Part I

Description, explanation, and agency

2 Problems, methods, and theories in the study of politics, or: what's wrong with political science and what to do about it

Ian Shapiro

Donald Green and I have previously criticized contemporary political science for being too method driven, not sufficiently problem driven (Green and Shapiro 1994; 1996). In various ways, many have responded that our critique fails to take full account of how inevitably theory laden empirical research is. Here I agree with many of these basic claims, but I argue that they do not weaken the contention that empirical research and explanatory theories should be problem driven. Rather, they suggest that one central task for political scientists should be to identify, criticize, and suggest plausible alternatives to the theoretical assumptions, interpretations of political conditions, and above all the specifications of problems that underlie prevailing empirical accounts and research programs, and to do it in ways that can spark novel and promising problem-driven research agendas.

My procedure will be to develop and extend our arguments for problem-driven over method-driven approaches to the study of politics. Green and I made the case for starting with a problem in the world, next coming to grips with previous attempts that have been made to study it, and then defining the research task by reference to the value added. We argued that method-driven research leads to self-serving construction of problems, misuse of data in various ways, and related pathologies summed up in the old adage that if the only tool you have is a hammer everything around you starts to look like a nail. Examples include collective action problems such as free riding that appear mysteriously to have been "solved" when perhaps it never occurred to anyone to free ride to begin with in many circumstances, or the concoction of elaborate explanations for why people "irrationally" vote, when perhaps it never occurred to most of them to think by reference to the individual costs and benefits of the voting act. The nub of our argument was that more attention to the problem and less to vindicating some pet approach would

An earlier version of this article was published in *Political Theory*, 30(4): 588–611.

be less likely to send people on esoteric goose chases that contribute little to the advancement of knowledge.

What we dubbed "method-driven" work in fact conflated theory-driven and method-driven work. These can be quite different things, though in the literature they often morph into one another, as when rational choice is said to be an "approach" rather than a theory. From the point of view elaborated here, the critical fact is that neither is problem driven, where this is understood to require specification of the problem under study in ways that are not mere artifacts of the theories and methods that are deployed to study it. Theory-drivenness and method-drivenness undermine problem-driven scholarship in different ways that are attended to below, necessitating different responses. This is not to say that problem selection is, or should be, uninfluenced by theories and methods, but I will contend that there are more ways than one of bringing theory to bear on the selection of problems, and that some are better than others.

Some resisted our earlier argument on the grounds that refinement of theoretical models and methodological tools is a good gamble in the advancement of science as part of a division of labor. It is sometimes noted, for instance, that when John Nash came up with his equilibrium concept (an equilibrium from which no one has an incentive to defect) he could not think of an application, yet it has since become widely used in political science (Nash 1950; Harsanyi 1986). We registered skepticism at this approach in our book, partly because the ratio of success to failure is so low, and partly because our instinct is that better models are likely to be developed in applied contexts, in closer proximity to the data.

I do not want to rehearse those arguments here. Rather, my goal is to take up some weaknesses in our previous discussion of the contrast between problem-drivenness and method- and theory-drivenness, and explore their implications for the study of politics. Our original formulation was open to two related objections: that the distinction we were attempting to draw is a distinction without a difference, and that there is no theory-independent way of characterizing problems. These are important objections, necessitating more elaborate responses than Green and I offered. My response to them in Parts I and II leads to a discussion, in Part III, of the reality that there are always multiple true descriptions of any given piece of social reality, where I argue against the reductionist impulse always to select one type of description as apt. This leaves us with the difficulty of selecting among potential competing descriptions of what is to be accounted for in politics, taken up in Part IV. There I explore the notion that the capacity to generate successful predictions is the appropriate criterion. In some circumstances this is the right answer, but it runs the risk of atrophying into a kind of method-drivenness that traps researchers into forlorn attempts to refine predictive instruments. Moreover, insisting on the capacity to generate predictions as the criterion for problem-selection risks predisposing the political science enterprise to the study of trivial, if tractable, problems. In light of prediction's limitations, I turn, in Part V, to other ways in which the aptness of competing accounts can be assessed. There I argue that political scientists have an important role to play in scrutinizing accepted accounts of political reality: exhibiting their presuppositions, both empirical and normative, and posing alternatives. Just because observation is inescapably theory laden, this is bound to be an ongoing task. Political scientists have a particular responsibility to take it on when accepted depictions of political reality are both faulty and widely influential outside the academy.

I. A distinction without a difference?

The claim that the distinction between problem- and theory-driven research is a distinction without a difference turns on the observation that even the kind of work that Green and I characterized as theory driven in fact posits a problem to study. This can be seen by reflecting on some manifestly theory-driven accounts.

Consider, for instance, a paper sent to me for review by the *American Political Science Review* on the probability of successful negotiated transitions to democracy in South Africa and elsewhere. It contended, *inter alia*, that as the relative size of the dispossessed majority grows the probability of a settlement decreases for the following reason: members of the dispossessed majority, as individual utility maximizers, confront a choice between working and fomenting revolution. Each one realizes that, as their numbers grow, the individual returns from revolution will decline, assuming that the expropriated proceeds of revolution will be equally divided among the expropriators. Accordingly, as their relative numbers grow they will be more likely to devote their energy to work than to fomenting revolution, and, because the wealthy minority realizes this, its members will be less inclined to negotiate a settlement as their numbers diminish since the threat of revolution is receding.

One only has to describe the model for its absurdity to be plain. Even if one thought dwindling minorities are likely to become increasingly recalcitrant, it is hard to imagine anyone coming up with this reasoning as part of the explanation. In any event the model seems so obtuse with respect to what drives the dispossessed to revolt and fails so obviously to take manifestly relevant factors into account (such as the changing likelihood of revolutionary success as the relative numbers change), that it is impossible to take seriously. In all likelihood it is a model that was designed for some other purpose, and this person is trying to adapt it to the study of democratic transitions. One can concede that even such manifestly theory-driven work posits problems, yet nonetheless insist that such specification is contrived. It is an artifact of the theoretical model in question.

Or consider the neo-Malthusian theory put forward by Charles Murray to the effect that poor women have children in response to the perverse incentives created by AFDC and related benefits (Murray 1984). Critics such as Katz (1984) pointed out that on this theory it is difficult to account for the steady increase in the numbers of children born into poverty since the early 1970s, when the real value of such benefits has been stagnant or declining. Murray's response (in support of which he cited no evidence) has only to be stated for its absurdity to be plain: "In the late 1970s, social scientists knew that the real value of the welfare benefit was declining, but the young woman in the street probably did not" (Murray 1994: 25). This is clearly self-serving for the neo-Malthusian account, even if in a palpably implausible way. Again, the point to stress here is not that no problem is specified; Murray is interested in explaining why poor women have children. But the fact that he holds on to his construction of it as an attempt by poor women to maximize their income from the government even in the face of confounding evidence suggests that he is more interested in vindicating his theory than in understanding the problem.

Notice that this is not an objection to modeling. To see this, compare these examples to John Roemer's (1999) account of the relative dearth of redistributive policies advocated by either political party in a two-party democracy with substantial *ex-ante* inequality. He develops a model which shows that if voters' preferences are arrayed along more than one dimension – such as "values" as well as "distributive" dimensions – then the median voter will not necessarily vote for downward redistribution as he would if there were only a single distributive dimension. His model seems to me worth taking seriously (leaving aside for present purposes how well it might do in systematic empirical testing), because the characterization of the problem that motivates it is not forced as in the earlier examples. Trying to develop the kind of model he proposes in order to account for it seems therefore to be worthwhile.[1]

In light of these examples we can say that the objection that theory-driven research in fact posits problems is telling in a trivial sense only. If the problems posited are idiosyncratic artifacts of the researcher's

[1] It should be noted, however, that the median voter theorem is eminently debatable empirically. For discussion, see Green and Shapiro (1994).

theoretical priors, then they will seem tendentious, if not downright misleading, to everyone except those who are wedded to their priors. Put differently, if a phenomenon is characterized as it is so as to vindicate a particular theory rather than to illuminate a problem that is specified independently of the theory, then it is unlikely that the specification will gain much purchase on what is actually going on in the world. Rather, it will appear to be what it is: a strained and unconvincing specification driven by the impulse to vindicate a particular theoretical outlook. It makes better sense to start with the problem, perhaps asking what the conditions are that make transitions to democracy more or less likely, or what influences the fertility rates of poor women. Next see what previous attempts to account for the phenomenon have turned up, and only then look to the possibility of whether a different theory will do better. To be sure, one's perception of what problems should be studied might be influenced by prevailing theories, as when the theory of evolution leads generations of researchers to study different forms of speciation. But the theory should not blind the researcher to the independent existence of the phenomenon under study. When that happens appropriate theoretical influence has atrophied into theory-drivenness.

II. All observation is theory laden?

It might be objected that the preceding discussion fails to come to grips with the reality that there is no theory-independent way to specify a problem. This claim is sometimes summed up with the epithet that "all observation is theory laden." Even when problems are thought to be specified independently of the theories that are deployed to account for them, in fact they always make implicit reference to some theory. From this perspective, the objection would be reformulated as the claim that the contrast between problem-driven and theory-driven research assumes there is some pre-theoretical way of demarcating the problem. But we have known since the later Wittgenstein, J. L. Austin, and Thomas Kuhn that there is not. After all, in the example just mentioned, Roemer's specification of problem is an artifact of the median voter theorem and a number of assumptions about voter preferences. The relative dearth of redistributive policies is in tension with that specification, and it is this tension that seems to call for an explanation. Such considerations buttress the insistence that there simply is no pretheoretical account of "the facts" to be given.

A possible response at this juncture would be to grant that all description is theory laden but retort that some descriptions are more theory laden than others. Going back to my initial examples of democratic transitions and welfare mothers, we might say that tendentious or contrived theory-laden descriptions fail on their own terms: one doesn't need to challenge the theory that leads to them to challenge them. Indeed, the only point of referring to the theory at all is to explain how anyone might come to believe them. The failure stems from the fact that, taken on its own terms, the depiction of the problem does not compute. We have no good reason to suppose that revolutionaries will become less militant as their relative numbers increase or that poor women have increasing numbers of babies in order to get decreasingly valuable welfare checks. Convincing as this might be as a response to the examples given, it does not quite come to grips with what is at stake for social research in the claim that all description is theory laden.

Consider theory-laden descriptions of institutions and practices that are problematic even though they do not fail on their own terms, such as Kathleen Bawn's (1999) claim that an ideology is a blueprint around which a group maintains a coalition or Russell Hardin's (2000) claim that constitutions exist to solve coordination problems. Here the difficulty is that, although it is arguable that ideologies and constitutions serve the designated purposes, they serve many other purposes as well. Moreover, it is far from clear that any serious investigation of how particular ideologies and constitutions came to be created or are subsequently sustained would reveal that the theorist's stipulated purpose has much to do with either. They are "just so" stories, debatably plausible conjectures about the creation and/or operation of these phenomena.[2]

The difficulty here is not that Bawn's and Hardin's are functional explanations. Difficult as functional explanations are to test empirically, they may sometimes be true. Rather, the worry is that these descriptions might be of the form: trees exist in order for dogs to pee on. Even when a sufficient account is not manifestly at odds with the facts, there is no reason to

[2] I leave aside, for present purposes, how convincing the debatable conjectures are. Consider the great difficulties Republican candidates face in forging winning coalitions in American politics that keep both social and libertarian conservatives on board. If one were to set out to define a blueprint to put together a winning coalition, trying to fashion it out of these conflicting elements scarcely seems like a logical place to start. Likewise with constitutions viewed as coordinating devices, the many veto points in the American constitutional system could just as arguably be said to be obstacles to coordination. See George Tsebelis (2002). This might not seem problematic if one takes the view, as Hardin does, that the central purpose of the US constitution is to facilitate commerce. On such a view institutional sclerosis might arguably be an advantage, limiting government's capacity to interfere with the economy. The difficulty with going down that route is that we then have a theory for all seasons: constitutions lacking multiple veto points facilitate political coordination, while those containing them facilitate coordination in realms that might otherwise be interfered with by politicians. Certainly nothing in Hardin's argument accounts for why some constitutions facilitate more coordination of a particular kind than do others.

suppose that it will ever get us closer to reality unless it is put up against other plausible conjectures in such a way that there can be decisive adjudication among them. Otherwise we have "Well, in that case what are lamp posts for?" Ideologies may be blueprints for maintaining coalitions, but they also give meaning and purpose to people's lives, mobilize masses, reduce information costs, contribute to social solidarity, and facilitate the demonization of "out-groups" – to name some common candidates. Constitutions might help solve coordination problems, but they are often charters to protect minority rights, legitimating statements of collective purpose, instruments to distinguish the rules of the game from the conflicts of the day, compromise agreements to end or avoid civil wars, and so on. Nor are these characterizations necessarily competing; ideologies and constitutions might well perform several such functions simultaneously. Selecting any one over others implies a theoretical commitment. This is one thing people may have in mind when asserting that all observation is theory laden.

One can concede the point without abandoning the problem-driven/theory-driven distinction, however. The theory-driven scholar commits to a sufficient account of a phenomenon, developing a "just so" story that might seem convincing to partisans of her theoretical priors. Others will see no more reason to believe it than a host of other "just so" stories that might have been developed, vindicating different theoretical priors. By contrast, the problem-driven scholar asks, "why are constitutions enacted?" or "why do they survive?" and "why do ideologies develop?" and "why do people adhere to them?" She then looks to previous theories that have been put forward to account for these phenomena, tries to see how they are lacking, and whether some alternative might do better. She need not deny that embracing one account rather than another implies different theoretical commitments, and she may even hope that one theoretical outcome will prevail. But she recognizes that she should be more concerned to discover which explanation works best than to vindicate any priors she may happen to have. As with the distinction-without-a-difference objection, then, this version of the theory-ladenness objection turns out on inspection at best to be trivially true.

III. Multiple true descriptions and aptness

There is a subtler sense in which observation is theory laden, untouched by the preceding discussion though implicit in it. The claim that all observation is theory laden scratches the surface of a larger set of issues having to do with the reality that all phenomena admit of multiple true descriptions. Consider possible descriptions of a woman who says "I do" in a conventional marriage ceremony. She could be:

- Expressing authentic love
- Doing (failing to do) what her father has told her to do
- Playing her expected part in a social ritual
- Unconsciously reproducing the patriarchal family
- Landing the richest husband that she can
- Maximizing the chances of reproducing her genes

Each description is theory laden in the sense that it leads to the search for a different type of explanation. This can be seen if in each case we ask the question *why*?, and see what type of explanation is called forth.

- *Why does she love him?* predisposes us to look for an explanation in terms of her personal biography
- *Why does she obey (disobey) her father?* predisposes us to look for a psychological explanation
- *Why does she play her part in the social ritual?* predisposes us to look for an anthropological explanation
- *Why does she unconsciously reproduce patriarchy?* predisposes us to look for an explanation in terms of ideology and power relations
- *Why does she do as well as she can in the marriage market?* predisposes us to look for an interest-based rational choice explanation
- *Why does she maximize the odds of reproducing her genes?* predisposes us to look for a sociobiological explanation

The claim that all description is theory laden illustrated here is a claim that there is no "raw" description of "the facts" or "the data." There are always multiple possible true descriptions of a given action or phenomenon, and the challenge is to decide which is most apt. Indeed, the idea of a complete description of any phenomenon is chimerical. We could describe the activity in her brain when she says "I do," the movements of her vocal chords, or any number of other aspects that are part of the phenomenon.[3]

From this perspective theory-driven work is part of a reductionist program. It dictates always opting for the description that calls for the explanation that flows from the preferred model or theory. So the narrative historian who believes every event to be unique will reach for personal biography; the psychological reductionist will turn to the psychological determinants of her choice; the anthropologist will see the constitutive role of the social ritual as the relevant description; the feminist will focus

[3] Recognizing that there can never be a complete description of any phenomenon does not, however, commit us to relativism. Any particular description, while incomplete, is nonetheless true or false in virtue of some state of affairs that either holds or does not hold in the world. See Bernard Williams (2002).

on the action as reproducing patriarchy; the rational choice theorist will reach for the explanation in terms of maximizing success in the marriage market; and for the sociobiologist it will be evolutionary selection at the level of gene reproduction.

Why do this? Why plump for any reductionist program that is invariably going to load the dice in favor of one type of description? I hesitate to say "level" here, since that prejudges the question I want to highlight: whether some descriptions are invariably more basic that others. Perhaps one is, but to presume this to be the case is to make the theory-driven move. Why do it?

The common answer rests, I think, on the belief that it is necessary for the program of social science. In many minds this enterprise is concerned with the search for general explanations. How is one going to come up with general explanations if one cannot characterize the classes of phenomena one studies in similar terms? This view misunderstands the enterprise of science, provoking three responses, one skeptical, one ontological, and one occupational.

The skeptical response is that whether there are general explanations for classes of phenomena is a question for social-scientific inquiry, not one to be prejudged before conducting that inquiry. At stake here is a variant of the debate between deductivists and inductivists. The deductivist starts from the preferred theory or model and then opts for the type of description that will vindicate the general claims implied by the model, whereas the inductivist begins by trying to account for particular phenomena or classes of phenomena, and then sees under what conditions, if any, such accounts might apply more generally. This enterprise, might, often, be theory influenced for the reasons discussed in Parts I and II, but is less likely to be theory driven than the pure deductivist's one because the inductivist is not determined to arrive at any particular theoretical destination. The inductivist pursues general accounts, but she regards it as an open question whether they are out there waiting to be discovered.

The ontological response is that although science is in the second instance concerned with developing general knowledge claims, it must in the first instance be concerned with developing valid knowledge claims. It seems to be an endemic obsession of political scientists to believe that there must be general explanations of all political phenomena, indeed to subsume them into a single theoretical program. Theory-drivenness kicks in when the pursuit of generality comes at the expense of the pursuit of empirical validity. "Positive" theorists sometimes assert that it is an appropriate division of labor for them to pursue generality while others worry about validity. This leads to the various pathologies Green and I wrote about, but the one we did not mention that I emphasize here is that it invites tendentious characterizations of the phenomena under study because the selection of one description rather than another is driven by the impulse to vindicate a particular theoretical outlook.

The occupational response is that political scientists are pushed in the direction of theory-driven work as a result of their perceived need to differentiate themselves from others, such as journalists, who also write about political phenomena for a living, but without the job security and prestige of the professoriate. This aspiration to do better than journalists is laudable, but it should be unpacked in a Lakatosian fashion. When tackling a problem we should come to grips with the previous attempts to study it, by journalists as well as scholars in all disciplines who have studied it, and then try to come up with an account that explains what was known before – and then some. Too often the aspiration to do better than journalists is cashed out as manufacturing esoteric discourses with high entry costs for outsiders. All the better if they involve inside-the-cranium exercises that never require one to leave one's computer screen.

IV. Prediction as a sorting criterion?

A possible response to what has been said thus far is that prediction should be the arbiter. Perhaps my skepticism is misplaced, and some reductionist program is right. If so, it will lead to correct predictions, whereas those operating with explanations that focus on other types of description will fail. Theory-driven or not, the predictive account should triumph as the one that shows that interest maximization, or gene preservation, or the oppression of women, or the domination of the father figure, etc., is "really going on." On this instrumentalist view, we would say, with Friedman: deploy whatever theory-laden description you like, but lay it on the line and see how it does in predicting outcomes. If you can predict from your preferred cut, you win (Friedman 1953).

This instrumental response is adequate up to a point. Part of what is wrong with many theory-driven enterprises, after all, is that their predictions can never be decisively falsified. From Bentham through Marx, Freud, functionalism, and much modern rational choice theory, too often the difficulty is that the theory is articulated in such a capacious manner that some version of it is consistent with every conceivable outcome. In effect, the theory predicts everything, so that it can never be shown to be false. This is why people say that a theory that predicts everything explains nothing. If a theory can never be put to a decisive predictive test, there seems little reason to take it seriously.

Theories of everything to one side, venturing down this path raises the difficulty that prediction is a tough test that is seldom met in political science. This difficulty calls to mind the job applicant who said in an interview that he would begin a course on comparative political institutions with a summary of the field's well-tested empirical findings, but then had nothing to say when asked what he would teach for the remaining twelve weeks of the semester. Requiring the capacity to predict is in many cases a matter of requiring more than can be delivered, so that if political science is held to this standard there would have to be a proliferation of exceedingly short courses. Does this reality suggest that we should give up on prediction as our sorting criterion?

Some, such as MacIntyre (1984), have objected to prediction as inherently unattainable in the study of human affairs due to the existence of free will. Such claims are not convincing, however. Whether or not human beings have free will is an empirical question. Even if we do, probabilistic generalizations might still be developed about the conditions under which we are more likely to behave in one way rather than another. To be sure, this assumes that people are likely to behave in similar ways in similar circumstances, which may or may not be true, but the possibility of its being true does not depend on denying the existence of free will. To say that someone will probably make choice *x* in circumstance *q* does not mean that they cannot choose not-*x* in that circumstance or, that, if do they choose not-*x*, it was not nonetheless more likely *ex-ante* that they would have chosen *x*. In any event, most successful science does not proceed by making point predictions. It predicts patterns of outcomes. There will always be outliers and error terms; the best theory minimizes them *vis-à-vis* the going alternatives.

A more general version of this objection is to insist that prediction is unlikely to be possible in politics because of the decisive role played by contingent events in most political outcomes. This, too, seems overstated unless one assumes in advance – with the narrative historian – that everything is contingent. A more epistemologically open approach is to assume that some things are contingent, others not, and try to develop predictive generalizations about the latter. For instance, Courtney Jung and I developed a theory of the conditions that make negotiated settlements to civil wars possible involving such factors as whether government reformers and opposition moderates can combine to marginalize reactionaries and revolutionary militants on their flanks. We also developed a theory of the conditions that are more and less likely to make reformers and moderates conclude that trying to do this is better for them than the going alternatives (Jung and Shapiro 1995). Assuming we are right, contingent triggers will nonetheless be critical in whether such agreements are successfully concluded, as can be seen by reflecting on how things might have developed differently in South Africa and the Middle East had F. W. DeKlerk been assassinated in 1992 or Yitzhak Rabin had not been assassinated in 1995. The decisive role of contingent events rules out *ex-ante* prediction of success, but the theory might correctly predict failure – as when a moderate IRA leader such as Jerry Adams emerges but the other necessary pieces are not in place, or if Yassir Arafat is offered a deal by Ehud Barak at a time when he is too weak to outflank Hamas and Islamic Jihad. Successful prediction of failure over a range of such cases would suggest that we have indeed taken the right descriptive cut at the problem.[4]

There are other types of circumstances in which capacity to predict will support one descriptive cut at a problem over others. For instance, Przeworski, Alvarez, Cheibub, and Limongi (2000) have shown that although level of economic development does not predict the installation of democracy, there is a strong relationship between level of per capita income and the survival of democratic regimes. Democracies appear never to die in wealthy countries, whereas poor democracies are fragile, exceedingly so when per capita incomes fall below $2,000 (in 1985 dollars). When per capita incomes fall below this threshold, democracies have a one in ten chance of collapsing within a year. Between per capita incomes of $2,001 and $5,000 this ratio falls to one in sixteen. Above $6,055 annual per capita income, democracies, once established, appear to last indefinitely. Moreover, poor democracies are more likely to survive when governments succeed in generating development and avoiding economic crises. If Przeworski *et al.* are right, as it seems presently that they are, then level of economic development is more important than institutional arrangements, cultural beliefs, presence or absence of a certain religion, or other variables for predicting democratic stability. For this problem the political economist's cut seems to be the right sorting criterion.[5]

These examples suggest that prediction can sometimes help, but we should nonetheless be wary of making it the criterion for problem selection in political science. For one thing, this can divert us away from the study of important political phenomena where knowledge can advance

[4] What is necessary in the context of one problem may, of course, be contingent in another. When we postulate that it was necessary that Arafat be strong enough to marginalize the radicals on his flank if he was to make an agreement with Barak, we do not mean to deny that his relative strength in this regard was dependent on many contingent factors. For further discussion, see Courtney Jung, Ellen Lust-Okar, and Ian Shapiro (2002). In some ultimate – if uninteresting – sense, everything social scientists study is contingent on factors such as that the possibility of life on earth not be destroyed due to a collision with a giant meteor. To be intelligible, the search for law-like generalizations must be couched in "if . . . then" statements that make reference, however implicitly, to the problem under study.

[5] For Przeworski *et al.*'s (2000) discussion of other explanatory variables, see pp. 122–37.

even though prediction turns out not to be possible. For instance, generations of scholars have theorized about the conditions that give rise to democracy (as distinct from the conditions that make it more or less likely to survive once instituted, just discussed). Alexis de Tocqueville (1966 [1832]) alleged it to be the product of egalitarian mores. For Seymour Martin Lipsett (1959) it was a byproduct of modernization. Barrington Moore (1966) identified the emergence of a bourgeoisie as critical, while Rueschemeyer, Stephens, and Stephens (1996) held the presence of an organized working class to be decisive. We now know that there is no single path to democracy and, therefore, no generalization to be had about which conditions give rise to democratic transitions. Democracy can result from decades of gradual evolution (Britain and the United States), imitation (India), cascades (much of Eastern Europe in 1989), collapses (Russia after 1991), imposition from above (Spain and Brazil), revolutions (Portugal and Argentina), negotiated settlements (Bolivia, Nicaragua, and South Africa), or external imposition (Japan and West Germany) (Przeworski 1991; Przeworski *et al.* 2001; Huntington 1991; Shapiro 1996).

In retrospect this is not surprising. Once someone invents a toaster, there is no good reason to suppose that others must go through the same invention processes. Perhaps some will, but some may copy it, some may buy it, some may receive it as a gift, and so on. Perhaps there is no cut at this problem that yields a serviceable generalization, and as a result no possibility of successful prediction. Political scientists tend to think they must have general theories of everything as we have seen, but looking for a general theory of what gives rise to democracy may be like looking for a general theory of holes.[6] Yet we would surely be making an error if our inability to predict in this area inclined us not to study it. It would prevent our discovering a great deal about democracy that is important to know, not least that there is no general theory of what gives rise to it to be had. Such knowledge would also be important for evaluating claims by defenders of authoritarianism who contend that democracy cannot be instituted in their countries because they have not gone through the requisite path-dependent evolution.

Reflecting on this example raises the possibility I want to consider next: that making a fetish of prediction can undermine problem-driven research via wag-the-dog scenarios in which we elect to study phenomena because they seem to admit the possibility of prediction rather than because we

[6] Perhaps one could develop such a theory, but only of an exceedingly general kind such as: "holes are created when something takes material content out of something else." This would not be of much help in understanding or predicting anything worth knowing about holes.

have independent reasons for thinking it worthwhile to study them. This is what I mean by method-drivenness, as distinct from theory-drivenness. It gains impetus from a number of sources, perhaps the most important being the lack of uncontroversial data concerning many political phenomena. Predictions about whether or not constitutional courts protect civil rights run into disagreements over which rights are to count and how to measure their protection. Predictions about the incidence of war run into objections about how to measure and count the relevant conflicts. In principle it sounds right to say let's test the model against the data. In reality there are few uncontroversial datasets in political science.

A related difficulty is that it is usually impossible to disentangle the complex interacting causal processes that operate in the actual world. We will always find political economists on both sides of the question of whether cutting taxes leads to increases or decreases in government revenue, and predictive tests will not settle their disagreements. Isolating the effects of tax cuts from the other changing factors that influence government revenues is just too difficult to do in ways that are likely to convince a skeptic. Likewise, political economists have been arguing at least since Bentham's time over whether trickle-down policies benefit the poor more than do government transfers, and it seems unlikely that the key variables will ever be isolated in ways that can settle this question decisively.

An understandable response to this is to suggest that we should tackle questions where good data are readily available. But taking this tack courts the danger of self-defeating method-drivenness, because there is no reason to suppose that the phenomena about which uncontroversial data are available are those about which valid generalizations are possible. My point here is not one about curve-fitting – running regression after regression on the same data-set until one finds the mix of explanatory variables that passes most closely through all the points to be explained. Leaving the well-known difficulties with this kind of data-mining to one side, my worry is that working with uncontroversial data because of the ease of getting it can lead to endless quests for a holy grail that may be nowhere to be found.

The difficulty here is related to my earlier discussion of contingency, to wit, that many phenomena political scientists try to generalize about may exhibit secular changes that will always defy their explanatory theories. For instance, trying to predict election outcomes from various mixes of macro political and economic variables has been a growth industry in political science for more than a generation. But perhaps the factors that caused people to vote as they did in the 1950s differ from those forty or fifty years later. After all, this is not an activity with much of a track record of success in political science. We saw this dramatically in the 2000 election where all of the standard models predicted a decisive Gore victory.[7] Despite endless *post-hoc* tinkering with the models after elections in which they fare poorly, this is not an enterprise that appears to be advancing. They will never get it right if my conjecture about secular change is correct.

It might be replied that if that is really so, either they will come up with historically nuanced models that do a better job or universities and funding agencies will pull the plug on them. But this ignores an occupational factor that might be dubbed the Morton Thiokol phenomenon. When the Challenger blew up in 1986 there was much blame to go around, but it became clear that Morton Thiokol, manufacturer of the faulty O-ring seals, shouldered a huge part of the responsibility. A naïve observer might have thought that this would mean the end of their contract with NASA, but, of course, this was not so. The combination of high-entry costs to others, the dependence of the space program on Morton Thiokol, and their access to those who control resources mean that they continue to make O-ring seals for the space shuttle. Likewise with those who work on general models of election forecasting. Established scholars with an investment in the activity have the protections of tenure and legitimacy, as well as privileged access to those who control research resources. Moreover, high methodological entry costs are likely to self-select new generations of researchers who are predisposed to believe that the grail is there to be found. Even if their space shuttles will never fly, it is far from clear that they will ever have the incentive to stop building them.

To this it might be objected that it is not as if others are building successful shuttles in this area. Perhaps so, but this observation misses my point here: that the main impetus for the exercise appears to be the ready availability of data which sustains a coterie of scholars who are likely to continue to try to generalize on the basis of it until the end of time. Unless one provides an account, that, like the others on offer, purports to retrodict past elections and predict the next one, one cannot aspire to be a player in this game at which everyone is failing. If there is no such account to be found, however, then perhaps some other game should be being played. For instance, we might learn more about why people vote in the ways that they do by asking them. Proceeding instead with the macro models risks becoming a matter of endlessly refining the predictive instrument as an end in itself – even in the face of continual failure and the absence of an argument about why we should expect it to be

[7] See the various postmortem papers in the March 2001 issue of PS: *Political Science and Politics* 34(1): 9–58.

successful. Discovering where generalization is possible is a taxing empirical task. Perhaps it should proceed on the basis of trial and error, perhaps on the basis of theoretical argument, perhaps some combination. What should *not* drive it, however, is the ready availability of data and technique.

A more promising response to the difficulties of bad data and of disentangling complex causal processes in the "open systems" of the actual world is to do experimental work where parameters can be controlled and key variables more easily isolated.[8] There is some history of this in political science and political psychology, but the main problem has been that of external validity. Even when subjects are randomly selected and control groups are included in the experiments (which often is not done), it is far from clear that results produced under lab conditions will replicate themselves outside the lab.

To deal with these problems Alan Gerber and Donald Green (2000; 2002) have revived the practice of field experiments, where subjects can be randomized, experimental controls can be introduced, and questions about external validity disappear. Prediction can operate once more, and when it is successful there are good reasons for supposing that the researcher has taken the right cut at the problem. In some ways this is an exciting development. It yields decisive answers to questions such as which forms of mobilizing voters are most effective in increasing turnout, or what the best ways are for partisans to get their grassroots supporters to the polls without also mobilizing their opponents.

Granting that this is an enterprise that leads to increments in knowledge, I want nonetheless to suggest that it carries risks of falling into a kind of method-drivenness that threatens to diminish the field experiment research program unless they are confronted. The potential difficulties arise from the fact that field experiments are limited to studying comparatively small questions in well-defined settings, where it is possible to intervene in ways that allow for experimental controls. Usually this means designing or piggy-backing on interventions in the world such as get out the vote efforts or attempts at partisan mobilization. Green and Gerber have shown that such efforts can be adapted to incorporate field experiments.

To be sure, the relative smallness of questions is to some extent in the eye of the beholder. But consider a list of phenomena that political scientists have sought to study, and those drawn to political science often want to understand, that are not likely to lend themselves to field experiments:

[8] For discussion of the difficulties with prediction in open systems, see Roy Bhaskar (1975: 63–142) and (1979: 158–69).

- The effects of regime type on the economy, and vice versa
- The determinants of peace, war, and revolution
- The causes and consequences of the trend toward creating independent central banks
- The causes and consequences of the growth in transnational political and economic institutions
- The relative merits of alternative policies for achieving racial integration, such as mandatory bussing, magnet schools, and voluntary desegregation plans
- The importance of constitutional courts in protecting civil liberties, property rights, and limiting the power of legislatures
- The effects of other institutional arrangements, such as parliamentarism *vs.* presidentialism, unicameralism *vs.* bicameralism, federalism *vs.* centralism on such things as the distribution of income and wealth, the effectiveness of macroeconomic policies, and the types of social policies that are enacted
- The dynamics of political negotiations to institute democracy

I could go on, but you get the point.

This is not to denigrate field experiments. One of the worst features of methodological disagreement in political science is the propensity of protagonists to compare the inadequacies of one method with the adequacies of a second, and then declare the first to be wanting.[9] Since all methods have limitations and none should be expected to be serviceable for all purposes, this is little more than a shell game. If a method can do some things well that are worth doing, that is a sufficient justification for investing some research resources in it. With methods, as with people: if you focus only on their limitations you will always be disappointed.

Field experiments lend themselves to the study of behavioral variation in settings where the institutional context is relatively fixed, and where the stakes are comparatively low, so that the kinds of interventions required do not violate accepted ethical criteria for experimentation on human subjects. They do not obviously lend themselves to the study of life or death and other high stakes politics, war and civil war, institutional variation, the macropolitical economy or the determinants of regime stability and change. This still leaves a great deal to study that is worth studying, and creative use of the method might render it deployable in a wider array of areas than I have noted here. But it must be conceded that it also leaves out a great deal that draws people to political science, so that

[9] For discussion of an analogous phenomenon that plagues normative debates in political theory, see Shapiro (1989).

if susceptibility to study via field experiment becomes the criterion for problem selection then it risks degenerating into method-drivenness.

This is an important caution. I would conjecture that part of the disaffection with 1960s behaviorism in the study of American politics that spawned the model mania of the 1990s was that the behaviorists became so mindlessly preoccupied with demonstrating propositions of the order that "Catholics in Detroit vote Democrat" (Taylor, 1985: 90). As a result, the mainstream of political science that they came to define seemed to others to be both utterly devoid of theoretical ambition as well as detached from consequential questions of politics; frankly, boring. To paraphrase Kant, theoretical ambition without empirical research may well be vacuous, but empirical research without theoretical ambition will be blind.

V. Undervaluing critical reappraisal of what is to be explained

The emphasis on prediction can lead to method-drivenness in another way: it can lead us to undervalue critical reappraisals of accepted descriptions of reality. To see why this is so, one must realize that much commentary on politics, both lay and professional, takes depictions of political reality for granted that closer critical scrutiny would reveal as problematic. Particularly, though not only, when prediction is not going to supply the sorting device to get us the right cut, political scientists have an important role to play in exhibiting what is at stake in taking one cut rather than another, and in proposing alternatives. Consider some examples.

For over a generation in debates about American exceptionalism, the United States was contrasted with Europe as a world of relative social and legal equality deriving from the lack of a feudal past. This began with de Tocqueville, but it has been endlessly repeated and has became conventional wisdom, if not a mantra, when restated by Louis Hartz (1955) in *The Liberal Tradition in America*. But as Rogers Smith (1995) showed decisively in *Civic Ideals*, it is highly misleading as a descriptive matter. Throughout American history the law has recognized explicit hierarchies based on race and gender whose effects are still very much with us. Smith's book advances no well-specified predictive model, let alone tests one, but it displaces a highly influential orthodoxy that has long been taken for granted in debates about pluralism and cross-cutting cleavages, the absence of socialism in America, arguments about the so-called end of ideology, and the ideological neutrality of the liberal tradition.[10]

[10] Indeed, Smith's argument turns out to be the tip of an iceberg in debunking misleading orthodoxy about American exceptionalism. Eric Foner (1984) has shown that its assumptions about Europe are no less questionable than its assumptions about the United States.

Important causal questions are to be asked and answered about these matters, but my point here is that what was thought to stand in need of explanation was so misspecified that the right causal questions were not even on the table.

Likewise with the debate about the determinants of industrial policy in capitalist democracies. In the 1970s it occurred to students of this subject to focus less on politicians' voting records and campaign statements and look at who actually writes the legislation. This led to the discovery that significant chunks of it were actually written by organized business and organized labor with government (usually in the form of the relevant minister or ministry) in a mediating role. The reality was more of a "liberal corporatist" one, and less of a pluralist one, than most commentators who had not focused on this had realized (Schmitter 1974; Panitch 1977). The questions that then motivated the next generation of research became: under what conditions do we get liberal corporatism, and what are its effects on industrial relations and industrial policy? As with the Tocqueville–Hartz orthodoxy, the causal questions had to be reframed because of the ascendancy of a different depiction of the reality.[11]

In one respect, the Tocqueville–Hartz and pluralist accounts debunked by Smith and the liberal corporatists are more like those of democratic transitions and the fertility rates of welfare mothers discussed in Part I than the multiple descriptions problem discussed in Part III. The difficulty is not how to choose one rather than another true description, but rather that the Tocqueville–Hartz and pluralist descriptions fail on their own terms. By focusing so myopically on the absence of feudalism and the activities of politicians, their proponents ignored other sources of social hierarchy and decisionmaking that are undeniably relevant once they have been called to attention. The main difference is that the democratic transition and welfare mother examples are not as widely accepted as the Tocqueville–Hartz and pluralist orthodoxies were before the debunkers came along. This should serve as a salutary reminder that orthodox views can be highly misleading, and that an important ongoing task for political scientists is to subject them to critical scrutiny. This involves exhibiting their presuppositions, assessing their plausibility, and

[11] It turns out that joint legislation writing is a small part of the story. What often matters more is ongoing tri-partite consultation about public policy and mutual adjustment of legislation/macroeconomic policy and "private" but quasi-public policy (e.g., wage increases, or multiemployer pension and health plans). There is also the formalized inclusion of private interest representatives in the administration and implementation process where *de facto* legislation and common law like adjudication takes place. The extent of their influence in the political process varies from country to country and even industry to industry, but the overall picture is a far cry from the standard pluralist account.

proposing alternatives when they are found wanting. This is particularly important when the defective account is widely accepted outside the academy. If political science has a role to play in social betterment, it must surely include debunking myths and misunderstandings that shape political practice.[12]

Notice that descriptions are theory laden not only in calling for a particular empirical story, but often also in implying a normative theory that may or may not be evident unless this is made explicit. Compare the following two descriptions:

- The Westphalian system is based on the norm of national sovereignty
- The Westphalian system is based on the norm of global Apartheid

Both are arguably accurate descriptions, but, depending on which of the two we adopt, we will be prompted to ask exceedingly different questions about justification as well as causation. Consider another instance:

- When substantial majorities in both parties support legislation, we have bipartisan agreement
- When substantial majorities in both parties support legislation, we have collusion in restraint of democracy

The first draws on a view of democracy in which deliberation and agreement are assumed to be unproblematic, even desirable goals in a democracy. The second, antitrust-framed, formulation calls to mind Mill's emphasis on the centrality of argument and contestation, and the Schumpeterian impulse to think of well-functioning democracy as requiring competition for power.[13]

Both *global apartheid* and *collusion in restraint of democracy* here are instances of problematizing redescriptions. Just as Smith's depiction of American public law called the Tocqueville–Hartz consensus into question, and the liberal corporatist description of industrial legislation called then conventional assumptions about pluralist decisionmaking into question, so do these. But they do it not so much by questioning the veracity of the accepted descriptions as by throwing their undernoticed benign normative assumptions into sharp relief. Describing the Westphalian system as based on a norm of global Apartheid, or political agreement among the major players in a democracy as collusion in restraint of democracy, shifts the focus to underattended features of reality, placing different empirical and justificatory questions on the table.

But are they the right questions?

To answer this by saying that one needs a theory of politics would be to turn once more to theory-drivenness. I want to suggest a more complex

[12] For a discussion of the dangers of convergent thinking, see Charles E. Lindblom (1990: 118–32).

[13] For discussion of differences between these models, see Shapiro (2002).

answer, one that sustains problematizing redescription as a problem-driven enterprise. It is a two-step venture that starts when one shows that the accepted way of characterizing a piece of political reality fails to capture an important feature of what stands in need of explanation or justification. One then offers a recharacterization that speaks to the inadequacies in the prior account.

When convincingly done, prior adherents to the old view will be unable to ignore it and remain credible. This is vital, because it will, of course, be true that the problematizing redescription is itself usually a theory-influenced, if not a theory-laden endeavor. But if the problematizing redescription assumes a theory that seems convincing only to partisans of her priors, or is validated only by reference to evidence that is projected from her alternative theory, then it will be judged tendentious to the rest of the scholarly community for the reasons I set out in Parts I and II of this chapter. It is important, therefore, to devote considerable effort to making the case that will persuade a skeptic of the superiority of the proffered redescription over the accepted one. One of the significant failings of many of the rational choice theories Green and I discussed is that their proponents fail to do this. They offered problematizing redescriptions that were sometimes arrestingly radical, but their failure to take the second step made them unconvincing to all except those who agreed with them in advance.

VI. Concluding comments

The recent emphases in political science on modeling for its own sake and on decisive predictive tests both give short shrift to the value of problematizing redescription in the study of politics. It is intrinsically worthwhile to unmask an accepted depiction as inadequate, and to make a convincing case for an alternative as more apt. Just because observation is inescapably theory laden for the reasons explored in this essay, political scientists have an ongoing role to play in exhibiting what is at stake in accepted depictions of reality, and reinterpreting what is known so as to put new problems onto the research agenda. This is important for scientific reasons when accepted descriptions are both faulty and influential in the conduct of social science. It is important for political reasons when the faulty understandings shape politics outside the academy.

If the problems thus placed on the agenda are difficult to study by means of theories and methods that are currently in vogue, an additional task arises that is no less important: to keep them there and challenge the ingenuity of scholars who are sufficiently open-minded to devise creative ways of grappling with them. It is important for political scientists to throw their weight against the powerful forces that entice scholars to embroider fashionable theories and massage methods in which they are professionally invested while failing to illuminate the world of politics. They should remind each generation of scholars that unless problems are identified in ways that are both theoretically illuminating and convincingly intelligible to outsiders, nothing that they say about them is likely to persuade anyone who stands in need of persuasion. Perhaps they will enjoy professional success of a sort, but at the price of trivializing their discipline and what one hopes is their vocation.

REFERENCES

Bawn, Kathleen. 1999. "Constructing 'Us': Ideology, Coalition Politics and False Consciousness." *American Journal of Political Science* 43(2): 303–34.

Bhaskar, Roy. 1975. *A Realist Theory of Science*. Sussex: Harvester Wheatsheaf.

1979. *The Possibility of Naturalism*. Sussex: Harvester Wheatsheaf.

de Tocqueville, Alexis. 1966 [1832]. *Democracy in America*, J. P. Mayer (ed.) and George Lawrence (trans.). New York: Harper Perennial.

Foner, Eric. 1984. "Why Is There No Socialism in the United States?" *History Workshop Journal* 17: 57–80.

Friedman, Milton. 1953. "The Methodology of Positive Economics," in *Essays in Positive Economics*. Oxford: Oxford University Press.

Gerber, Alan and Donald Green. 2000. "Do Phone Calls Increase Voter Turnout? A Field Experiment." *Public Opinion Quarterly* 65: 75–85.

2002. "Reclaiming the Experimental Tradition in Political Science," in Ira Katznelson and Helen Milner (eds.). *Political Science: The State of the Discipline*, 3rd edn. New York: W. W. Norton.

Green, Donald and Ian Shapiro. 1994. *Pathologies of Rational Choice Theory: A Critique of Applications in Political Science*. New Haven: Yale University Press.

1996. "Pathologies revisited: Reflections on our critics," in Jeffrey Friedman (ed.). *The Rational Choice Controversy: Economic Models of Politics Reconsidered*. New Haven: Yale University Press. 235–76.

Hardin, Russell. 2000. *Liberalism, Constitutionalism, and Democracy*. Oxford: Oxford University Press.

Harsanyi, John. 1986. "Advances in understanding rational behavior," in Jon Elster (ed.). *Rational Choice*. New York: New York University Press. 92–4.

Hartz, Louis. 1955. *The Liberal Tradition in America*. New York: Harcourt Brace.

Huntington, Samuel P. 1991. *The Third Wave: Democratization in the Late Twentieth Century*. Norman: University of Oklahoma Press.

Jung, Courtney and Ian Shapiro. 1995. "South Africa's Negotiated Transition: Democracy, Opposition, and the New Constitutional Order." *Politics and Society* 23: 269–308.

Jung, Courtney, Ellen Lust-Okar, and Ian Shapiro. 2002. "Problems and prospects for democratic transitions: South Africa as a model for the Middle East and Northern Ireland?" Mimeo. New Haven: Yale University.

Katz, Michael B. 1989. *The Undeserving Poor: From the War on Poverty to the War on Welfare*. New York: Pantheon.

Lindblom, Charles E. 1990. *Inquiry and Change: The Troubled Attempt to Understand and Shape Society*. New Haven: Yale University Press.

Lipsett, Seymour Martin. 1959. "Some Social Requisites of Democracy: Economic Development and Political Legitimacy." *American Political Science Review* 53: 69–105.

MacIntyre, Alasdair. 1984. *After Virtue*. Notre Dame, IN: University of Notre Dame Press.

Moore, Barrington. 1966. *The Social Origins of Dictatorship and Democracy: Lord and Peasant in the Making of the Modern World*. Boston: Beacon Press.

Murray, Charles. 1984. *Losing Ground – American Social Policy, 1950–1980*. New York: Basic Books.

1994. "Does Welfare Bring More Babies?" *The Public Interest* (Spring): 17–30.

Nash, John. 1950. "The Bargaining Problem." *Econometrica* 18: 155–62.

Panitch, Leo. 1977. "The Development of Corporatism in Liberal Democracies." *Comparative Political Studies* 10: 61–90.

Przeworski, Adam, Michael Alvarez, Jose Cheibub, and Fernando Limongi. 2000. *Democracy and Development: Political Institutions and Well-Being in the World, 1950–1990*. Cambridge: Cambridge University Press.

Przeworski, Adam. 1991. *Democracy and the Market*. Cambridge: Cambridge University Press.

Political Science and Politics (PS). 2001. "Al Gore and George W. Bush's Not-So-Excellent Adventure" 34(1): 9–58.

Roemer, John. 1999. "Does Democracy Engender Justice?," in Ian Shapiro and Casiano Hacker-Cordón (eds.). *Democracy's Value*. Cambridge: Cambridge University Press. 56–68.

Rueschemeyer, Dietrich, Evelyne Huber Stephens, and John D. Stephens. 1992. *Capitalist Development and Democracy*. Cambridge: Polity Press.

Schmitter, Philippe. 1974. "Still the Century of Corporatism?" *Review of Politics* 36(1): 85–121.

Shapiro, Ian. 1989. "Gross Concepts in Political Argument." *Political Theory* 17(1): 51–76.

1996. *Democracy's Place*. Ithaca: Cornell University Press.

2002. "The State of Democratic theory," in Ira Katznelson and Helen Milner (eds.). *Political Science: The State of the Discipline*, 3rd edn. New York: W. W. Norton.

Smith, Rogers. 1995. *Civic Ideals: Changing Conceptions of Citizenship in US Law*. New Haven: Yale University Press.

Taylor, Charles. 1985. "Neutrality in Political Science," in *Philosophical Papers II: Philosophy and the Human Sciences*. Cambridge: Cambridge University Press.

Tsebelis, George. 2002. *Veto Players: How Political Institutions Work*. Princeton: Princeton University Press.

Williams, Bernard. 2002. *Truth and Truthfulness: An Essay in Genealogy*. Princeton: Princeton University Press.

3 The politics of identities and the tasks of political science

Rogers M. Smith

Those of us in political science who subscribe to the view that our research should be "problem driven," with methods chosen according to what is needed to explore those problems, bear the burden of defining and defending the substantive political problems that we think we should address, and explaining why certain methods are suitable for doing so. This chapter contends that contemporary political science should give high (though certainly not exclusive) priority to studies of the processes, especially the *political* processes, through which senses of political membership, allegiance, and identity are formed and transformed. To do so, we will need to rely more than the discipline collectively has done on approaches that provide empathetic interpretive understandings of human consciousnesses and values, and on identification of historical and contextual differences in identities and values. We cannot rely solely or even predominantly on approaches that seek to improve our formal grasp of instrumental rationality, or even on efforts to seek to identify timeless, enduring regularities in political behavior that are relatively independent of specific historical contexts. Though valuable for other purposes, in regard to processes of political identity formation and transformation, those methods generally contribute most when they are but components of projects that rest extensively on contextually informed interpretive accounts.

Yet even if, in regard to such problems, we must often move away from striving to develop a single thin, universal theory of all politics, we should still be able to formulate somewhat less abstract theoretical frameworks that can help us identify both near-universal patterns of political conduct, and some necessarily suprahistorical theories about the means and mechanisms of historical transformations. And even in our interpretive and contextual characterizations, we also still must conform as rigorously as we can to what King, Keohane and Verba have urged to be a unified "logic of scientific inference" – though we should not equate that logic with whatever particular statistical techniques, all always imperfect, that happen to be widely used to approximate it in any particular phase of social science research (King, Keohane, and Verba 1994: 36–41). If we are to judge, for example, to what sorts of conceptions of their identities and interests particular political actors are giving priority, we need to form some notions or hypotheses based on what we know about these actors, and then look for observable data about their lives that we can use to falsify some of those hypotheses. That logic is constant, though the techniques of falsification will vary with the content of the hypotheses, the types of problems particular data present, and the tools currently at our disposal.

But if the challenge of drawing reliable inferences is one universal dimension of social science research, the most crucial work in analyzing political identities must often be done via immersing ourselves in information about the actors in question, and using both empathy and imagination to construct credible accounts of their senses of identity and interest. In many instances, if that work is well done, the potential of one interpretive framework to explain more than others will be reasonably evident even prior to more quantified inferential testing. Once James Scott elaborated the concepts of the "hidden transcript" and "arts of resistance," for example, it was clear that analyses that included cognate hypotheses about peasant political behavior were far more credible than those that automatically coded anything except outright defiance as acquiescence, and that much peasant conduct clearly did represent more subtle forms of resistance, even if we still need to do careful empirical work to decide how much "much" is in particular contexts (Scott 1985, 1990).

If we accept that political science should take the illumination of political processes of political identity making and remaking as one of its core "problems," then we will be compelled to acknowledge that many research methods often deemed less "scientific," such as comparative historical case studies, ethnographies, in-depth interviewing, textual interpretation and the like, are as vital to our scientific enterprise as more formal or quantitative ones. We will need to place as much emphasis on preparation for, and execution of, illuminating concept formation as on our necessary but often highly limited efforts at inferential testing. And, more trivially but not inconsequentially, we will also be well advised to redefine the basic fields that comprise, and often constrain, the discipline of political science.

I. Why the "politics of identity," and why now?

Let us start with the reflections that lead me to suggest that the political processes of identity formation and transformation need to become a more central substantive concern for political scientists than they have been in our discipline's past. I concede at the outset that throughout human history, we can find people thinking and often writing about virtually every dimension of human political existence. Hence all generalizations about the "central tendencies" of political analysis in different eras are subject to innumerable exceptions. Even so, there are certain topics and problems that appear to loom particularly large in the bulk of the political writing in distinct historical periods.

Given the limited pecuniary rewards, I suspect that most political scholars have always been motivated, at least to begin with, by concerns about what they were experiencing as major political problems in their time. If so, then we should expect that the major external political developments, changes, and conflicts of different eras would shape political writing in those periods, even as some of those writings have sometimes had impacts on those developments.

A casual glance at the history of Western political thought suggests that prevailing political issues have indeed provided the main agendas for political writers in different epochs. In imperial democratic Athens, the rise of a relatively large class of free citizens with leisure (derived from slave labor) made lives of reflection more feasible, though they were controversial. Consequently, the thought of Plato and Aristotle turned centrally on the merits of alternative ways of life – political, poetic, religious, above all the potentially subversive philosophic path they had chosen for themselves. After the Roman Empire became Christian and then fell as the Catholic Church began to grow in power, for many centuries reflection focused instead on the comparative claims of religious and civil authorities, in the works of writers like Augustine, Aquinas, and Marsilius (Islamic writers like Alfarabi and Averroes had similar concerns). When secular rulers of various sorts increasingly broke away from clerical constraints, though with much violence, and sought to consolidate authority in themselves, Western political thought turned chiefly to debates over centralized "sovereign" power, as in Bodin, Grotius, and Hobbes. But first in Renaissance Italy, then in Britain, its North American colonies, France, and elsewhere, dissatisfactions with monarchical rulers led to a new focus on issues of popular government and personal rights versus aristocratic and autocratic rule, developments foreshadowed by Machiavelli and carried to diverse fruitions in the seventeenth-and eighteenth-century writings of Locke, Hume, Montesquieu, Rousseau, the *Federalist* authors, and Kant. The successes of efforts to create modern popular or "republican" governments also led to new emphases on "nations" as bodies of self-governing peoples, though initially as a subordinate theme.

The consequences of the commercial and then industrial revolutions that arose with modern commercial republics increasingly shifted the center of attention for political writers in the nineteenth and twentieth centuries. Questions of political economy, ultimately of *laissez-faire* capitalism versus more intensely regulated market systems versus socialism and communism, came to predominate in the writings of early figures like Adam Smith, then Karl Marx and most of the utilitarians, late nineteenth-century social Darwinists like William Graham Sumner and Herbert Spencer, and early twentieth-century democratic theorists like John Dewey and Harold Laski, followed by a host of other "liberal" and "socialist" thinkers. Some of these writers, like John Stuart Mill and V. I. Lenin, wrote not only of representative government and free markets versus socialism and dictatorship, but also of nationalism – though again as a subordinate theme. More purely "nationalist" theorists like Giambattista Vico and Johann Gottfried Herder had an important place in Western intellectual and political endeavors, but not a dominant place.

The Cold War period in the second half of the twentieth century only heightened all the more the tendencies of American political analysts, in particular, to believe that the "big issues" ultimately were questions of democratic capitalism versus authoritarian socialism – fascism having been crushed after its brief but destructive burgeoning. The many dimensions of that opposition were analyzed on the limiting assumption that democratic capitalism and authoritarian socialism were political and economic systems constructed within particular states, usually thought of as nation-states, and that those nation-states allied for and against each other on behalf of these politicoeconomic systems, thereby defining the terrain of "international relations." Such "national" identities were the political identities that generally received attention if identity questions were raised at all.

There are certainly many other currents and themes visible in all these periods and in all the writers whom I have named, as well as their numerous and varied contemporaries. Still, it seems accurate to assert that the central concerns of leading political writers historically indeed have been, and quite understandably have been, rooted in what seemed then the great issues of the day. The bulk of contemporary American political science has been shaped by the paths thus begun, especially concerns with democracy versus alternative political systems, and market capitalism versus alternative economic systems, all conceived of as structures within particular "nation-states."

These, at least, are my impressions after working in the field of American political science for many years and reading most of its major disciplinary histories. These characterizations seem to me confirmed by our discipline's very structure. The "fields" of American political science emerged during the course of the twentieth century – chiefly, it appears, through the diffusion of definitions adopted by particular departments – in ways that expressed an American take on a political world conceived in these substantive terms. By the early 1960s, in the heart of the Cold War, sources such as department directories, professorial self-descriptions in surveys, and the American Political Science Association all identified the discipline's chief fields as "American government and politics," "international relations," "comparative politics," and "political theory," with declining numbers of political scientists still practicing "public law," "public administration," and "general politics and political processes" (Somit and Tanenhaus 1964: 52–54). Though distinct subfields such as "political economy" and "formal theory" have since emerged, the grand quartet of American politics, international relations, comparative politics, and political theory still provides the most prevalent set of field definitions in the discipline (Schwartz-Shea 2003). And this structure still articulates a perspective in which politics takes place essentially within and between nation-states, with the United States as the paradigmatic nation-state, exceptionally exemplary of both democracy and capitalism. Our disciplinary fields label all other nation-states as having "comparative" significance, or else they matter in their roles as international allies or adversaries of America, democracy, and capitalism.

But the political world changed during the Cold War years in ways that have pushed new problems on to intellectual as well as political agendas (yes, even among stodgy political scientists). The Cold War era saw not only the great dualistic rivalry between the US and the USSR, the "free world" and the "Communist world," that so captured scholarly imaginations. It also witnessed the gradual dismantling of most of the Western European empires and the rise of "new nations" in Africa and Asia from the end of World War II through the 1990s. That ending of empires eventually included the collapse of the Soviet Union itself and, with it, many other communist regimes, a world-historical turn of events that has precipitated a still unfolding further wave of nation-building, but also other, more novel forms of political reconstruction, in many parts of the world. The post-Cold War years have also displayed the proliferation of transnational economic entities, such as the old European Common Market, the North American Free Trade Association, and the World Trade Organization; the subsequent development of regional political bodies, notably the European Union and now, perhaps, the African Union; heightened roles for international institutions like the United Nations, its International Court of Justice, and, it seems, the new

International Criminal Court; the growth of politically potent and active multinational corporations; and the development of many transnational "movement" organizations, such as environmental, labor, and human rights groups. All of these have generated not only new forms of political community, but also some novel political identities.

The ending of formal European imperial rule was also accompanied by and complexly linked to the demise of formal domestic systems of racial hierarchy in many regimes. Those officially renounced include the Jim Crow system in the US and its related legal denials of equal status to Asian, Latino, and Native Americans; many Latin American governmental measures to promote the "whitening" of their societies; the "white Australia" policy and numerous doctrines denying rights to Aboriginal peoples there and in other British Commonwealth nations; and eventually, even the apartheid system in South Africa. As all those forms of domestic legal discrimination fell, they sparked further efforts to change other unequal statuses. The successes of the African-American civil rights movement in the 1950s and early 1960s, in particular, helped inspire a range of other liberation movements in the US and elsewhere – for women's rights; for cultural, linguistic, and ancestral ethnic groups and national minorities; for gays, lesbians, and others with non-traditional life styles; for religious minorities; for the disabled. Most readers can continue this list.

Yet its familiarity should not mask the fact that these events have gradually put on the agenda of political science a great range of topics that were absent or much more marginal before; and they have done so for good reason. All these developments wrought fundamental transformations in existing forms of political membership, status, and identity that were in themselves tremendously important for billions of human beings. They also set in train further sweeping changes that are far from being completed. Stirred by this series of historic events, popular writers, political activists, and scholars in many disciplines, again, eventually including our own, have in the past two decades devoted increasing attention, both normative and empirical, to what are widely termed issues of "identity politics" or (if that suggests purely expressive rather than constitutive processes) the "politics of identity." Work on "racial and ethnic politics," "women's politics," "politics and religion," "immigrant politics," "indigenous peoples," "GLBT politics," along with "globalization," "cosmopolitan citizenship," "transnational social movements," and so on, may not have been entirely absent before these events. Such work may, moreover, seem still remote from the pinnacle of prestigious political science endeavors today. Yet most of these topics were far less visible up to the 1980s than they have since become.[1] However one assesses their place now, they clearly have been and almost certainly still are in the ascendant in receiving scholarly attention.

Ascendant – but still neither central nor secure. Though scholars from across our ideological and methodological spectra now engage in such work, an equally varied group is deeply disturbed by these developments. To many, concerns about "identity" merely represent a politically fashionable academic "fad," one all too often pursued by methods derived from the humanities that lack the rigor of true social science. Others worry that attention to such "faddish" topics can distract attention from earlier sets of political, economic, and social issues that remain crucially important; and certainly topics of democratic and authoritarian systems and economic regulation and distribution matter today as much as ever.

In political science as in other disciplines, a number of lively debates are therefore proceeding over whether "globalization" and "transnationalism" are really making "nation-states" less central to political life, or only changing the contexts within which they still play dominant roles (e.g. Held 1995; Newman 2000; Slaughter and Mattli 1995; Garrett 1995). Probably even more scholars are concerned that, analytically, an undue focus on racial, ethnic, religious, cultural, and gender "identities" misses deeper, often economic causes of political action. Many also argue more normatively that concentration on such political "identities" operates to deflect attention from persistent material inequalities that need to be addressed. Many more fear that such work strengthens false and politically divisive essentialist understandings of the character of political actors. Hence one can find many sharp critiques, on the left as well as the right, of scholarly turns to "identity politics" (e.g., Bloom 1987; D'Souza 1991; Bromwich 1992; Rorty 1998; Reed 2000; Barry 2001). Most such critics contend that the increased concern with "identity politics" has been at best intellectually and politically undesirable, at worst deeply immoral.

Though I too have concerns about many varieties of the work that can be grouped under the "politics of identity" heading, I believe that the events of the latter part of the twentieth century have made greater focus on both empirical and normative questions about the politics of memberships and identities inescapable as we enter the twenty first. The problem, I submit, is not that we are paying attention to such issues; it is that we as a profession still have not taken them seriously enough. Again,

[1] See e.g., Walton *et al.* 1995, 1997, documenting how from their inception through 1990, only about 2 percent of the articles published in the *American Political Science Review* and *Political Science Quarterly* addressed the political experience of African-Americans.

we have long had work devoted to "nation-building" and nationalism, never more so than in the past two decades.[2] Post-World War II political science revived "group theory," with some attention to group formation. The political upheavals of the 1960s, especially, also generated work on the construction of gender, racial, and ethnic identities and social movements. But much of the latter work has followed in the wake of studies in other disciplines – sociology, anthropology, social psychology, as well as literary, linguistic, and postmodernist philosophic theorizing – stressing the "social" construction of identities, rather than formulating explicitly political accounts of political identity formation.[3]

Furthermore, though there has been some cross-fertilisation, most of these accounts have been rather "identity-specific." There have been few thoroughgoing efforts to unify theories of national identity formation with accounts of the creation of groups, social movements, and gender, racial, cultural, and religious identities, though there have been many persuasive arguments for specific linkages of one sort or another.

That may be quite sensible. The many historical varieties of political identities may be so different from one another that trying to generalize across them could represent a hopeless "apples and oranges" exercise. A more general theory of the formation of political identities may just be too general to be of real use.[4] And if we are talking about a general

[2] See e.g., A. D. Smith 1983: 191; Turner 1986: 18–19; Brubaker 1992: 35.

[3] For a similar critique, see Schnapper 1998: 12, 145–9. To be sure, there are outstanding exceptions to these generalizations. Unlike Schnapper, I see John A. Armstrong's *Nations before Nationalism* (1982), as stressing the political origins of ethnic and national identities, though how far is unclear as he explicitly eschews much general theory-building (3). The resistance of political scientists to portraying identities as deeply politically constructed may have reflected the wariness of many post-World War II Western liberal scholars toward perspectives that appeared to dismiss individual autonomy, in ways associated with both fascist and communist abuses. In the years of Marxism's greatest prestige, moreover, many writers on the left generally took economically determined class identities to be most important and treated other identities as epiphenomenal, as have many social scientists influenced by liberal forms of political economy. Thus no theory of political identity formation beyond Marxism or economic theory more broadly seemed necessary. If so, it is not surprising that theoretical explorations of the politics of identity have proliferated after the fall of Communism.

In one such effort, Alexander Wendt (1999) offers an insightful "general, evolutionary" model of collective identity formation from which I have greatly benefited. Wendt does so, however, out of recognition that we scholars with "constructivist" views of social identities do not clearly possess a general theory of their creation (317), and his effort, too, is a partial one. He is specifically concerned with the formation of "state" identities, not all political identities.

[4] For a historian's sharp and reasonable critique of exactly the sort of general theorizing I am discussing here, see Breuilly (1982: 2). Yet in an attempt to avoid too general an approach, Breuilly and others have to define their topics, in his case "nationalism," in plausibly but eminently contestable ways that generate lengthy and often unfruitful debates over what "nationalism" (or "ethnicity," or "class," or some other category) "really" is.

theory of phenomena that should not centrally occupy our attention in the first place, we might well conclude that the discipline's priorities should remain where they long have been or revert back further.

I believe to the contrary that the breakdown of older imperial systems, the subsequent waves of new nation-making, the range of "identity politics" movements, and the emergence of new forms of international and transnational political associations have not only made it clear that "nationalism" cannot be our only "identity" theme. They have brought to the surface a whole fundamental dimension of political life that should have been seen as a basic concern of political analysis all along – as basic as the questions of whether people should see themselves as governed by God and his deputies or by themselves; whether rule should be by the one, the few, or the many; whether politics is driven by economic interests, and whether governance should be weighted toward the interests of the rich, the middle class, or the poor, toward those who control capital or those who provide labor. Just as foundational is the issue of how human beings come to see themselves as having certain identities that are "political" in the following sense: some identities and memberships define persons' trumping allegiances in cases where the demands of those memberships conflict with ones advanced on behalf of other human associations, groups, or societies (whether those are "nation-states," different levels of government, religious bodies, racial or ethnic communities, corporate, worker, or other class organizations, or other groups).

And though it may in the end prove to be the case that such identities emerge rather seamlessly from what we may conclude to be extrapolitical sources – perhaps inherited languages, ancestral groups, geographical clusters, or economic structures – it is far from obvious that this is the case. It seems more plausible to assume that, out of the multiple possible identities that human existence presents to most people, political activities of various sorts play important roles in determining which become salient political identities. Certainly, political scientists ought to explore possible political explanations for such identities thoroughly and intensively before concluding that their origins lie outside their disciplinary domains.

I take more seriously the objection that we do not have more explicit efforts to generate a "Unified Field Theory of Political Identities" that seeks to explain their creation and transformations because such a theory cannot be achieved. Any all-encompassing account of such diverse phenomena may have to be so abstract and thin as to be useless. It may turn out to be the case that what we need is a set of explanations of different sorts of political identities that cannot be tightly connected to one another, which is more or less what we currently seem to be developing. Even if this so, I still contend that we ought to treat the development of such explanations as one of the central endeavors of political science, much more than we do.

But on this point, too, we should not give up too easily. There are undoubtedly ways in which the crafting of different sorts of political identities occurs through contrasting processes, but there may also be illuminating ways in which these processes and patterns of identity formation have elements in common. That, again, is a possibility that political scientists have an intellectual duty to take seriously. For, if it is so, then these elements are surely fundamental for both empirical and normative political analyses. There can hardly be any topic more richly significant for explanatory political inquiries than questions of how people come to have the senses of political affiliation and allegiance that they possess and how those senses of belonging change. There can hardly be any topic more important to normative political debates than the question of the forms of political membership and identity we ought to embrace as our own.

For though the relationship of political identities to human interests is complex, there is good reason to think that it is reciprocal – that just as economic interests influence our affiliations, so those affiliations shape our sense of our economic interests. The same person might conceive of her most politically salient identity as simply "worker," or as a "white worker," or as an "American worker," or as a "female worker. She is likely to define her economic interests differently and to pursue very different political courses depending on which conception she favors. The same sort of list can easily be compiled for "capitalists" and other economic identities. Similarly, it is plausible to think that our memberships reflect but also help define our interests in personal physical security and political power. The same person might seek protection and representation primarily as a Jew, or as a Brooklyn resident, or as a member of a radical socialist party, conceptions that again have quite different implications for political conduct. If it is credible to think that political identities can play such decisive roles, then the problem of how they are made and remade cannot be ignored or marginalized in a genuinely adequate political science.

But accepting these questions as fundamental to our discipline, especially today, quickly makes clear that our conventional field structure is ill suited to accord such work proper recognition. That structure treats "nations" as the only forms of political identity worthy of *structural* recognition – the fields, again, are governance in the American nation-state, other "comparative" nation-states, relations between nation-states, and political theorizing, which tends to be about nation-state politics. The field structure also "recognizes" nation-states as, for the most part, somehow already "given" or established prior to "politics," though to be fair the emergence of "new nations" is a "comparative politics" theme.

But political life everywhere has always involved many political identities beyond nation-state memberships and they have always been subjects of political contestation and change, as they so vividly are now. If we have begun to conclude that most of those identities – racial, ethnic, gendered, religious – are worthy of study, and if we also see the development of at least *some* new transnational forms of political association and politically engaged economic association as important, then we need to recognize that the old field structure is, at best, a particular artifact of certain past historical processes of political identity construction. That artifact has long been under pressure; most political science departments have often heard and often granted requests to create "special fields" on topics like "racial and ethnic politics," "gender politics," "religion and politics" and the like. As these demands accumulate, they provide evidence that the time when the prevailing field structure served the discipline's analytical purposes well is passing, if indeed it ever really did so. By reifying a certain subset of political identities as all-important, that structure now stands in the way of even asking, much less answering, all the questions that should concern us in the contemporary world. We need to study problems and processes of political identity formation and transformation more broadly than our traditional focus on "nations" permits. We also need to focus on the explicitly political dimensions of these processes more than we have done. And we should at least try to do so via more general theorizing, if we can. The existing field structure is, in short, a Cold War relic that we should consign to that era, while we experiment with better ways to define and structure the diverse sorts of work we must now do.[5]

II. The limited utility of rational choice theorizing

If we are to make questions of political identity formation and transformation as central to our discipline as issues such as constitutional design, electoral representation, legislative behavior, global security and political economy long have been, that shift in our agenda has implications for the methods we employ (I leave aside questions of what methods are really most appropriate for these other sorts of substantive topics). Many important aspects of the politics of identity cannot be adequately probed without methods that are richly "interpretive," that involve grasping discursively the consciousnesses and senses of value and meaning that

[5] I will not venture to offer a personal alternative field mapping here, but both the departments with which I have been affiliated, at Yale University and the University of Pennsylvania, have developed distinct and promising formulations in their recent "Initiatives."

identities involve for the human beings who possess them. More formal methods that reduce identity choices to points on hypothetical preference functions are valuable for many purposes, including many kinds of empirical "identity" studies. They cannot, however, go very far in helping us to comprehend the substantive appeal and normative significance of particular identities, features important for understanding how those identities are likely to shape conduct. Formal models that are so "thin" that they treat sense of identity and interest as exogenously "given," moreover, simply do not seek to shed any light on how identities are formed and changed.

It is on these topics that, I contend, interpretive textual analyses, ethnographic fieldwork, biographical studies, in-depth interviews, individual and comparative case studies, participant observation research, and other methods generally labeled "qualitative" are likely to yield more insights. Each of these varied methods has a growing literature analyzing their distinctive elements, strengths, and limitations.[6] No method, however, provides a failsafe "cook book" listing steps to take in grasping the phenomena of political identities. As suggested above, it is a combination of contextual immersion, psychological empathy, and creative imagination, utilized in a variety of research techniques, that enables researchers like James Scott (1985), Cathy Cohen (1999), Claire Kim (2000), and Courtney Jung (2000) to offer alternative conceptions of the identities they study, whether their subjects have previously been conceived as passive peasants, politically homogeneous African-Americans, irrational racial nationalists, or ethnic tribespeople. Again, I think even the partly ineffable process of concept formation always involves at least a rough method of inference testing; researchers like these invariably check preliminary hypotheses formed on the basis of a few observations against many others made in the course of their fieldwork before elaborating them as positive arguments. Such conceptions can, in turn, often be used to construct better large-N surveys that can be more easily subjected to systematic, if always imperfect, statistical analyses to test particular claims. That work is valuable, but it is dependent on the strength of concepts that are generally formed through other means.

And not only the means through which concepts are generated, but also all the methods used to formalize, operationalize, and test them, must be to some minimal degree historically and contextually sensitive. The political identities that prevail in one location in one era may well be very different than in another place and time, prompting "local" patterns

[6] For a valuable collection of materials, see the Consortium for Qualitative Research Methods website at http://www.asu.edu/clas/polisci/cqrm/.

in political behavior that we can understand only by grasping the substantive characteristics and distinctions among ways of life in those contrasting periods. A large team of anthropologists and economists has recently provided support for this claim. In a set of field experiments, they asked members of many different societies on several continents to participate in rational-choice-style games involving decisions either to offer or to accept cash drawn from a researcher-provided "pie." These researchers contend that standard economic models of wealth-maximizing behavior were not borne out by actual patterns of conduct in most of these societies – and that the societies varied widely from each other in the sorts of "rationality" they displayed. Both to conduct the games successfully and to find explanations for these variations, the research team relied heavily on ethnographic studies of the social and economic practices of each society, and they argued that these distinctive patterns of group behavior did much to explain individual choices.[7] Their work shows how essential such ethnographic analysis is to insightful investigation of how distinct identities and values are formed and maintained. But like most research in the other social sciences, their study does not much explore whether political struggles, and the social policies and practices thus established, have played a major role in constructing the different kinds of group behaviors and identities that the researchers found to characterize different societies. Nor, arguably, should such scholars have to do so. Analyzing the contributions of politics is the job for political science.

But are we not doing it well enough? Let me consider a prime counterexample. In apparent refutation of everything I have said so far, one of the most celebrated recent works in political science is David Laitin's *Identity in Formation: The Russian-Speaking Populations in the Near Abroad* (1998). It not only takes processes of identity formation and transformation as its subject; it begins with a chapter-long "theory of political identities." And it presents, as the heart of that theory, an abstract rational choice "tipping model," which Laitin suggests does the key explanatory work in illuminating his topic, the question of whether Russian-speaking persons in states that were formerly part of the Soviet Union will learn the dominant language of those newly independent nations or remain "Russian," at least in linguistic terms (Laitin 1998: 24–32, 243–60).

Let me underline at the outset that I find most aspects of Laitin's work exceedingly impressive. After its introductory theory chapter, it presents a highly informative, analytical recounting of historical and institutional developments that help explain why so many parts of what became the

[7] Henrich *et al.* 2003. I am grateful to the editorial staff of *Perspectives on Politics* and to James Robins for independently calling this body of work to my attention.

Soviet Union never became Russian speaking, under either centralized Tsarist or Soviet rule. Laitin then draws on four ethnographies conducted in the newly independent republics he studies to analyze recent trends, and he personally lived in Estonia for seven months as part of this fieldwork. He and his colleagues also conducted an extensive large-N survey of Russian and "native language" speakers in those republics that he uses in a variety of quantitative analyses. And Laitin adds both interviews and documentary analyses to identify pertinent state policies in each republic, along with a Russian-language discourse analysis based on articles in the Russian press and the presses of the republics. The result is a landmark work that exhibits the great benefits of multiple methods of analysis for political science research, even if Laitin claims "to have provided only a descriptive account" (1998: x).

Still, one must wonder why he makes this surprisingly modest claim in a book that, again, begins by trumpeting a rational choice theory of identity formation that it seeks to test. In fact, his work is more theoretically valuable than this characterization suggests; but Laitin's modesty is nonetheless understandable. Even on its own terms, Laitin's rational choice model of identity formation contributes little to the understanding of processes of identity formation that his book provides. Instead, most of the explanatory power in his analysis comes from features he treats as less theoretically significant, especially his analysis of historical institutional patterns and current state policies, as well as his evidence of some group attitudes for which he does, indeed, have no real theoretical account. In the end the book does much more to make the case for the limitations of rational choice models as responses to the kinds of questions he is addressing rather than the reverse.

Laitin seeks to build on Thomas Schelling's justly renowned "tipping model" analysis, which clarified, among other things, why discriminatory real estate practices do not have to be massive to produce sharply segregated residential patterns. I wholeheartedly affirm the usefulness of Schelling's model for demonstrating the consequences of racially inflected (rather than strictly wealth-maximizing) economic choices. Laitin, however, uses his "tipping model" to support a somewhat more obvious claim. Russian speakers in societies where Russian is not the dominant language will learn that dominant language, he suggests, when they find it advantageous to do so. But when is that? The "tipping model" suggests that at least one important element is their perception that most other Russian speakers in the region are learning the dominant language and shifting to it. If most are, then presumably it will be harder in a whole variety of respects to remain one of the few monolingual Russian speakers left (1998: 21–4). We thus need to know what sort of factors make first some, then many, decide to pay the costs of learning the dominant language, eventually creating a "cascade" in which it becomes understandable why most of those remaining go ahead and take that step. We also want to know why Russian speakers in some republics make the shift more readily and massively than in others.

To answer, Laitin elaborates his theoretical choice model. People might learn a new language under three conditions – when it is economically advantageous to do so; when members of their own "in-group" do not punish and perhaps even encourage such adaptation (again, perhaps by doing it themselves); and when members of the regionally dominant language group do not punish and perhaps even welcome and reward this learning (1998: 28–9). Though these elaborations are presented in fairly formal, abstract terms, they quickly move the bulk of the work done by Laitin's analysis to its non-formal, extramodel elements. If choices are greatly shaped by the attitudes and behavior of "in-group" and "out-group" members, then we need to know how those senses of group identity came to be, and why those groups are punishing and encouraging members and outsiders in the ways that they are. Yet the "tipping model" tells us nothing at all about those decisive issues.

Perhaps the model might seem more useful if economic calculations proved to be most of the story in explaining personal choices, as in Schelling's real estate example (though there, too, we ultimately need to grasp the different reasons that lead people to place a high value on residential racial homogeneity). Yet when Laitin goes to test his model against data gathered in his surveys of the four republics, he is compelled to conclude that if "the tipping model relied solely on economic returns and probabilities for occupational mobility, these data present an insurmountable challenge" (1998: 254). The expected economic returns for language acquisition in the four republics simply do not map on to the findings about the willingness to engage in such learning expressed by Russian speakers in the different regions. The economic returns of change are, for example, worst in Latvia; yet it is the republic where Russian speakers are most eager to undertake linguistic assimilation. Sadly, the economic results do not display a nice pattern of inverse correlation, either. They just do not do much explanatory work at all.

What works best is a sense of "social distance"; in republics where Russians share a similar language and religion with the dominant group, as in Latvia (Indo-European language and Christian in religion), the Russians are most open to change. Where that social distance is greatest, as in Kazakhstan (non-Indo-European language and Islamic), they are least open. But how do we judge whether a language or religion is "distant" or "close" to another? We can do so only by learning about the particularities of the different languages and religions in question and determining how far, on balance, they seem to resemble each other in decisive respects, or to be distinct. Those determinations are prior to the construction of a data set measuring "social distance" and must be performed persuasively if any analysis of that data set is to be convincing. It appears, then, that on Laitin's evidence, the things we can only grasp by interpretively characterizing the identities in question – our accounts and judgments of substantive religious and linguistic similarities and contrasts – are much more crucial to explaining patterns of change than anything at all in the tipping model.

But Laitin sloughs off this evidence, saying it applies to conditions during the Soviet era, when there were few other reasons to learn the regional language other than personal interest. He argues that now, "Russian-speakers need to calculate more consciously the potential payoffs for learning" the local language, and so the tipping model will be of more help in the future (1998: 253). The focus has to be on this speculative future vindication, because Laitin's other data do not provide any strong corroboration for the tipping model, either. For example, Russian speakers face relatively low friendship losses among their "in-group" for learning the dominant language in Kazakhstan, and also have the best prospects for friendships among the Kazakh "out-group" (though those prospects are still negative); yet they are the least open to doing so. But the data do allow Laitin to contend that where out-group opposition to Russians learning the local language is low, as in Latvia, Russians are more likely to assimilate, and where losses of in-group respect for doing so are high, as in Kazakhstan, they are less likely to do so (252, 255–6).

These results are, however, weak in their own terms; there are contradictory findings, such as the unwillingness of Kazakhs to support inter-group marriages; and, in any case, they do little to explain either the *power* of in-group and out-group identities or the reasons for their *variations* in disapproval and receptivity. For those questions, Laitin turns in part to historical patterns he identified in the first part of his book, in part to current state policies to which he thinks those patterns have contributed. His analysis suggests that in those republics whose local elites had extensive opportunities for mobility within the Soviet Union, such as Ukraine, the rewards for learning the local language are not so high, in part because those elites are long accustomed to intermingling with Russian speakers and often speak Russian themselves, even though they do not wish it to predominate. Conversely, in republics such as Kazakhstan, where the Soviets governed on a more colonial model, with relatively little meaningful power of any sort for local elites, many Russians remain hostile to embracing what they still see as an inferior Kazakh identity. The rewards for learning the local language are also not great, perhaps because resentful opposition to Russian assimilation and mobility remain strong as well (though again, the data only ambiguously support that claim). In places like Latvia and Estonia, where local elites had real internal power but not many prospects to rise in the Soviet Union outside their borders, there tend to be more benefits to learning the dominant language and otherwise assimilating. Elites are comparatively receptive to Russians who do so, but they do expect business to be conducted in the local tongue. These different patterns of receptivity, reflecting identity-inflected power structures under Soviet and Tsarist rule, may help explain why in some places enough Russians learn the locally dominant language to launch a "cascade" (though Laitin has no data showing any actual cascades) (1998: 59–82, 353–61).

However much weight should be given to such explanations – again, it is "social distance" variables that really fare best in Laitin's data set – they clearly gain whatever power they possess not from the "tipping model" or any other formal theory. They rest instead on reasoning about the results of different historically shaped patterns of institutional structures and state policies, patterns that have both expressed and reinforced distinctive senses of political identity and allegiance. That reasoning, if borne out by empirical evidence more fully than Laitin's, would be sufficient to explain the prevailing patterns of linguistic assimilation without any reliance on the tipping model. If state institutions and policies and the behavior they foster make it sensible for most people to change, then most people will change; the intervening variable of how many others are changing when one does so is likely to prove more intercorrelated than independently causal. There does not, in the end, seem to be much "value added" provided to Latin's historically, institutionally, ethnographically, and survey-based analysis by his invocation of this rational choice model.[8]

I therefore conclude from Laitin's justly acclaimed example that grasping identity formation and change often requires ethnographically based interpretive understandings of different identities, in ways that can enable us to identify and comprehend phenomena like "social distance." It may also be advanced by an understanding of historical political processes of institution-building and power-structuring that have strengthened and modified certain existing identities, sometimes fostered novel ones, and often played strong roles in defining the relationships of those identities to various others.

[8] For a related critique from which I have benefited, see Motyl 2002 (237–41). I am grateful to Yitzhak Brudny for calling this article to my attention.

At this point, my fellow "interpretivists" and "historical institutionalists" among political science readers may feel cheered; but all political scientists may also have a mounting sense of anxiety. Political science may not be the best place for such "qualitative" work. Sociological and anthropological ethnographers, literary critics, and narrative historians may do it better. And even if we do it as well as it can be done, it may not satisfy our aspirations to achieve social scientific explanatory theories. Those theories are supposed to be, if not universal, at least meaningfully and usefully generalizable over some middle range of cases. They should not be "just" thickly descriptive case studies or narrative accounts of unique historical trajectories. Again, though I do not disparage any of these sorts of work, and indeed think they must play indispensable roles in any adequate analyses of the formation of political identities, I do not think political scientists need to despair of their theoretical ambitions so entirely.

III. The role of social science explanatory theories

Let me reiterate that though questions of political identity-making have not been central to the American political science profession's agenda through much of the twentieth century, they have come much more to the fore in recent decades. We have valuable political analyses of the formation of various identities, including nations, races, ethnic social movements, gender identities, religions affiliations, and others. These works show that a focus on "identity politics" topics need not necessarily be limited to ethnographies or narrative histories, useful as those are (see, e.g., Armstrong 1982; Pateman 1988; Marx 1998; Jung 2000; Marquez 2001). My message is in part that our discipline ought to be further encouraging and rewarding all these sorts of projects, because they productively address what ought to be among our most fundamental concerns.

Cathy Cohen's *Boundaries of Blackness* (1997), for example, uses a variety of research methods, including participant observation research, to develop a novel theory of how modern politics has reconstructed the identities of African-Americans, one that other researchers are finding applicable to different long-suppressed groups. She argues that in the wake of the victories of the civil rights era, black Americans find themselves placed in a position of "advanced marginalization" – included as civic equals to a greater degree than ever before, yet still confronted by a variety of more subtle forms of discrimination. Those circumstances create lamentable incentives for more prosperous and socially conforming members of American black communities to distance themselves from blacks who, because of their poverty, drug use, sexual orientation, or other traits, are seen as threats to the still-fragile civic status of African-Americans generally. The better off still define themselves as black Americans, but as ones who are now closer to being "normal middle-class Americans." The latter identity creates heightened pressures to limit the inclusiveness of the former. Such attention to the complex processes of political identity construction enacted both upon and by different groups of black Americans remains an infant enterprise in political science, but it is essential if we are to understand central dynamics in contemporary American politics.

Yet though Cohen's theory of "advanced marginalization" may well serve to illuminate processes affecting many groups in the US and elsewhere, it is still focused on those who have long faced official discrimination. I suggest that we seek to do even more. Grandiose as the notion of a "Unified Field Theory of Political Identities" may prove to be, it seems worth exploring whether we can generate such a theory or, better yet, rival theories that we can test against each other and then improve. I make this recommendation because my own studies of the power of racism and nativism in American history have compelled me to move to higher levels of theoretical generality in the quest for satisfying explanations of that power (Smith 2001, 2003).

A sketch of the framework I now use to think about political identities may help clarify, not what the "right" theoretical approach to questions of identity formation is, but rather the sorts of theoretical endeavors that I am urging. In various recent writings, I contend that senses of membership in a political community, a political "people," to which one is commonly understood to owe allegiance, are indeed political creations. They do not emerge semi-automatically from economic, demographic, sociological, geographic, linguistic, ancestral, biological, religious, or cultural characteristics. Rather, drawing on and constrained by such features of human life, elites and would-be elites craft many forms of political "peoplehood," many kinds of "imagined community" (not just Benedict Anderson's nations), by winning the support of a critical mass of constituents for their visions of political identities and memberships (Anderson 1983).

That crafting of senses of political identity takes place through two fundamental means: coercive force and persuasive stories. Many Native Americans are Americans because some of their ancestors were conquered; many African-Americans are Americans because some of their ancestors were enslaved; and all Americans are Americans because a critical mass of colonists wielded sufficient force to make the more powerful British imperial authorities decide that it was easier to let them go than to coerce them into continued subjectship. But mainstream political science is well equipped to grasp the role of coercive force in crafting political identities; that seems hard, solid, often measurable stuff. So in my recent work I have concentrated on "stories" or "narratives" of peoplehood and the role they play in winning some constituent allegiances voluntarily. That role is necessary, I assert, because no leaders ever have enough force to make all members of their community subscribe to their vision through coercion alone.

What kinds of stories inspire persons to embrace certain senses of imagined political community, memberships in particular political peoples? Like Laitin's, my model has three components. "Economic" stories offer material benefits for membership – ideologies espousing both capitalist and socialist political communities are stories of this sort. "Political power" stories promise personal protection and a share in great collective power – both republicanism and fascism are instances here. "Ethically constitutive" stories suggest that membership in a particular community is somehow intrinsic to who its members are, in ways that are ethically valuable. I view most racial, religious, ethnic, cultural, linguistic, historical, and gendered senses of community membership, among others, as such "ethically constitutive" accounts.

I argue next for the existence of an empirically falsifiable behavioral regularity; not all "real-world" visions of political membership capable of attracting significant support, from "Americanism" to "radical Islam" to those offered by the leaders of communal "eco-villages" (Jackson 2000), always blend together distinctive versions of all these three types of stories. All political actors and movements advance an account of the community they wish to shape and lead that includes identifiable economic, political power, and ethically constitutive stories. The emphases vary – some more pronouncedly economic, some more predominantly "ethically constitutive," and so forth – and of course the substantive contents of their stories differ as well; but the three types are always present. Furthermore, it is vanishingly rare for any political actor, party, or movement to succeed in institutionalizing their vision of political peoplehood in unqualified fashion, without any concessions to any opposing views. Instead, all political societies are and always have been made up by the compromised, aggregated results of contests among proponents of different visions of that community, each offering accounts that include some version of all three story types, in ways that are always undergoing contestation and change. That is the "politics of political identity formation" in a nutshell. Or so I claim.

Here, my concern is not to show that this framework is correct or even helpful, though I hope it is. Rather, I use it to suggest the sort of theory building in regard to political identities that seems advisable to undertake now. This is a framework, theory, or account that meets social science aspirations to heuristically useful simplification – since it relies as much as possible on elaboration of a few basic ideas; and universality – since it purports to identify fundamental elements through which political identities are constructed in all times and places. It is therefore necessarily an abstract theory; but it is not as abstract as models of pure instrumental rationality. Rather, it postulates that people do have basic, recognizable types of substantive interests: in material well-being; in some forms of political protection and therefore, necessarily, political power; and in senses of ethically constitutive identity. It also presumes that these needs can be understood and met in a great many ways.

In other writings I try to explain why, in particular, my relatively novel category of "ethically constitutive" stories is not only a coherent one, but also one that highlights discourses capable of playing vital roles in human political life that the other types of stories cannot play so well. Among other things, ethically constitutive stories can support beliefs that political memberships are morally valuable more easily than economic and political power stories can. Ethically constitutive stories are also harder to discredit via empirical evidence than economic or power ones, meaning that they can sustain loyalty even in materially bad times. For me, the framework helps answer my longstanding question about why racial conceptions have so often been politically potent in the US, even under circumstances where they were not clearly helping to rationalize economic exploitation and were costly to enforce. Ugly as they are to all but true believers, American racial stories have long conferred to many whites a sense of superior moral worth that no evidence could dispel, but that racial equality could threaten. In so arguing, I advance a specific understanding of W. E. B. Du Bois's famous claim for the existence of a "sort of public and psychological wage" for whiteness (Du Bois 1969: 700).

This "theory of people-making" also generates a number of other hypotheses about the circumstances in which such ethically constitutive stories are likely to come to the political forefront, thereby altering dominant senses of political identity, and when instead economic and political power themes are more prominent. It is reasonable, for example, to suppose that when leaders can hope to offer quite concrete economic and power benefits, they will stress them, and that they will turn to ethically constitutive accounts when they cannot, or when a people can only obtain material benefits by means that would seem illegitimate if not sanctioned by an ethically constitutive story. Furthermore, by arguing that we can always find political leaders blending the three types of stories with a content that is explicable in terms of what will appeal to the core constituents whose support they must win, this framework provides a general guide to identifying and analyzing the mechanisms of political identity formation in different locales (Smith 2002). Its categories can structure empirical inquiries, and those inquiries can then empirically falsify the key claims I am advancing; it may be that in many instances, some or all of these types of stories cannot be found at all, or cannot be plausibly seen to play anything like the roles I am suggesting.

My admittedly elementary framework for analyzing the "politics of people-making" is, then, a kind of general, transhistorical theory about some important dimensions of political identity formation. Because it has more specific substantive content, because it promises to help us understand the distinct political contributions of different types of political identity, because it can spark ideas about their strengths and limitations, both politically and normatively, it seems to me to do more work than Laitin's tipping model does in his analysis. But it is at the same time necessarily quite limited in what it can contribute to full understandings of the politics of identity making and remaking.

On its own terms, it requires us to engage in historical, empirical, and often ethnographic and interpretive work to grasp the kinds of economic arrangements that render certain sorts of economic stories more probable and successful; the contextual traditions and structures of power that make specific political power stories capable of inspiring allegiance; and the existing array of demographic, cultural, linguistic, and other identities that make the formulation of certain sorts of resonant ethically constitutive stories, and not others, possible. Those investigations are likely often to confirm the belief that people have conceived of their "economic interests," "political power interests," and certainly "ethically constitutive" interests, very differently in different times and places, and that different populations have done so within the same times and places. The patterns, in other words, are likely to show the same sorts of variations found in the ethnographically informed work of the researchers on game-playing, and in Laitin's own ethnographically and historically shaped arguments. These variations mean, again, that though the same general underlying processes may be operating, the specific explanations for and predictions of political conduct that "work" in those contexts will differ from those that "work" in others. Many political behavioral regularities that we can accurately and usefully identify in one historical and geographical context will not be applicable to times and places where identities and senses of interests have been constituted in ways that are substantively quite distinct. Because that is likely to prove to be a truth about political behavior, we need to use theoretical ideas and investigative approaches that can accommodate and develop that truth by explicitly leading us to undertake many kinds of detailed contextual investigation. We will not be well served by approaches that treat this apparent hard fact of political life as an impediment to scientific progress, to be evaded whenever possible.

We need, in short, to study processes of political identity formation in large part through interpretive, ethnographic, and historical methods of various sorts. But there is hope, I think, of developing theoretical frameworks that can enable us to knit those studies together to a greater degree than we are now doing. A final example: my colleague Ian Lustick has been exploring "agent-based modeling," which seeks to wed the precision of formal models with more contextual knowledge to illuminate processes of identity change in just the ways I am urging. He seeks to show, for example, how political actors draw on repertoires of identity available to them in particular contexts to "remake" themselves in circumstantially more optimal ways (Lustick and Miodwinik 2000). Here, too, I doubt that most of the work is being done by the formal elements of the analysis, as opposed to the documenting of the available identities and the circumstances that make the embrace of some new sense of self attractive. But this work is a clear instance of the somewhat less thin but still theoretically ambitious endeavors that seem appropriate today.

How much transhistorical insight and explanatory theorizing can be developed from such efforts remains to be seen. Perhaps we will eventually be able to generate plausible transhistorical theories of patterns in the historical evolution of all political identities. Perhaps we will simply be able to make sense of political identity formation at some more middle-range theoretical levels. But through such work we will, I think, be able to advance significantly our insights into these key features of political life.

If, that is, we choose to try to do so. My aim here has been to provide some ideas that may be of assistance to those already predisposed to work in these directions, and to move some who are not so disposed a little closer toward thinking that we should attempt such labors. If I have had any success at all in modifying anyone's sense of professional identity and purpose even a little bit, then, given how highly politicized our professional identities are, I will further contend that this chapter is actually a rough kind of admittedly non-randomized "field experiment" – one that has, through its impact of its professional story on its readers, in contrast to unaltered masses of non-readers, provided empirical support for its theoretical claims.

And if it turns out that no one is persuaded in the least by any of this, then, in the identity-preserving words of the late Gilda Radner, never mind.

REFERENCES

Anderson, B. 1983. *Imagined Communities: Reflections on the Origin and Spread of Nationalism*. New York: Verso.

Armstrong, J. A. 1982. *Nations before Nationalism*. Chapel Hill: University of North Carolina Press.

Barry, B. 2001. *Culture and Equality: An Egalitarian Critique of Multiculturalism*. Cambridge: Polity Press.

Bloom, A. 1987. *The Closing of the American Mind*. New York: Simon and Schuster.

Breuilly, J. 1982. *Nationalism and the State*. Manchester: Manchester University Press.

Bromwich, D. 1992. *Politics by Other Means: Higher Education and Group Thinking*. New Haven: Yale University Press.

Brubaker, W. R. 1992. *Citizenship and Nationhood in France and Germany*. Cambridge, MA: Harvard University Press.

Cohen, C. J. 1999. *The Boundaries of Blackness: AIDS and the Breakdown of Black Politics*. Chicago: University of Chicago Press.

D'Souza, D. 1991. *Illiberal Education: The Politics of Race and Sex on Campus*. New York: Free Press.

Du Bois, W. E. B. 1992 [orig. 1935]. *Black Reconstruction in America*. New York: Atheneum.

Garrett, G. 1995. "The Politics of Legal Integration in the European Union." *International Organization* 49: 171–81.

Held, D. 1995. *Democracy and the Global Order: From the Modern State to Cosmopolitan Governance*. Stanford: Stanford University Press.

Henrich, J., R. Boyd, S. Bowles, C. Camerer, E. Fehr, H. Gintis, R. McElreath, Michael Alvard, A. Barr, J. Ensminger, K. Hill, F. Gil-White, M. Gurven, F. Marlowe, J. Q. Patton, N. Smith, and D. Tracer. 2003. "'Economic Man' in Cross-cultural Perspective: Behavior Experiments in 15 Small-scale Societies." Available at http://www.hss.caltech.edu/~camerer/overviewpaper.pdf.

Jackson, J. T. R. 2000. *And We Are Doing it: Building an Ecovillage Future*. San Francisco: Robert T. Reed Publishers.

Jung, C. 2000. *Then I Was Black: South African Political Identities in Transition*. New Haven: Yale University Press.

Kim, C. J. 2000. *Bitter Fruit: The Politics of Black–Korean Conflict in New York City*. New Haven: Yale University Press.

King, G., R. O. Keohane, and S. Verba. 1994. *Designing Social Inquiry*. Princeton: Princeton University Press.

Laitin, D. 1998. *Identity in Formation: The Russian-Speaking Populations in the Near Abroad*. Ithaca: Cornell University Press.

Lustick, I. and D. Miodwinik. 2000. "Deliberative Democracy and Public Discourse: The Agent Based Argument Repertoire Model." *Complexity* 5: 13–30.

Marquez, B. 2001. "Choosing Issues, Choosing Sides: Constructing Identities in Mexican–American Social Movement Organizations." *Ethnic and Racial Studies* 24: 218–35.

Marx, A. 1998. *Making Race and Nation: A Comparison of South Africa, the United States, and Brazil.* New York: Cambridge University Press.

Motyl, A. J. 2002. "Imagined Communities, Rational Choosers, Invented Ethnies." *Comparative Politics* 233–50.

Newman, S. 2000. "Nationalism in Post-Industrial Societies: Why States Still Matter." *Comparative Politics* 33: 21–41.

Pateman, C. 1988. *The Sexual Contract.* Stanford: Stanford University Press.

Reed, A. 2000. *Class Notes: Posing as Politics and Other Thoughts on the American Scene.* New York: The New Press.

Rorty, R. 1998. *Achieving Our Country: Leftist Thought in Twentieth-Century America,* Cambridge, MA: Harvard University Press.

Schnapper, D. 1998. *Community of Citizens: On the Modern Idea of Nationality,* Séverine Rosée (trans.). New Brunswick, NJ: Transaction Publishers. Orig. 1994. *La Communauté des Citoyens: Sur l'idée moderne de nation.* Paris: Gallimard.

Schwartz-Shea, P. 2003. "Is This the Curriculum We Want? Doctoral Requirements and Offering in Method and Methodology." *PS: Political Science and Politics* 36: 379–86.

Scott, J. C. 1985. *Weapons of the Weak: Everyday Forms of Peasant Resistance.* New Haven: Yale University Press.

1990. *Domination and the Arts of Resistance: Hidden Transcripts.* New Haven: Yale University Press.

Slaughter, A.-M. B. and W. Mattli. 1995. "Law and Politics in the European Union: A Reply to Garrett." *International Organizations* 49: 183–93.

Smith, A. D. 1983. *Theories of Nationalism,* 2nd ed. New York: Holmes & Meier.

Smith, R. M. 2001. "Citizenship and the Politics of People-Building." *Journal of Citizenship Studies* 5: 73–96.

2003. *Stories of Peoplehood: The Politics and Morals of Political Membership.* Cambridge: Cambridge University Press.

Somit, A. and J. Tanenhaus. 1964. *American Political Science: A Profile of the Discipline.* New York: Atherton Press.

Turner, B. S. 1986. *Citizenship and Capitalism: The Debate over Reformism.* Boston: Allen & Unwin.

Walton, Hanes, Jr. and Joseph P. McCormick II. 1997. "The Study of African American Politics as Social Danger: Clues from Disciplinary Journals," in Georgia A. Persons (ed.), *National Political Science Review* 6. New Brunswick: Transaction Publishers.

Walton, Hanes, Jr., Cheryl M. Miller, and Joseph McCormick, II. 1995. "Race and Political Science: The Dual Traditions of Race Relations and African-American Politics," in James Farr, John S. Dryzek, and Stephen T. Leonard (eds.), *Political Science in History: Research Programs and Political Traditions.* Cambridge: Cambridge University Press.

Wendt, A. 1999. *Social Theory of International Politics.* Cambridge: Cambridge University Press.

4 Political science as a vocation

Anne Norton

Perhaps rigorous science makes for totalitarian politics, but democratic politics permits rigorous science. Perhaps democratic politics makes errors necessary, but discoveries possible. Perhaps ethics are fulfilled in the abandonment of principle for practice. Perhaps science is advanced less by duty than by desire. Perhaps an uncertain politics can provide better guidance than a certain science. These are the possibilities this chapter advances.

The ethic in which I was trained suggested that if we were careful, if we were responsible, if we took certain precautions, if we confined politics to a carefully demarcated sector (rather like the ghetto or the West Bank) we could do science honestly and honorably. The ethical conduct of political science required discipline. The ethical conduct of political science required the evacuation of politics or rather, that politics come into political science only as an object, never as a source of imperatives. This ethic, for which I have a lingering respect, argued that "the professor has other fields for the diffusion of his ideals" and "should not demand the right as a professor to carry the marshal's baton of the statesman or reformer in his knapsack"[1] (Weber 1949: 5).

This ethic directed us to science as a discipline, science as a duty. We were required to subordinate our political to our scientific ideals. We were impelled to go where our investigations led us. If our work led us to conclusions that we found politically distasteful, we were obliged to accept them. Our paths were marked out for us not by politics but by science. Method was discipline, an ascetism of the intellect, a means of maintaining the strict suppression of longing, of desire. Whether that desire was for a higher salary or a better world, it was corrupting. Discipline would make us neutral, and in that neutrality we could fulfill our duty.

Another, contending, ethic suggested that to seek such neutrality was to make political science a job, that we ought to be scholar activists,

[1] The ethic I was taught did not, however, do justice to Weber's nuanced understanding of the demands of politics and scholarship, most notably his appreciation of passionate scholarly advocacy.

conscious of our political roles, and performing those roles conscientiously, in accordance with our political principles. We ought to hone our political understandings, develop political strategies and tactics, and pursue particular political objectives. We ought to keep the faith. We ought to serve.

Method-driven political science is an instance of this first ethic, though not, perhaps, the ethic in its highest or most honorable form. In method-driven political science, we see how the desire for neutrality betrays itself. The experience of politics has shown us that doing one's duty and following an ethic of official neutrality – following orders – can be forms of corruption and betrayal. The experience of political science is equally revealing. A method-driven political science tends toward an incorporative totalitarianism in principle, and a series of anathemas, excisions and exclusions in practice. Methods require judgment and exclusions. One's method directs one to particular problems, to particular sources of information and techniques of analysis. One's method directs one not only to particular problems and practices, but also to particular people: to a distinct research community. The choice of one subject entails the neglect of others. The choice of one technique requires the refusal of others: a post-structuralist must refuse the exegesis of intent; a rational choice theorist must refuse psychoanalytic techniques. The conviction that one (that is to say, one's own) method is correct leads to proselytizing, imperialism, and coercive assimilation. If we believe that politics requires pluralism, if we look for ethical forbearance, then we must foreclose the exclusionary judgments methods (any method) will impel. Methods will not provide a justification of pluralism. On the contrary, they will provide imperatives for the elimination of other methods.

These objections to method are widely acknowledged, and, I believe, widely shared. There are few, I think, who retain their faith that methodological rigor can ensure political neutrality. Those who do have betrayed their principles, for the proof is an easy one, and the evidence overwhelming and often replicated.

Problem-oriented political science is an instance of the second ethic. The "problem-oriented political scientist" is the tamed domestic form of the "scholar activist." Both believe that the vocation of the political scientist is to act in the world for the improvement of the world. Both have no small confidence that they know what a good world would look like, and that they know what the advancement of a better world requires. Both believe (in defiance of Foucault, Derrida, history, and common sense) that their acts will follow their will, that they are able to determine the consequences of their actions, and control the uses of their work. Both move with the good will, the generous heart, the parochial judgment, and the intrusive magnanimity of the nineteenth-century missionary.

Mindful of the history of earlier missionary efforts, I have come to distrust the scholar activist. I am sympathetic to the view that political science is too abstruse and inaccessible, too specialized, too inclined to narcissism and navel-gazing; that it too often fails to examine that power which is its avowed object, too often abdicates engagement with issues for a vacuous professionalism. Yet it is exactly because I think political scientists should look more to the operation of power, that they should speak truth to power, that I oppose the primacy of problem-oriented political science.

Political science is a discipline in history without a history; a discipline ignorant of its own genealogy and unconscious of its place in a world historical context. Recovering the history of political science, and the place of political science in history, should make us suspect the utility and the ethics of problem-oriented political science. The history of that enterprise does not inspire confidence. The history of efforts, throughout the social sciences, to find solutions to social and political problems serves as a cautionary tale. Those moments when scholars have been called, and disciplines devoted, to the solving of problems have not been the moments when they showed themselves best. I intend to begin by looking briefly at three of these moments: Reconstruction, colonialism, and the Cold War.

Reconstruction is prior to political science as a nominal field but, as Robert Vitalis has shown, it is also a crucial site for the development of the discipline, particularly for the subfield of international relations. The early twentieth century saw the conscious adaptation of models of development created for, and enacted in, the American South (Vitalis 1998: 21; West 1992: 371–88). Solving the problem of poverty, whether at a national or global level, was understood by early social scientists as a question intimately tied to race. The inaugural issue of the *Journal of Race Development* declared that the United States "has as fundamental an interest in races of a less developed civilization as have the powers of Europe. The key to the past seventy-five years of American history is the continuing struggle to find some solution to the negro problem – a problem still unsolved" (Vitalis and Markovits 2002: 6). This was problem-solving at its most self-conscious. This also, as Vitalis has demonstrated, placed problem-solving at the originary site of the discipline's efforts to work in the world. The *Journal of Race Development* "was renamed the *Journal of International Relations* in 1919. Three years later it became *Foreign Affairs*" (Ross 1972: 414).

In many respects, Reconstruction is the most heartening of the moments I will put before you. It was impelled, at least in its first or best instance, by a commitment to right wrongs done to former slaves, to integrate them legally, socially, economically, into the American nation. The government committed significant resources of land and labor to the enterprise. For a moment, African-Americans entered the House, the Senate, the polity, as equals. For a moment, white and black Americans worked together to make a more perfect Union. For a moment, inequities of race and class were addressed together. There is great merit in the view Du Bois accepted, that "no approximately correct history of civilization can ever be written which does not throw out in bold relief, as one of the great landmarks of political and social progress, the organization and administration of the Freedmen's Bureau" (Du Bois 1996: 24–5).

In chapter 2 of *The Souls of Black Folk*, Du Bois gives a poignant account of the ambitions, hopes, and efforts of the Freedmen's Bureau. This was an effort to address, honorably, effectively, and systematically the problem of "that dark cloud that hung like remorse on the rear of those swift columns" (Du Bois 1996: 19). He reports the diverse, passionate, and practical efforts of the military, the missionaries, and all the "fifty or more active organizations, which sent clothes, money, schoolbooks, and teachers southward." He also reports the establishment of "strange little governments like that of General Banks in Louisiana, with its ninety thousand black subjects," the damage done by the failure of the Freedmen's Bank, the hazards of "paternalism" and most of all the complex of trials, constraints and distortions that leave us with "the large legacy of the Freedmen's Bureau, the work it did not do because it could not" (Du Bois 1996: 19, 37, 42). Some of the defects of Reconstruction were due to the ordinary defects of men and institutions: "too much faith in human nature," "narrow-minded busybodies and thieves," and the dilemmas and difficulties of land redistribution (Du Bois 1996: 24, 30, 26). Others, more fundamental in my view, are the defects of American liberalism and those of the emerging discipline of political science: an arrogant, evangelical imperialism, and an unrecognized parochialism.

Reconstruction aimed, as Saidiya Hartman's work has shown, not only at the reconstruction of the Union, nor the reconstruction of the South, but at the reconstruction of the freed slave as well[2] (Hartman 2002). Rather than an acknowledgment of rights, Reconstruction extended a series of instructions that would, if undertaken, make from the unpromising material of the freed slave a democratic subject. Manner, mores, and morals had to be remade. They were to be modeled not on an imagined American, or African-American of the future, but in the image of the white, Northern, Americans of the present. The freed slaves had to learn

[2] While I find Reconstruction the most poignant instance of this errant evangelism, examples can be found in virtually every moment (and at every ideological site) of American politics and science. Jane Addams and the settlement movement, the planned communities of Lowell and Pullman, and the labor-saving practices of Taylor and Ford are impelled by the same imperatives of universal improvement.

not only (indeed not primarily) to read, but also how to keep house, what they ought to wear, and the bodily habitus preferred by those who taught them. This was, as W. E. B. Du Bois, wrote, "the crusade of the New England school ma'am" (Du Bois 1996: 27). This instruction was well intentioned, generously offered, and sometimes well received, but it was assimilation at its most coercive. It established the norms and structural parameters for a new racial order, one in which the currency of freedom was the enactment of whiteness. Like the Mizrachim an Ashkenazi-dominated Israel met with Lysol spray and language lessons, those newly freed remained to be redeemed (Oz 1983: 34; Segev 1988: 215). As this suggests, these enterprises of redemptive assimilation are not peculiar to the United States.

As DuBois famously noted, the problem of the twentieth century was the problem of the color line, a line that extended not only across the United States, but around the globe. The desire to make real the freedom and progress which appeared only as announced goals manifested itself in both the domestic and international dimensions of the question of race. As Reconstruction had sought to establish the institutions necessary for the practice of freedom and the progress of development so too did the early proponents of development and modernization theory. In each case, universalist ideals were accompanied and undermined by the easy assumption of cultural superiority. Solutions to the problems of dependence and poverty presented themselves in the image of the beneficent modernizers. The modernizers not only brought the solution, they were the solution; for the standards by which progress was to be measured mirrored their understanding of themselves (Adas 1989; Coronil 1997; Escobar 1995).

The discipline of anthropology has recognized, as political science has not, the extent of its involvement in colonial enterprises, the degree to which anthropological research produced the institutions, tools, and mentalities that impelled and facilitated colonization, and the impact of these ethical and political failures on the discipline. Other social sciences, especially in the United States, have been slower to acknowledge their implication in colonial rule. Yet in this historical moment, the effects of a disposition to – and involvement in the structures of – problem-solving show themselves in a particularly vivid and troubling form. Consider the case of political theory. I am sympathetic to the view that political theory is, as a field, too often given to what Ian Shapiro has called navel-gazing, too preoccupied with presenting an appearance of erudition and inaccessibility, too much enamoured of arcane literatures, too prone to abstraction. The era of colonialism shows political theorists in another guise. In the government of India we find political theorists actively engaged with the world; developing solutions, as they saw it, to problems of governance, surveillance, poverty, and civilizational decline. They present no good object for emulation. "As a general matter," Uday Mehta has written, "it is liberal and progressive thinkers such as Bentham, both the Mills, and Macaulay, who, notwithstanding – indeed, on account of – their reforming schemes, endorse the empire as a legitimate form of political and commercial governance; who justify and accept its largely undemocratic and nonrepresentative structure" and who develop in that enterprise repressive techniques and institutions[3] (Mehta 1999: 2).

If the involvement of political science in the later stages of colonialism was limited by disciplinary parameters that placed "the native" beyond the discipline's boundaries, the Cold War offered a virtually unlimited license. Funding for research was supplied, in unprecedented amounts, by the CIA, the Department of Defense, and, perhaps more importantly, by myriad nominally private foundations with agendas tied to nationalist and anti-communist ideologies. The state permeated the academy. Research always reaches into the classroom, but here the reach of state and corporate interests was more direct. Many have condemned (often too easily) those scholars who joined forces with the Department of Defense and the Central Intelligence Agency in their efforts to forestall the spread of communist totalitarianism, or to profess and propound the ideals and practices of democracy. I will not rehearse the arguments against these, for they draw too heavily on persistent prejudices. Instead, I invite you to consider policies more likely to appeal to our ideals and preferences, those moments on the periphery of the Cold War, more celebrated in the academy, where the state turned its strength to efforts to alter access to education. One might consider, in this regard, the still-sacrosanct GI Bill and the Reserve Officer Training Corps (ROTC). These are not simple cases. The GI Bill countered, for some, the effects of class and race, extended the benefits of education and opened the academy to men once excluded from it. I have seen ROTC do the same. Yet the GI Bill also gave soldiers a privileged access to higher education that reflected (if it did not enhance) the subordination of women (Ritter 2002). The effects of this licensing of social privilege and reinforcement of gender hierarchies are still visible throughout the academy.

One might argue that these are far from the only examples of problem-oriented political science. One might set other examples against these, or argue that political scientists performed their work critically, with an eye

[3] We are now in the midst of another imperial moment, exacting an increasingly burdensome corvée in expressions of allegiance, and offering an extravagant recompense for willing clients. With the example of Bentham and the Mills before us, the unlovely spectacle of political theorists willing to solve the problem of Islam may be less surprising.

not only to the interests and ends of the state. I would argue, in response that however carefully the academy protects dissidents, however large the field of academic freedom it grants to researchers, political science will always serve the state and corporate interests best. It will do so because research requires money and these institutions have money in hand. The temptation to serve power will be most powerful, and most seductive, where the enemy is clear and the problem evident. Yet it is exactly at those moments that we ought most to distrust our judgment. Problem-oriented political science serves established interests best. The presence of heresy was a problem that called for inquiry, indeed, for an Inquisition. That Jerusalem remained in the hands of the Muslims was a problem that called for – and to our lasting cost, gained – global (that was to say, European) cooperation. Mad cow disease is a problem. Is meat-eating a problem? The oil cartel is a problem. Are SUVs a problem?

What are the problems with problem-oriented political science? It is bad for politics and bad for science. It encourages arrogance; persuading the young and uneducated (and occasionally the old and the erudite) that they can solve problems beyond their reach; that they can answer questions they do not fully understand. For those with a better sense of the depth of the problems and the profundity of the questions, it encourages quietism. Do not take up these problems, for they are, if not insoluble, beyond your reach. The foolish are encouraged, the wise disheartened. A political science preoccupied with problems suggests that the aim of study is to find answers. With these imperatives, under such conditions, it is easy for the arrogant and the uneducated to believe they have found them, and easy for those who see farther to feel undone. Quick conclusions are encouraged; study, consideration, reflection, and debate are not. Science is not advanced in this economy. Politics is harmed by it.

Think of the question Du Bois posed. "What does it feel like to be a problem?"[4] (Du Bois 1996: 3, 4; Prashad 2001). The discourse of problem-solving places a thin veneer of abstraction over a series of material relations. Politics is shielded behind this veneer. Problem-solving presents the political effects of material structures as technical problems, concealing the work of power. The discourse of problem-solving entails and enhances particular relations of power. The researcher is the subject who answers, others are made the objects of research, subjected to analyses in which they are spoken of and perhaps spoken for, but never speaking. The problem speaks to the researcher. The individual subjects who embody or enact or experience the problem are merely instances, and consequently not adequate to the problem which appears in its wholeness

[4] One might also ask Vijay Prakash's question, "What does it feel like to be a solution?"

only in and through the analysis of the problem-solving social scientist. One might paraphrase Gayatri Spivak's question, "Can these problematic subjects speak?" and answer no, these subjects cannot speak. The imperatives of problem-solving require that they be read, and read only in one way: as a problem that must be solved.

There are other troubling problems with problem-oriented political science. Weber recognized them in his essays on ethical neutrality and on the uneasily coupled vocations of politics and science. He wrote movingly of the ethical implications of the selective silencing of speech. "In view of the fact that certain value-questions which are of decisive political significance are permanently banned from university discussion, it seems to me to be only in accord with the dignity of a representative of science *to be silent* as well about such value-problems as he is allowed to treat"[5] (Weber 1949: 8).

Subsequent thought on these matters has made these responses seem less effective than Weber had hoped. The silence of the professor did not serve, as Weber had hoped, as a wall to keep the power out, but rather as a buttress to established authority. Neither advances in the freedom of the universities, nor advances in scholarship, have set the problems aside. On the contrary, ethical hazards became all the more evident when we realized the fraught relation of knowledge and power.

What have we learned? All science is political. All science is conducted in language, or, more precisely, in languages. These languages are collective and conventional. Their content conveys contemporary preferences, prejudices, norms, standards, and assumptions. Their structure carries, less visibly, more constitutively, and finally inevitably, the conventions of the group to whom it belongs. There are things we cannot say, and things we must say, if we are to be understood. This is politics. There are things we cannot do, and things we must do, if we are to be understood as scientific, as logical, as sane. These imperatives maintain the parameters of the political. But politics is not confined to this fundamental level.

Science comprises institutions and discourses. We have studied institutions. We have learned that institutions call identities and interests into being. The presence of funding for particular projects, the absence of funding for others, will ensure that in some (if not in many) cases, individuals will undertake research projects not because they think these are the most important, but because these are the projects that can be accomplished, or even because these are the projects that bring the greatest rewards. The power of the state is evident here, but that of private funding is no less to be deprecated. As Weber recognized, such a system

[5] The italics are Weber's.

"gives the advantage to those with large sums of money and to groups which are already in power" (Weber 1996: 7 n. 1).

Academics are arguably moved less by financial incentives than by intellectual passion and the ambition for recognition. These motives also operate within institutions. Reputation, honor, recognition are also parceled out within institutions, and it is within universities that intellectual passions are (for most of us) most easily pursued.

Foucault has shown how institutions can call not only categories but subjects into being. We are all familiar with Foucault's account of how the study of sexuality created new identities, new subjects, new pathologies, and new crimes. Knowledge worked as a form of power, productive power, that brought into being the invert, the fetishist, the sexual deviant. Foucault's very extensive historical researches give us compelling reasons, historical and theoretical, for distrusting a problem-oriented political science. Efforts to solve the problem of punitive cruelty and recidivism issued in new institutions, new cruelties, new tactics, new subjects, new criminals, and new problems: the prison, the panopticon, tactics of surveillance, delinquents, and delinquency (Foucault 1973). Efforts to solve the problems of deviant sexualities created them; proliferating categories of deviance and pathology, identifying deviant and pathological subjects and developing modes of treatment, categories of transgression, criminality and punishment. Whether one came to it through Foucault or by another means, the recognition that in solving problems institutions proliferate them is not new to us. Why ignore it?

Problem-oriented political science doesn't simply solve problems – it makes them. If political scientists choose to solve problems they will be wise to leave some of their colleagues in reserve to consider, more thoughtfully, and carefully, the problems they create.

Problem-oriented political science is prey not only to the effects of institutions, but to the effects of practice, persons, and particularity. Science is practiced by people who have political convictions, political principles, political objects, political strategies. Science is practiced in particular situations, particular nations, in the context of particular events. The particular products of science – ideas, inventions, insights, measures, techniques, theories, taxonomies – have particular political utilities, particular uses and salience in particular times, with regard to particular political events.

If we turn away from a problem-oriented political science, what do we turn to? If our disciplines, our institutions, our languages, our colleagues, and ourselves are imbricated with established structures of power, how are we to escape or amend them? If the ends and ideals, the immediate objects and the fantastic aspirations of politics are not to guide our work, what is?

First we must cease the pretense that we can work outside politics. If we continue to act as if objectivity and methodological neutrality were possible we are not acting ethically but acting the ethical: performing a fictional character. This seems to indicate a curious, shamanic faith in Baudrillard's description of the simulacrum: "the map that precedes the territory." By this formula, acting as if we were ethical produces a performance that may bring its anticipated referent into being. I would argue, however, that this is not simulation but dissimulation. Both Franklin and Machiavelli, those very rational actors, thought the discipline of virtue was an expensive, if ascetic, luxury. For less effort one might have the advantages of reputation and retain the flexibility, the room for maneuver, that ethical elasticity could offer. For those reluctant to resolve this ethical problem by evading it, recognition of our political character and effects would seem to be necessary at the outset.

We can no longer rely on Weber's guidance in the essays "Politics as a Vocation" and "Science as a Vocation." We have learned that science is political, and that politics opens itself to inquiry. For us, the imperatives of politics and science are fused. We are neither politicians nor scientists, but scholars impelled by the uneasily coupled imperatives of politics and science. What is political science as a vocation? How do we learn to learn for power? How do we learn to learn against power? Where do we seek guidance on how to study science, on how to learn? Where are we to find ethical guidance on where and when and how to serve power, and when to oppose it? Where are we to find the guidance necessary to navigate a science now visible as politics?

The work that gives the most daunting caveats also gives some reassurance to those who believe that science can act in and for politics. The recognition that power and knowledge are fused; that power breeds knowledge, that knowledge brings power and serves power, does not come alone (Foucault: 1980; 1972). Power incites its own resistance. The knowledge derived from and allied to that power incites the acquisition of critical and dissident knowledge. Systems of research developed to advance one research agenda will suggest others competitive with, and perhaps opposed to, the first. Power/knowledge thus produces not only knowledge from power, and knowledge that serves power, but also knowledge that undermines power, knowledge against power, and fissures within power.

What we have learned then (I learned it from Foucault and Ibn Khaldun), is that power incites resistance, and that structures fall impelled by that which serves as the engine of their expansion.

We cannot be too much reassured by this recognition. This set of effects is as inevitable as the first, but it is not quite equal to it. Look first at the knowledge that comes from power. This is the expected knowledge sought by institutions – by the state, by corporations, by foundations and universities. Much, indeed, most, critical and dissident knowledge remains within these institutions, constrained within their normative parameters, governed by the standards of the disciplinary regimes (scientific and political) in which it is produced. Even the knowledge that emerges to question these regimes remains under their sway, dependent on their validation for advancement, reputation, recognition, and employment[6] (Bourdieu 1987). Critical and resistant knowledge that places itself outside the normative parameters of these disciplinary regimes (that dispenses with the standards it challenges, rather than adhering to them while challenging them) will be rejected not only by those who adhere to those standards but also by most of those who question, challenge, and reject them. It will also, of course, find itself without funding, without employment, and with only stigmatized venues for its dissemination.

This condition is not a particularly happy one for power or knowledge, politics or scholarship, but it has its advantages. The slow pace of reform may be politically, if not scientifically, salutary. Gradually, yet inevitably, structures produce the conditions for their own correction. Change is conducted largely within existing institutions, through established methods and in accordance with extant norms and standards. Reform is gradual, disruption is slight. For scholars, scarcity might produce, of necessity, a commendable independence of mind, fortitude, and perhaps a higher degree of solidarity among scholars more than usually obliged to rely upon one another (Ibn Khaldun: 1967).

If problem-solving makes problems, if attempts to improve the world inevitably transform themselves into attempts at domination, if science tends toward the trivial and the totalitarian, where are we to turn for guidance? What might counter the tendency of both method-driven and problem-solving political science to domination?

In a famous essay aimed at people, like myself, who read the work of Lacan, Derrida, Irigaray, and Cixous, Habermas contrasted two enterprises that he saw as fundamentally at odds: problem-solving and world disclosure. Habermas sees problem-solving as the aim of philosophy (a perception that may surprise many readers of philosophy in general and many readers of Habermas in particular). World disclosure was the province of the literary; not only of literature, but of those he called "literary theorists" and sought to exclude from the province of philosophy (Habermas 1993: 185–210).

[6] One might describe this position using the term Bourdieu coined for a larger position with similar claims and temptations: "the dominated fraction of the dominant class."

This might be taken simply as a low moment in the work of an eminent theorist, the employment of a petty and all too familiar strategy – exclusion by fiat. We have all heard "that's not political science" used as an excommunicatory anathema meant to send enemies into exile, protect positions of privilege, and diminish the competition. In this instance the exclusionary binary has surprisingly revelatory effects. First, it reminds us how much would be lost if we turned from world-disclosure to problem-solving. Would we be obliged to forgo problems that admitted of no answer? Would we be obliged to silence the old questions: what is being? what is thinking? what is justice?

I might argue, for those of you still committed to problem-solving, that world-disclosure may be of some assistance in solving certain problems that living in the world presents. For feminists, the world disclosures of Beauvoir and Friedan, Butler and bell hooks, made politics possible. As a woman – not as a feminist, but simply as a woman living under circumstances that are occasionally hostile – I have found more of use to me in Lacan than in all the problem-oriented efforts of the "Women in Politics" section. The world Fanon and Memmi disclosed in the psychology of colonizer and colonized, the world Said disclosed in Orientalism, the worlds subaltern studies disclosed, have enabled entire peoples to begin to resolve problems that have plagued them for centuries.

Yet like W. E. B. Du Bois, faced with a problem-solving rival in Booker T. Washington, I am not content to let the argument rest on claims of utility alone (Du Bois 1996: 83). I am not prepared to trade Plato for policy statements, Wittgenstein for the Wilson Center, no, not Bhabha for the Brookings Institution, not even with thirty pieces of silver thrown in.

This is, after all, tantamount to betrayal – a betrayal of earlier, more daring, and more dangerous commitments. To abandon open fields and wild speculation for what? Forced labor? Are we all to be harnessed to the plow of problem-solving? Is each political scientist to be chained to an oar in the ship of state? Must we all do what some of us think best?

The call for a problem-solving political science is after all, exactly that, an effort to enlist all political scientists in a common ethic, if not a common effort. In this we see the all-too-imperial desire for a rule, a single standard that can measure all, a maxim that applies in every case. Problem-solving here reveals its likeness to the method-driven political science it condemns.

Whatever form it takes, this is an errant enterprise. There is more than one ethical life. There may be more than we presently recognize. Endorsement of this view does not require endorsement of an all-embracing relativism. On the contrary, it is a recognition built into the most intransigent and intolerant of religions. The priest and the father of a family, the mother and the mother superior, live ethical lives guided by radically different rules and imperatives.

We scholars are accustomed, in practice, to this ethical diversity. We belong to a more expansive world, with more demanding imperatives for inclusion. Scholars are often said to live apart, "in an ivory tower" isolated from the world. This is not so. Scholars live in more worlds than most people. We live in the ordinary world of commerce and media, of grocery stores and banks, day care centers, elementary schools, and gyms. We live in the world of the academy, and move, albeit on the periphery, in the worlds of our students. We live in other worlds as well – in the worlds of those we study. We know more lives and lifeworlds, in reading and in experience. Our worlds are large, and we know they differ from one another. Within these worlds we see different ethics made real.

These not all consonant with one another. They cannot be held to the same rule. There are radical, and sometimes dangerous differences in the practice of these ethics. There are those who testify to their faith, like some of my old colleagues, shocking students with stories of their conversions or how God spoke to them, or the certainty of their condemnations and their prescriptions for living. There are more secular preachers, whom some have called sanctimonious, who believe they know the good and what will bring us to it. There are activists and partisans, always persuading, recruiting new troops and sending them out into the field. There are, rarely and beautifully, the cool ascetics of scholarship, whose politics one never knows, who lay before you elegant and dispassionate accounts in which even the most diligent exegete can find no trace of partisan politics. There are the critics, the skeptics, the questioners, the determined agnostics of scholarship who live in a world foreign to them, and never fully find their feet there, though they swim well. We need all of these, not only because of the value they have in themselves, but because more grows in this fertile matrix. My university said *Crescat scientia, vita excolatur*. One might also say let lives grow, that knowledge may be enriched. The alternative is that which Weber rightly rejected. If we say there can be no scholarship in the class room, no expressions of political passion, no advocacy, then we say there can be "no Marxists and no Manchesterites," and soon after, no Marx, and no politics in Manchester[7] (Weber 1949: 7).

This approach is not without dangers. There are those who, if they succeed, will see (insofar as they can) that there are none unlike themselves,

7 Here Weber rejects the position of his teacher, Schmoller. For my part, I am indebted to my student, Arthur Bochner, whose critique of claims of political infallibility directed the conclusion of this chapter.

that the broad world that nourished them is folded away. If the Christians win, we will have a Christian academy. If Martha Nussbaum succeeds we will cease our reading of Judith Butler. They are a hazard, these zealots. I shelter them, and I ask you to do the same, not because I like them or I wish their projects well, but because I see possibilities in democracy and scholarship.

Democracy requires that we encompass the possibility (even the enactment) of our own annihilation. Democracy requires that at some moments, and in some respects always, one will cease to be. One will be not only ruled and overruled, but made absent. One must not only live with others, one must accept – in principle and quite often in practice – the sacrifice of one's own imperatives to theirs (Norton 2001). One must accept, even submit to those others whose ethics and imperatives one finds alien. This is a duty for democrats, it is a drive and a desire for scholars. Science is impelled by the drive to know the alien, the desire to comprehend the other.

A miscegenate, polyglot world is the world most consistent with the fundamental impulse and the ordinary practice of scholarship. All should be studied, all should be available for study. This is the world most consistent with the work of teaching, for knowledge grows best when error gets a hearing. This is the world most consistent with research – in the desire for the other, the alien, the unknown. This is, I think, the most fertile, the most productive world. And, as al Farabi said of the democratic city: this is the most beautiful.

> On the surface it looks like an embroidered garment full of colored figures and dyes. Everybody loves it and loves to reside in it, because there is no human wish or desire that this city does not satisfy. The nations emigrate to it and reside there and it grows beyond measure. People of every race multiply in it, and this by all kinds of copulations and marriages, resulting in children of extremely varied dispositions, with extremely varied education and upbringing. Consequently this city develops into many cities, distinct yet intertwined, with the parts of each scattered throughout the parts of the others. Strangers cannot be distinguished from the residents. All kinds of wishes and ways of life are to be found in it . . . the bigger, the more civilized, the more populated, the more productive, and the more perfect it is, the more prevalent and the greater are the good and the evil it possesses.[8] (al Farabi 1963: 51)

But it is in this city alone that knowledge and virtue may appear. The risks are too good to pass up.

[8] al Farabi 1963: 51. This is only the best of the ignorant cities, but ignorance is the condition of philosophy. I am indebted to Charles Butterworth for conversations on al Farabi, and for his translations of al Farabi's *Selected Aphorisms* (2001).

Scientific methods, and the priority of methodological rigor in scholarship mandate a world of common standards, uniform procedures, and unquestioned principles. The desire to solve problems and the attendant imperative to progress to a better world advance those who claim to know the contours of that world in advance. Both method-driven and problem-solving political science are driven to totalitarianism. Neither can foster an environment of contest, agon, and argument. Neither can support a politics of difference, division, and diversity. Without this politics, neither can be true to its own internal imperatives. The desire of science is to know the other, to pursue the alien, to exceed the limits of method and theory. The practice of problem-solving requires a constant attention to the demands of the particular. The practice of scholarship is the accommodation of difference. These practices and this desire permit the politics that makes them possible. In the shadows cast by the old Weberian rule, we can see that politics is served better by practice than principle, and that drive and desire may advance an ethic greater than duty.

REFERENCES

Adas, Michael. 1989. *Machines as the Measure of Men: Science, Technology, and Ideologies of Western Dominance*. Ithaca, NY: Cornell University Press.

Bourdieu, Pierre. 1987. *Distinction*. Cambridge, MA: Harvard University Press.

Coronil, Fernando. 1997. *The Magical State: Nature, Money and Modernity in Venzuela*. Chicago: University of Chicago Press.

Du Bois, W. E. B. 1996. *The Souls of Black Folk*. New York: Modern Library.

Escobar, Arturo. 1995. *Encountering Development: The Making and Unmaking of the Third World*. Princeton: Princeton University Press.

al Farabi, Abu Nasr. 1963. *The Political Regime*, Fauzi Najjar (trans.), in *Medieval Political Philosophy*, Ralph Lerner and Muhsin Mahdi (eds.): New York: Free Press.

2001. *Selected Aphorisms*, Charles Butterworth (trans.). Ithaca, NY: Cornell University Press.

Michel Foucault. 1972. *Archaelogy of Knowledge*, A. M. Sheridan Smith (trans.). New York: Pantheon.

1973. *The Birth of the Clinic*, A. M. Sheridan Smith (trans.). New York: Pantheon.

1980. *Power/Knowledge: Selected Interviews and Other Writings*, Colin Gordon (ed.). New York: Pantheon.

Habermas, Jurgen. 1993. "Excursus on Levelling the Genre Distinction Between Philosophy and Literature," in *The Philosophical Discourse of Modernity*, Frederick Lawrence (trans.). Cambridge, MA: MIT Press.

Hartman, Saidiya. 2002. *Scenes of Subjection*. Oxford: Oxford University Press.

Ibn Khaldun. 1967. *Muqaddimah*, Franz Rosenthal, (trans.), N. J. Dawood (ed.). Princeton: Princeton University Press.

Mehta, Uday Singh. 1999. *Liberalism and Empire*. Chicago: University of Chicago Press.

Norton, Anne. 2001. "Evening land," in *Democracy and Vision*, Aryeh Botwinick and William Connolly (eds.). Princeton: Princeton University Press.

Oz, Amos. 1983. *In the Land of Israel*. New York: Harcourt Brace and Co. 34.

Prashad, Vijay. 2001. *The Karma of Brown Folk*. Minneapolis: University of Minnesota Press.

Ritter, Gretchen. 2002. "Social Identity and Civic Membership in the American Constitutional Order: Some Lessons from the 1940s," paper delivered at American Political Science Annual Meeting, August.

Ross, Dorothy. 1972. *G. Stanley Hall: The Psychologist as Prophet*. Chicago: University of Chicago Press.

Segev, Tom. 1988. *1949: The First Israelis*. New York: Henry Holt. 215.

Vitalis, Robert. 1988. "Crossing Exceptionalism's Frontiers to Discover America's Kingdom." *Arab Studies Journal* 6 (Spring).

Vitalis, Robert and Marton Markovits. 2002. "The Lost World of Development Theory in the United States: 1865–1930." Unpublished ms., February.

Weber, Max. 1949. "The Meaning of Ethical Neutrality," in *Methodology of the Social Sciences*, Edward Shils and Henry Finch (trans.). New York: The Free Press.

West, Michael O. 1992. "The Tuskegee model of development in Africa." *Diplomatic History* 16 (Summer): 371–88.

5 The politics of policy science

Frances Fox Piven

No matter our growing sophistication about the manifold ways in which social and political influences penetrate and shape what is seen as scientific inquiry, the ideas of the policy science establishment seem fixed in an earlier era of unbridled confidence that science is science, even social science is science, and politics is politics. As for how science should be used in the formulation of social policy, the view is easy to summarize, based as it is on a simple model of rational decisionmaking. Presumably the objective is to alter some condition deemed a social problem, whether poverty, or insecurity, or dependency, or social disorganization. That objective is determined not by policy science, but by the political process. Policy science rather has to do with identifying the key variables or chains of variables that bear on this politically determined objective, and with identifying and measuring the effects of manipulating those variables. The effort is of course susceptible to great complication, as researchers try to cope with multiple intervening and interactive variables and with difficult problems of measurement. But complexities aside, the social scientist uses multivariate analysis in the effort to identify and measure ostensible cause-and-effect relations which are then the basis for scientifically informed interventions by government to alleviate social problems.

This dominant model in turn generated a dominant critique. For several decades, thoughtful social thinkers have agreed that this view of the uses of social science rested on an impossibly simplified view of real-world policy decisionmaking. In brief, the core of this criticism was that the simplification demanded by empirical research into cause-and-effect relations was inevitably incommensurate with the complexity and fluidity of the factors that are taken into account in real-world policy decisions (Merton 1957). The point is perhaps especially apt in matters of social policy, where complexity is compounded by the importance of informal and discretionary processes of implementation (Brodkin 1995). More recently, other commentators have taken up the thread of the dominant critique and proposed that abstract models of cause and effect, particularly economic models, should be shelved in favor of something called "local knowledge" that eschews generalization and quantification in favor of contingent and complex assessments tied to action in concrete settings (Schram 2002; Toulmin 2002; Flyvbjerg 2001; Scott 1998).

I am sympathetic with this critique. Still, the difficulty or even futility of the policy science effort as rational decisionmaking strategy is not to my mind the sole or even the main problem distorting the use of social science in social policy. Indeed, this discussion is often beside the point because social science is far less important as a decisionmaking tool in the rational pursuit of politically agreed upon goals, than as a resource employed by different participants, but especially by powerful participants, in the politics of social policy. I want to illustrate this by discussing the uses of policy science in the process of "welfare reform," specifically the passage of the Personal Responsibility and Work Opportunity Reconciliation Act (PRWORA) of 1996. The Act terminated open-ended federal funding for cash assistance, imposed time limits on the receipt of assistance, required work activity of those assisted and, by guaranteeing the states a lump sum for the program no matter how many families were aided, created a strong incentive for the states to administer the program in ways that would drive down the numbers of people receiving welfare.

Research as political performance

Decisions about social policy are of great moment to many groups, not only because they involve the expenditure of billions of dollars with large implications for taxation and government borrowing, and not only because those expenditures have widespread reverberations on capital and labor markets, on communities, and on the multiple providers of social services, but also because the policy process has become a kind of political theater, played to wide audiences, and increasingly important to politicians trying to sway voting publics.[1]

The dominant policy science model usually treats all of this as mainly external to policy science investigations.[2] Politics happens when the goals of a policy are decided, when the desirability or undesirability of the consequences of different interventions are assessed. To be sure, policy

[1] Davis B. Bobrow makes a similar argument when he says that "Demand for policy analysis . . . features purchase of rationalization and justification at least as much as reform" so that most policy analysis "amounts to competitive advocacy argumentation." Dobrow seems to think that all sorts of advocates generate policy analysis, and ignores the reality of differential political power and its effects on the market for policy analysis (2002: 28).

[2] Or wishes away the dependent political relationship, as when Dryzek calls for a "post-positivist policy analysis" that fosters "communicative rationality" and (citing Fischer and Forester: 1993) the "argumentative turn" in policy analysis (2002: 32–3).

scientists are often advocates of one or another course of intervention. But the basis for their advocacy is presumably their knowledge of cause and effect, which is what policy science is about.

However, this sort of separation of means and goals, of policy science and politics, does not match real-world policy politics where the goals of major contenders are typically unacknowledged, as well as conflictual and ambiguous, and in any case different from the proclamations which count as officially articulated goals. Moreover, key actors may be less interested in the ultimate consequences of policy alternatives, than in the multiple consequences of actions undertaken in the process of promoting and implementing the policy. In other words, in real-world politics, means are also goals, and goals are always also means.

These rather obvious observations about the complex and often unacknowledged factors involved in policymaking are part of the reason for the usual skepticism about the uses of social science as a tool of rational decisionmaking. But neither politicians nor social scientists are deterred. Policy science is, if anything, more prominent than ever in the policy process. The reason has less to do with social science research as a tool of rational decisionmaking than as a tool of politics. The variable relationships which are treated by the social scientist as causal sequences bearing on valued goals or outcomes are strategies in the play of power and performance that shape policy. The cause-and-effect relations between government interventions and desired outcomes which are framed as questions by the researcher become the stories told to promote a particular policy direction. In other words, policy science, the investigation of cause and effect, is not mainly a strategy to aid rational decisionmaking in the manner the policy scientist imagines. Rather, it comes to be used as political strategy, providing arguments or stories to elicit support from among those in a wider public whose participation can influence the outcome in a political contest.[3] In the process, the cause and effect arguments provided by the policy analyst become political claims, used for some audiences under some conditions, and perhaps not for others, and concealing the intentions of particular political contenders as much as they reveal intentions.

The human endeavor to understand the social world, whether scientifically or otherwise, is always deeply and pervasively shaped by social influences. That acknowledged, we can try to specify some of the distinctive ways that policy research is shaped to buttress the argument of a dominant regime. The contemporary reliance on multivariate analysis to

[3] It was Schattschneider 1960 who pointed out that the outcome of political conflicts is often determined by who is drawn into the audience.

explore causal relations is important, not because multivariate analysis is more susceptible to political use and abuse than other forms of inquiry, but because it is susceptible in distinctive ways. For one thing, while all social analysis is selective, multivariate analysis demands a sharply selective use of data. A narrow range of variables is identified and investigated, treated as independent or intervening or dependent variables, and represented by selected empirical measures. Meanwhile, most other "variables," or other facets of social life, are ignored, including those that are difficult to observe and measure, but including also those that do not suit the dominant political argument of the regime.

Another part of the problem is that while everyone knows that correlations do not establish causality, policy scientists, and the politicians who call on the authority of policy science findings, tend to overlook this elementary truth. Indeed, the whole point of the policy science exercise seems to be to proclaim correlations as causes. This is especially significant in the area of welfare policy because the paradigmatic case against social spending for the poor usually gains force in the wake of a rise in such spending, itself the result of a chain of events which includes economic dislocation and distress, and the social disorganization which often follows in the wake of profound dislocation. With sufficient methodological sophistication, everything is available to correlate with everything.[4] Of course, all efforts to understand the social world are inherently limited, and multivariate analysis is probably not worse than any other. My point is only to make explicit the ways in which these limitations facilitate the bending of research to suit the dominant political argument of the era.

Finally, there is the data problem. Multivariate analysis relies on numbers that presumably count or measure some aspect of our reality. Many of these numbers are provided by government. Indeed, contemporary governments produce cascades of numbers that tend to be treated as rock-hard and utterly reliable measurements of whatever it is they are said to measure. The Census is, so to speak, truth. And the more policy scientists rely on such numbers, the greater the investment in them, and so the truer they seem to become. Just imagine the blow to the study of American politics were critical scrutiny to be made of historical voting statistics – which have in fact always been susceptible to manifold strategies of manipulation by the political organizations that control elections – as representations of the political preferences of the citizenry!

Governments produce statistics for any number of reasons. Statistics describing population and resources are indirect measures of state power;

[4] This point is discussed in regard to welfare policy analysis in Piven and Cloward 1987.

they also facilitate the apportionment of taxes and representation; and as statistical measures are multiplied to include descriptions of the incidence of pathology, for example, they are the stuff used by government agencies of social control and welfare state administration (Hacking 1990).

But government statistics are not simply objective descriptors. Rather they depend on often arbitrary measurement and enumeration decisions. Since politics and policy both draw upon statistics, the decisions made about the methods of producing them are also the stuff of political calculation. Only recall recent disputes about the enumeration methods of the 2000 Census, disputes triggered because different methods were expected to produce different statistics, in this case regarding just how many inner-city minorities would be counted in the allocation of federal grants, and for purposes of representation.

Political regimes change, and when they do they give us an opportunity to scrutinize the relationship of policy science to politics. As new power blocs assume control of the state, new arguments are launched about the relation between state, society, and economy, and about the social policies that government should therefore promote. The main role of social scientists who study social policy has been to legitimate these new policies and new arguments by spinning theories about the causes of social problems and the appropriate remedies. In the nineteenth century, for example, eminent English social thinkers developed the paradigmatic case to justify the elimination of outdoor poor relief, summoning learned arguments to show that giving bread to the poor does harm both to the larger society and to the poor themselves (Block and Somers 2003). And in the late twentieth century, policy scientists were helping to develop the paradigmatic case against public welfare.[5]

Of course, the social scientists who do "policy science" are, like other social scientists, a diverse lot, committed to quite different perspectives on social policy as on other matters. But the rise of new political regimes alters the pattern of status and financial rewards, inevitably influencing the intellectual arguments that flourish. New cohorts of policy scientists come of age under the influence of the new incentives, and older cohorts gradually bend. In any case, those who do not bend, lose status and resources, and therefore also lose voice.

Think, for example, of the shift in the social policy research establishment from the 1960s to the 1980s. In the earlier period, under the influence of a liberal Democratic regime sharply shaken by the social

[5] See the collection of essays in Rueschemeyer and Skocpol 1996, for a series of historical accounts of the intimate connection between the rise of contemporary social science and modern social policies. See also Schram 1996 for a sharp critique of this relationship.

movements of blacks, youth, and women, the policy scientists located the causes of poverty in larger institutional arrangements. They traced the consequences of agricultural displacement, structural unemployment and low wages, and an inadequate safety net, and produced arguments and findings that justified the expansion of social programs. In the 1970s and 1980s, under the influence of Republican administrations and a mobilized business community, poverty was ostensibly still the concern, but now policy scientists located the source of poverty in social welfare programs themselves. The shifting arguments of policy scientists paralleled the shifting orientations of different political regimes.

Thus, in the 1960s, most social policy researchers followed the prevailing political drift in searching for systemic or institutional causes for a variety of social problems, from poverty to ill health to educational deficits to social disorganization to crime to mental illness. Research projects were launched, books and articles published, which singled out labor market conditions, institutionalized discrimination, distortions in housing markets, and so on, as causes of poverty and associated social problems. When researchers did turn to social policies as causes or independent variables, it was to consider the ways in which the failure to provide services or income, or distorted methods of service provision, might be contributing to the social problems associated with poverty. Some researchers even began to examine the complex interplay between such government policies as military spending or highway construction, and patterns of industrial and urban growth, tracing out the implications of these developments for labor markets and housing patterns, and ultimately for poverty and social marginalization. But in the 1970s, as the social movements which had given the 1960s their liberal cast subsided, a combination of white backlash and a business mobilization caused the political wheel to turn. The liberalization of social policies came to a halt, and then, gradually, the cutbacks in social spending began. As this happened, so did the liberal direction of the policy science establishment first come to a halt, and then shifted into reverse.

To be sure, the policy science establishment in the universities and research institutes could not refashion itself on a dime; arguments and findings had been published, reputations established, and big research projects were underway. Nevertheless, the intellectual transition began, catalyzed in the early 1970s by the creation or rapid expansion of conservative think tanks funded by business. The American Enterprise Institute, the Hoover Institution, the Cato Institute, the Heritage Foundation, the Manhattan Institute, all were established or expanded in the 1970s. These think tanks then became the home base for new policy scientists like George Gilder or Charles Murray or Robert Rector or Marvin

Olasky, who proclaimed that it was the liberalization of social policy in the 1960s that was to blame for social problems ranging from poverty to ill health to crime and so on. The underlying argument was, of course, the famous hypothesis of perverse effects (Hirschman 1991). Instead of reducing economic hardship and sustaining families, welfare policies and practices were themselves important causes of poverty and family breakdown, mainly because welfare ostensibly generates disincentives to labor market participation by poor women, or to stable two-parent family formation. In other words, a new political regime was providing resources for a new group of policy scientists, and a new or newly fashionable argument. And soon the turn in the political wheel that allocates recognition and research funds would carry along the older liberal policy scientists as well, and younger liberal policy scientists coming of age. The work they produced was qualified and their findings ambiguous, as I will soon explain; nevertheless, and however indirectly, they contributed to the new consensus.

Fame and funds were surely not the only reason for the turn in policy science. Much of the research conducted by the liberal policy establishment as the politics of the 1960s faded was in a sense defensive, provoked by increasingly strident attacks on the more generous social policies the era had produced. To answer those attacks, liberal researchers focused on the bearing of different aspects of social policy on the decisions of poor women to bear children or to marry, on their employment behavior, on their income, on the well-being of their children, and so on. These questions were investigated with the aid of the enormous data sets now maintained by government and research institutes which record empirical measures of family organization, income, employment, school achievement, crime, and so on. These data, and especially the income data, were based on measurement decisions susceptible to politicization, as I will later show. For now I want to direct attention to the propositions that guided the research. As any graduate student knows, empirical data are not self-revealing. The theoretical framework that organizes the inquiry, that specifies variables and suggests their relationship, is what matters. And the theoretical framework, the hypothesis of perverse effects, had in effect been conceded to the new political regime, and every study reiterated it. In this way, the real debate about the difficult theoretical issues dealing with the causes of poverty and changes in family forms was simply avoided.

I should point out that a good many of the researchers involved in the literally hundreds and hundreds of studies that were undertaken, at least those located in universities or at the liberal research institutes, undertook this research with the expectation that their findings would prove the hypothesis substantially wrong, or at least overstated.[6] Their research findings would therefore constitute a sort of answer to the rising chorus of critics of welfare. In effect, the findings of policy science would be used to shore up the social programs in the political world, perhaps with minor reforms to remedy whatever disincentive effects were identified. And more than a decade ago, writing with Richard Cloward, I thought that the very existence of a research establishment tied to the welfare state would provide some reasoned defense for the social programs against the growing conservative assault (Piven and Cloward, 1987). But these expectations did not take sufficient account of the limits of multivariate analysis framed by politically motivated questions.

In any event, beginning with the guaranteed income experiments of the mid-1970s (Munnell 1986), a vast body of research was produced investigating the impact of existing welfare practices on family decisions, and particularly on the incidence of out-of-wedlock births and single-parent family formation. Presumably, the interest in such research was prompted by the broad coincidence between the expansion of welfare and the rise of out-of-wedlock births, although closer scrutiny showed no such straightforward relationship over time.[7] In fact, although the number of children living in out-of-wedlock households had been rising steadily, from 5.3 to 28 percent of all births between 1970 and 1990 (Acs 1995), the value of welfare benefits which presumably were the incentive for this trend had been falling steadily, dropping nearly 40 percent in the two decades (*Green Book* 1994: 324), during which time the fraction of those children living in AFDC households also fell (Ellwood and Summers 1983: 3; see also Greenstein 1985 and Parrot and Greenstein 1995). Moreover, comparisons between states with sharply varying benefit levels did not fare much better. While there was some evidence that divorce and separation rates were modestly influenced by welfare benefit levels, and even some evidence that out-of-wedlock births to white women were influenced by state benefit levels, overall out-of-wedlock birth rates did not appear to be influenced by welfare policy.[8] No matter. As these sorts of findings accumulated over a period of more than two decades, the failure to produce

6 The important research centers were the Institute for Research on Poverty at the University of Wisconsin, Institute for Social Research at the University of Michigan, the Urban Institute at Northwestern University, and the Urban Institute in Washington, DC.

7 See Piven and Cloward 1987 (54–8) for a discussion of the formidable measurement problems that have been largely ignored in counting single-parent families and out-of-wedlock births.

8 On out-of-wedlock births, see Cutright 1973; Fechter and Greenfield 1973; Winegarden 1974; Moore and Caldwell 1977; Ellwood and Bane 1986; Rank 1994; Blank 1995; Moffitt 1995. On divorce and separation rates, see Hoffman and Holmes 1976; Ross and Sawhill 1975; Lane 1981; Bishop 1980; Wilson and Neckerman 1986; Rank 1994.

definitive correlations between AFDC receipt and family formation decisions only prompted more research, and more methodologically sophisticated analysis of the same hypothesis of perverse effects. And so the main political story line, that welfare leads to out-of-wedlock births, lived on, in an ironic way perpetuated even by the research which purported to investigate it.

The political fracas over the New Jersey "family cap" experiment illustrates how much more powerful is the force of the storyline implicit in selective variables than the mere empirical findings of multivariate analysis. Several years ago, the New Jersey legislature pioneered one of the first proposals to deny AFDC families a benefit increase after the birth of a child conceived while the mother was on welfare. The Clinton Administration granted New Jersey the waiver that was required to implement this departure from the Social Security Act. This occurred with a good deal of press attention. Meanwhile, since waivers are supposed to be scientifically evaluated demonstrations, Rutgers University was given a contract to study the impact of the experiment on rates of out-of-wedlock births to AFDC recipients. In the time since, the Rutgers researchers have released several reports on their findings, which show family benefit reductions have little or no impact on rates of conception by welfare recipients (Laracy 1995). No matter. Negative findings had none of the force of the proposition that welfare benefits are an incentive for out-of-wedlock births and its logical corollary, that reducing benefits will discourage such births. Subsequently, the family cap was introduced in numerous other state waiver plans, and then explicitly permitted in the PRWORA. There were no comparable efforts to evaluate the impact of the resulting benefit cuts on the well-being of recipient families who are subject to this experiment.

Research on the impact of welfare incentives on the work behavior and earnings of welfare recipient families followed much the same pattern. The guaranteed income experiments of the 1970s had shown a modest work reduction effort, particularly among women in two-parent households, when cash payments were guaranteed. In a way, this was an almost common sense finding: if government cash benefits are increased and made secure while earnings from work are not, some people will work less, in this case women who were already working in the home. Subsequent studies went beyond this purely economistic argument to explore what might be called the dependency thesis, that welfare receipt psychologically disables recipients and especially their children for labor market participation. Findings tended not to support the thesis that welfare caused "intergenerational" welfare dependency (Hill and Ponza 1984; Duncan and Hoffman 1986; Freeman and Holzer 1986). A parallel line of research made use of large longitudinal data sets to study patterns of welfare utilization over time, with findings that showed most women were relatively short-term users of welfare, although repeat spells were common (Rein and Rainwater 1978; Coe 1979; Bane and Ellwood 1983; O'Neil *et. al.* 1984; Duncan et. al. 1984; Ellwood 1986).

Meanwhile, still another line of policy research focused on the consequences for labor market participation and earnings of manipulating welfare practices in a variety of reform schemes designed to move women from welfare to work. The package of interventions is familiar: supervised job search or training or workfare assignments on the one hand, and sanctions in the form of benefit reductions or cutoffs for non-compliance on the other hand. Much of this research was conducted under the impetus of a series of "workfare" programs inaugurated in the 1980s and 1990s as opposition to cash welfare escalated. Some studies conducted by the Manpower Development and Training Corporation actually attempted an experimental design, comparing recipients exposed to the "experimental" reforms with control groups who were not. A reasonable conclusion from this research was that the workfare programs had only trivial effects in public policy terms. To be sure, a handful of the programs show statistically significant effects, but it is hard to see how statistically significant aggregate increases of, say, 6 percent in earnings or hours worked among the experimental subjects made much of a difference in policy terms. Even the purportedly most successful workfare program of all, in Riverside, California, only brought the average earnings of participants to $286 a month (De Parle 1996). Moreover, as in the case of reforms designed to reduce out-of-wedlock births, there were few studies of the consequences of the benefit reductions or welfare terminations experienced by the women who were sanctioned by the programs (some proportion of whom, it would seem reasonable to presume, failed to show up for job clubs or workfare for the simple reason that they were already working).[9] Instead, the governors of the states that introduced the harshest programs were left free to crow that declining state caseloads proved the programs were succeeding.

The 1996 reversals in welfare policy illustrate another deep flaw in the policy science model, which arises from misunderstanding of the role of lofty public goals in the policy process. In politics, professed goals are not

[9] For a study that is an exception, see I. Wolock *et al.* 1985–86. Such efforts as there were to track the fate of recipients removed from the rolls were undertaken by advocacy organizations. See, for example, National Employment Law Project 1996; Carlson and Theodore 1995; Pawasarat and Quinn 1995. There were also some studies of the fate of those affected by earlier state level cutbacks in the general assistance program. See, for example, Danziger and Kossoudjii 1995; and Nichols and Porter 1995.

first principles guiding action, as they are in the policy science model. Rather, the articulation of goals is itself a political strategy, a maneuver in the effort to win support and deflect opposition in a complex and conflict-ridden terrain. The role of David Ellwood, eminent welfare expert and government policy advisor, illustrates this point.

Ellwood had for many years done research on the impact of welfare practices on work and family behavior. In the late 1980s, he undertook to move beyond the usual policy science study of cause and effect to formulate policy proposals. In doing so, he would draw both on the findings of his research and his estimation of what the American public wanted from a reformed welfare system, which he thought was to reduce poverty while also requiring wage work.

Ellwood's book, *Poor Support* (1988) outlined his recommendations and rationale. It put forward a series of policy reforms that would presumably eliminate the perverse disincentive effects of existing welfare policy, largely by making work pay. The proposals included universal medical protection, an increased minimum wage, comprehensive child support payments backed up by government, and guaranteed minimum wage jobs. He also advocated time limits on the receipt of cash benefits, presumably to eliminate the perverse incentive of welfare, and to make the policy package palatable to an American public that believed in work and self-reliance.

In other words, this was a proposal based not only on policy science, but on a sort of political science that policy scientists find compatible, perhaps because it matches their simplified model of government decision processes. Policy science justified Ellwood's proposals for expanded health care, wage and other (non-welfare) income improvements, and so on, because these proposals were derived from presumed causal relations between policy and the behavior of the poor. But it was a judgment about politics that justified wrapping these benign initiatives together with a proposal for time limits since the expectation was that tough time limits would generate the public support that would carry the entire package.

The real political world in which social policy is fashioned isn't, however, a dialogue between "the policy reformer" and "the public." For one thing, as social policies unfold, they have important economic consequences through their impact on labor markets and tax levels, and the anticipation of those economic consequences activate organized market actors. Moreover, policy initiatives are carried forward by politicians, not experts, and politicians bring to any reform effort a preeminent concern not with a particular social policy, but with the uses of a particular policy initiative in coping with the politician's central preoccupations – with appeasing organized interest groups, garnering campaign contributions, and appealing to voters. As for the voters themselves, their preferences are not firm and they do not drive policy. Rather, voter preferences are in most policy matters ambiguous and shifting, and susceptible to political manipulation, especially on policy matters of which people have little direct knowledge. At the same time, there is a deep reservoir of antipathy toward the poor and toward welfare in American culture, which is not necessarily revealed by polls asking whether people think the government should assist the poor. And that antipathy can be activated when doing so serves a political purpose.

This is why the foray of policy scientists into welfare reform turned out so badly, indeed calamitously. Reform turned into its opposite, as politicians competed to trumpet the message that the AFDC program was somehow a major and growing problem for American society, and proposed cutbacks in benefit levels, draconian sanctions for disapproved behaviors, the elimination of federal responsibility in favor of the states, and of course, time limits on the receipt of benefits. Sophisticated research into policy effects aside, none of this seemed to have to do with even the most obvious features of AFDC, which was a relatively small program, and consumed only a tiny proportion of federal or state budgets. When measured as a proportion of the poor, or as a proportion of single parent households, the numbers on the rolls were in fact declining. Nor did it have much to do with the volumes of policy science research accumulated over three decades regarding the consequences of welfare disincentives in shaping the behavior of the poor.

What it did have to do with was an escalating competition between Democrats and Republicans to claim leadership in welfare reform. Clinton, very likely influenced by *Poor Support*, fired the opening shot with his 1992 campaign promises to "end welfare as we know it" with "two years and off to work."[10] On taking office, the president recruited Ellwood, along with Mary Joe Bane, another welfare policy scientist from the Kennedy School, to his welfare reform team, and the talk was of new programs in job training and job creation, health care and day care, all of which would presumably smooth the path to work (and all of which would presumably be evaluated according to the most sophisticated canons of policy science). But services and jobs cost money, much more money than was being spent on AFDC. This, no one who mattered, intended. As the estimated costs mounted, the service and job provisions shrank. In the end, the important feature of the Clinton plan was the two-year lifetime limit on cash assistance.

[10] Splitting the difference, one presumes, in Ellwood's 1988 proposal for a time limit of between 18 months and three years.

This outcome may have seemed fortuitous from the perspective of the policy scientists, but events suggest that it was inevitable given the power constellations in the contest over welfare that emerged. Almost immediately after the 1994 congressional election, the victorious Republicans tried to take the welfare issue back from the Democrats, proposing to eliminate federal responsibility for cash assistance to poor families in favor of block grants, thus giving the states more latitude in deciding how the money should be spent, or whether it should be spent at all. This was coupled with strict time limits, rigid work requirements, and the outright denial of aid to young women who became pregnant before they were eighteen.

As the contest for front position on welfare escalated, the president maneuvered for position. True, he responded to liberal advocacy groups by vetoing the draconian block grant proposal that emerged from the Congress in the winter of 1995–96, claiming it did not do enough to ensure that recipients worked. But he exerted himself to regain the initiative. He had learned during his 1992 election campaign, if he had not already known, that bashing politically powerless welfare recipients while celebrating family values struck a chord with the electorate. And so, as the slogan "it's the economy, stupid" dissolved in the failure of a feeble economic stimulus package, as profits rose but wages stagnated, as the scandals and missteps of his administration piled one on the other, Clinton approved one state waiver request after the other for programs which essentially mimicked the Republican welfare cutbacks, and mimicked the rhetorical argument, of Clinton and the Republicans alike, that a too lenient welfare system was leading to worsening poverty and social pathology.

By mid-1996, Clinton had approved waiver programs in thirty-seven states. State initiatives included not only time limits, but benefit cuts and sharp sanctions for a variety of disapproved behaviors, quickly resulting in a 10 percent drop in the welfare rolls. As Douglas Besharov commented at an American Enterprise Institute meeting in April of 1996, "Based on what happened in the last year, President Clinton can justifiably claim he has ended welfare as we know it." He had accomplished, Besharov said, "welfare reform on the cheap" without an increase in spending for child care or "a penny for job training." The "revolutionary" result was "an end to personal entitlement" (Center for Law and Social Policy: May 16, 1996). Not surprisingly, in mid-1996 as the presidential election approached, Clinton signed a Republican-designed welfare bill little different from the one he had earlier vetoed.

The 1996 law expired in 2002. As the Congress debated proposals to modify the program, President Bush proposed to toughen the work requirements, raising the percentage of those on the rolls required to be in work activity from 50 percent to 70 percent and from 30 hours a week to 40 hours a week, while eliminating the earlier practice of allowing the states to count those who simply leave the rolls in meeting their required quota of working recipients.

In the intervening six years, research on the impact of the different elements in "welfare reform" has burgeoned, funded not only by the federal government, but by the foundations, including the more liberal foundations. Indeed, a convening of policy scientists shortly after the PRWORA became law celebrated the new research opportunities since, as a result of devolution, fifty-one distinct welfare systems would now be available for comparison. Most of the studies that were undertaken fastened on the causal arguments that had informed the earlier debate, and primarily on the impact of welfare reform on employment and income. As the Congress debated, and as of this writing continues to debate, proposals to toughen the PRWORA, the findings of the policy scientists again became news. Policy science was summoned to celebrate the success of the law; the rolls were down, and many more single mothers were working, and working more hours. No matter the questions that might be raised by careful scrutiny of the findings; as in the earlier period, the research was invoked to support the main story line. Tough love worked! "We have shattered the cycle of dependency and we are creating a cycle of opportunity for millions of American families, thanks to the welfare reform law of 1996," said Health and Human Services Secretary Tommy Thompson. Robert Rector of the Heritage Foundation announced that the reform "had been an enormous success" and warned against any modifications that would restore "perverse incentives" (2002).

In fact, the findings of the volumes of policy research now available are riddled with problems. On the one hand, by selecting for investigation the correlation between welfare reform, employment, and earnings, the research ignored other variable relationships that might reasonably be considered relevant. What other outcomes should be investigated, such as the material hardships experienced by families, or indicators of family stress, or the well-being of children? And if employment and earnings had increased among former welfare recipients, could that be the result of other, non-welfare, independent variables, such as the remarkably low unemployment rate of the late 1990s and the rising wages of unskilled workers that resulted, or the expanded availability of wage subsidies, mainly in the form of the Earned Income Tax whose costs now far outpaced welfare expenditures?

And then there were the numerous data problems. The positing of welfare and work as mutually exclusive alternatives by the policy science establishment distorted the real behavior of recipients. Some research, particularly ethnographic research, showed that many women on welfare were in fact working, and indeed, that perhaps most of them were working, albeit so erratically and at such low wages that they survived only with the aid of welfare benefits. At least 40 percent of adult recipients worked while they were on welfare, others cycled between welfare and work, and they worked on average as much as other women with children (Edin and Lein 1994; Lein *et al.*2002; Harris 1996; Spalter-Roth and Hartmann 1994).

At least as important in distorting impact findings was the focus of research on welfare leavers. Women had in fact always left welfare, only to be replaced by new applicants, or returning applicants. The larger part of the decline in the rolls resulted not from the numbers who left welfare, but from the various new practices of diversion that prevented women from getting on the rolls. But no one was tracking the diverted or measuring their circumstances, a task which admittedly would have been difficult because of the absence of state records (Kaplan 2002).

The data problems emerged starkly with the completion of Census 2000 which reported that poverty was down among mother-headed families, and employment was up. Not surprisingly, these findings were treated by politicians and the press as the ultimate vindication of welfare reform. There were many problems with Census 2000, as researchers knew. But the bold claim, that poverty was down, was the most egregious, for it rested on a poverty measurement, contrived in the 1960s, that calculated the cost of a market basket of essential foodstuffs and then applied a multiplier of three to account for all other subsistence costs, including transportation, fuel, housing, and medical care. Experts estimate that an accurate accounting of the inflation rate of these factors would in fact require that a multiplier of seven be employed to calculate a poverty line equivalent to the 1960s. I cannot recall a single press report that pointed this out.

I should pause to give credit where credit is due. Some policy researchers exerted themselves to broaden the inquiry, to shift political attention from the preoccupation with the impact of welfare utilization on labor market participation and income, to a focus on the impact of labor market conditions, especially conditions in the low wage labor market, on the utilization of welfare (Edin and Lein 1994, Harris 1996). In effect, they challenged the proposition underlying most impact studies, that welfare use led women to drop out of the labor market, by substituting the proposition that labor market conditions led to welfare use. The results of this modest but important shift of focus reveals the influence of the hypothesis of perverse effects. The presumably perverse incentive effect attributed to welfare could as well be labeled the perverse incentive effect of a low wage labor market that pays too little, too irregularly, with few work-related benefits such as health care, for families to survive. Years earlier, Ellwood and Bane had interpreted multiple spells of welfare as evidence of a dependency syndrome. But some researchers now interpreted the evidence that women move back and forth from work to welfare as evidence of the insecurity of much low wage work. Or, as Kathleen Harris said, "Work among poor women should be viewed as the problem rather than the solution" (1996: 424).

In fact, the welter of policy science findings now available can be seen to raise serious questions about the new policy directions: between one-quarter and half of those who left welfare remained unemployed; many were only intermittently employed; and the earnings of those who worked were low; many who remained eligible were not receiving food stamps; the overwhelming majority were not being helped with child-care costs (Mezey, Greenberg, and Schumacher 2002). Some studies reported rising food insecurity and hunger, stressed and depressed working mothers, reduced parental time spent with children and troubled adolescents (Cook *et al.* 2002; Burton, Hurt, and Avenilla 2002; Lein, Benjamin, McManus, and Roy 2002; Burton, Skinner, Matthews, and Lachicotte 2002; Hofferth 2002; Zedlewski 2002; Moffitt 2002, Boushey 2002). Nevertheless, the storyline that emerged from this research was that the "work first" thrust of the new regime was a success, most welfare leavers were working and earning paychecks.

Conclusion

Did policy science actually matter? Certainly, it did not provide empirical evidence of the consequences which could be expected from these new initiatives. If anything, the weight of the evidence argued that slashing benefits would have negligible influence on out-of-wedlock pregnancy rates, for example. (It is only another irony that some of that evidence had been provided earlier in much-cited work by Ellwood and Bane themselves.) What policy science provided was not evidence to justify the waiver proposals as serious policy experiments, but the cloak of evaluation research to justify the new policy direction. As for the main administrative thrust of these reforms, ceding administrative and fiscal responsibility to state governments, almost nothing was known, at least nothing was known that followed the policy science model of basing decisions on empirical knowledge of cause and effect relations (Lurie 1996). It should be pointed out, however, that a good deal was known (albeit not in the policy science mode of multivariate analysis) about the history of state administration of relief. Almost all of what was known was in fact alarming, including the history of state and county administration of poor relief programs prior to the passage of the Social Security Act, and subsequent state and county administration of residual programs for those not covered by federal assistance, known as "general assistance" or "home relief" programs. Scarcely any research had been done on the consequences of the cutoffs that would result from time limits, or on the consequences of benefit reductions entailed by sanctions for non-compliance in the new program requirements. No one, but no one, policy scientist or politician, had any empirical basis for predicting what would happen to the women and children affected. No matter, the reforms went forward. As for the policy scientists, it is hard to see what role they played, except for lending an aura of scientific authority to initiatives shaped by the partisan contest to claim leadership in the politics of scapegoating the benighted poor.

A number of important points seem to me to emerge from this experience. First, most of the policy research focused on narrowly selective policy variables as "causes" of the family and work behavior which were presumably of concern. But the justification for this preoccupation in the larger body of social science theory is extraordinarily weak. Why, given what we know about the historic persistence of family forms over time and throughout the world, would social scientists look mainly at a relatively small welfare program as the cause of the transformation of established family forms in a generation? Moreover, how can this narrow focus be justified when we also know that family forms are changing among all classes, and in all Western societies? At the very least, the policy establishment should have complemented their preoccupation with welfare policy with a search for other deep and enduring influences on family formation decisions arising from changes in the larger culture, in social organization, and in the economy.

Second, why did a methodologically sophisticated policy science establishment allow correlations to be treated as causes? True, there was a broad if crude correlation of welfare expansion and family change. But that correlation paralleled multiple other changes which social theory would suggest were more significant. And if correlations were to be treated as causes, then why assume, as in the case of the inquiry into the relationship between welfare availability and labor market behavior, that the direction of causation ran from social policy to the labor market and not the other way round?

Third, there is the question, in a sense the ultimate question, of whether research mattered, one way or the other. This is a difficult query. Overall, the findings were ambiguous or insignificant. The response to small or ambiguous findings was to call for more research, more complex multivariate analysis. In the process, the story told by the research question, and by the selective variables the question demanded, became what was politically important. Research was significant not because applied social science was a component of rational policy decisionmaking, but because the causes and consequences investigated by the policy scientists became in politics stories about who or what was to blame for family changes and poverty. Even weak and ambiguous findings seemed to confirm the main political storyline about the perverse effects of welfare (Schram 1995).

What if there had been theoretically broader and more complex research? Would it have made a difference? Well, no one can be sure. Still, maybe it would have mattered if the policy establishment had directed public attention to the broad changes in culture, community, and economy which were leading to deepening hardship and disorganization. As it was, advertently or inadvertently, the social researchers reinforced the argument that welfare actually harmed, perhaps a little, perhaps a lot, the people it was designed to help.

Finally, I am struck by the historical naivete of the entire research enterprise. After all, the preoccupations of this generation of social policy scientists were, good intentions aside, closely similar to those of the nineteenth-century social critics who, first in England and then later in the century in the United States, justified the draconian cutbacks of poor relief demanded by a new class of industrial capitalists by worrying endlessly and publicly about the perverse effects of a too generous poor relief on the family and work behavior of the poor. They also did not worry about the cruel consequences of the elimination of relief.

As for the question of why policy scientists go along, the answer is too obvious to bear telling. Perhaps an anecdote that shows what happens when they don't go along will make the point. When the state of Wisconsin introduced one of the first "learnfare" programs some years ago, before the passage of the PRWORA, they required a federal waiver since the program, which would cut family benefits if teenage children were truants, was a violation of the Social Security Act. A waiver in turn required the justification of a social policy experiment. Accordingly, the program was to be tested in several counties, and the rather feisty Employment and Training Institute at the University of Wisconsin at Milwaukee got the contract to evaluate the impact of family benefit cuts on the school absences of teenage children. When early findings showed that truancy in fact increased instead of decreasing, the state government simply terminated the research contract. Then, under the new provisions of the

PRWORA, the learnfare program was continued and expanded statewide (Pawasarat *et al.* 1992; Quinn 1995).

REFERENCES

Acs, Gregory. 1995. "Do Welfare Benefits Promote Out-of-Wedlock Childbearing?" in *Welfare Reform, An Analysis of the Issues*, Isabel V. Sawhill (ed.). Washington, DC: The Urban Institute Press.

Bishop, John H. 1980. "Jobs, Cash Transfers, and Marital Instability: A Review and Synthesis of the Literature." *Journal of Human Resources* 15(3): 301–34.

Bane, Mary Jo and David T. Ellwood. 1983. "The Dynamics of Dependence: The Routes to Self-Sufficiency." John F. Kennedy School of Government, Harvard University, Cambridge, MA. Mimeo.

Blank, Rebecca M. 1997. *It Takes a Nation: A New Agenda for Fighting Poverty.* Princeton: Princeton University Press.

Blank, Rebecca M. and Alan S. Blinder. 1986. "Macroeconomics, Income Distribution, and Poverty," in Sheldon H. Danziger and Daniel H. Weinberg (eds.). *Fighting Poverty: What Works and What Doesn't.* Cambridge, MA: Harvard University Press. 180–208.

Block, Fred and Margaret Somers. 2003. "In the Shadow of Speenhamland: Social Policy and the Old Poor Law." *Politics and Society,* June.

Bobrow, Davis B. 2002. "Knights, Dragons and the Holy Grail." *The Good Society, Journal of the Committee on the Political Economy of the Good Society* 2(1): 26–31.

Boushey, Heather. 2002. "The Needs of the Working Poor: Helping Working Families Make Ends Meet." Testimony before the US Senate Committee on Health, Education, Labor and Pensions, February 14, http://www.epinet.org/webfeatures/viewpoints/boushey_testimony_ 2002014.html.

Brodkin, Evelyn Z. 1995. "The State Side of the 'Welfare Contract': Discretion and Accountability in Policy Delivery." Working Paper no. 6, SSA Working Papers Series, University of Chicago School of Social Service Administration, November.

Burton, Linda M., Tera R. Hurt, and Frank R Avenilla. 2002. "The Three-City Ethnography: An Overview." Paper presented at the Annual Meeting of the American Sociological Association, Chicago.

Burton, Linda M., Debra Skinner, Stephen Matthews, and William Lachicotte. 2002. "Family Health, Economic Security, and Welfare Reform." Paper presented at the Annual Meeting of the American Sociological Association, Chicago.

Carlson, Virginia L. and Nikolas C. Theodore. 1995. "Are There Enough Jobs? Welfare Reform and Labor Market Reality." Job Gap Project, Office of Social Policy Research, Northern Illinois University.

Center for Law and Social Policy. May 16, 1996. CLASP Update, Washington, DC.

Center on Social Welfare Policy and Law. 1996. "Comment Opposing Granting of Waiver for Wisconsin Works ("W-2") Demonstration Projected, Submitted to Donna Shalala, Secretary, Department of Health and Human Services. July 9.

Coe, Richard D. 1979. "Participation in the Food Stamp Program Among the Poverty Population," in *Five Thousand American Families – Patterns of Economic Progress*. Vol. VII, Greg Duncan and James N. Morgan (eds.). Ann Arbor: Institute for Social Research, University of Michigan.

Cook, J. T. *et al.* 2002. "Welfare Reform and the Health of Young Children: A Sentinel Survey in Six United States Cities." *Archives of Pediatric and Adolescent Medicine* 156(7) (July): 678–84.

Cutright, Phillips. 1973. "Illegitimacy and Income Supplements." *Studies in Public Welfare,* Paper no. 12, prepared for the use of the Subcommittee on Fiscal Policy of the Joint Economic Committee, Congress of the United States. Washington, DC: Government Printing Office.

Danziger, Sandra K. and Kossoudji, Sherrie A. 1995. "When Welfare Ends: Subsistence Strategies of Former GA Recipients: Final Report of the General Assistance Project." University of Michigan School of Social Work, February.

De Parle, Jason. 1996. "The New Contract With America's Poor. *New York Times, Week in Review*. July 28, 1996.

Dryzek, John S. 2002. "A Post-Positivist Policy-Analytic Travelogue." *The Good Society, A Journal of the Committee on the Political Economy of the Good Society* 2(1): 32–6.

Duncan, Greg J., with Richard D. Coe, Mary E. Corcoran, Martha S. Hill, Saul D. Hoffman, and James N. Morgan. 1984. *Years of Poverty, Years of Plenty*. Ann Arbor: Institute for Social Research, University of Michigan.

Duncan, Greg J. and Saul D. Hoffman. 1986. "Welfare Dynamics and the Nature of Need." Institute for Social Research, University of Michigan, Ann Arbor. Mimeo.

Edin, Kathryn. 2002. "The Myths of Dependency and Self-Sufficiency: Women, Welfare, and Low-Wage Work." Working Paper no. 67. Center for Urban Policy Research, Rutgers University, Princeton.

Edin, Kathryn and Laura Lein. 1997. *Making Ends Meet: How Single Mothers Survive and Low-Wage Work*. New York: Russell Sage Foundation.

Ellwood, David T. 1986. "Targeting 'Would-Be' Long-term Recipients of AFDC." Mathematica Policy Research, Princeton. Mimeo.

1988. *Poor Support: Poverty in America*. New York: Basic Books.

Ellwood, David T. and Mary Jo Bane. 1986. "The Impact of AFDC on Family Structure and Living Arrangements." *Research in Labor Economics* 7: 137–207.

Ellwood, David T. and Larry H. Summers. 1986. "Poverty in America: is welfare the answer or the problem?" in Sheldon H. Danziger and Daniel H. Weinberg, eds., *Fighting Poverty: What Works and What Doesn't*. Cambridge, MA: Harvard University Press, pp. 78–105.

Fechter, A. and S. Greenfield. 1973. "Welfare and Illegitimacy: An Economic Model and Some Preliminary Results." Working Paper, Washington, DC: Urban Institute.

Fischer, Frank and John Forester, eds. 1993. *The Argumentative Turn in Policy Analysis and Planning*. Durham, NC: Duke University Press.

Flyvbjerg, Bent. 2001. *Making Social Science Matter: Why Social Science Fails and How it Can Succeed Again*. New York: Cambridge University Press.

Freeman, Richard B. and Harry J. Holzer. 1986. "The Black Youth Employment Crisis: Summary of Findings," in Richard B. Freeman and Harry J. Holzer (eds.). *The Black Youth Employment Crisis*. Chicago: University of Chicago Press. 3–20.

Green Book. 1994. Committee on Ways and Means, U.S. House of Representatives. Washington, DC: US Government Printing Office.

Greenstein, Robert. 1985. "Losing Faith in 'Losing Ground.'" *The New Republic*. March 25.

Hacking, Ian. 1990. *The Taming of Chance*. Cambridge: Cambridge University Press.

Harris, Kathleen Mullan. 1996. "Life After Welfare: Women, Work, and Repeat Dependency." *American Sociological Review*. 61 (June).

Hill, Martha S. and Michael Ponza. 1984. "Poverty Across Generations: Is Welfare Dependency a Pathology Passed on from One Generation to the Next?" Institute for Social Research, University of Michigan, Ann Arbor. Mimeo.

Hirshman, Albert O. 1991. *The Rhetoric of Reaction*. Cambridge, MA: Harvard University Press.

Hofferth, Sandra L. 2002. "Use of the Survey of Program Dynamics for Studying the Implications of Welfare Reform for Families and Children." University of Maryland. Mimeo.

Hoffman, Saul D. and John W. Holmes. 1976. "Husbands, Wives, and Divorce," in *Five Thousand American Families – Patterns of Economic Progress*. Vol. IV, Greg J. Duncan and James No Morgan (eds.). Ann Arbor: Institute for Social Research, University of Michigan.

Kaplan, Thomas. 2002. "TANF programs in nine states: Incentives, assistance, and obligation." *Focus* 22(2) (Summer): 36–40.

Lane, Jonathan P. 1981. "The Findings of the Panel Study of Income Dynamics about the AFDC Program." Assistant Secretary for Planning and Evaluation, US Department of Health and Human Services. Mimeo.

Laracy, Michael C. 1995. "If it Seems too Good To Be True, It Probably Is." Report by the Annie E. Casey Foundation, June 21.

Lein, Laura, Alan Benjamin, Monica McManus, and Kevin Roy. 2002. "Economic Roulette: When Is A Job Not A Job?" Paper presented at the Annual Meeting of the American Sociological Association.

Lurie, Irene. 1996. "The Impact of Welfare Reform: More Questions Than Answers." Symposium on American Federalism: The Devolution Revolution, Rockefeller Institute Bulletin.

Merton, Robert. 1957. "Role of the Intellectual in Public Bureaucracy," in Robert Merton, *Social Theory and Social Structure*, p. 217.

Mezey, Jennifer, Mark Greenberg, and Rachel Schumacher. 2002. "The Vast Majority of Federally Eligible Children Did not Receive Child Care in FY 2000 – Increased Child Care Funding Needed to Help More Families." Washington, DC: Center for Law and Social Policy. Updated October 2002.

Moffitt, Robert A. 1995. "The Effect of the Welfare System on Non-marital Fertility." In *A Report to Congress on Out-of-Wedlock Childbearing*. Washington, DC: US Department of Health and Human Services, National Center of Health Statistics.

2002. "From Welfare to Work: What the Evidence Shows." *Welfare Reform and Beyond*. Policy Brief 13. Washington, DC: The Brookings Institution: 28, January.

Moore, Kristin A. and S. B. Caldwell. 1976. "The Effect of Government Policies on Out-of-Wedlock Sex and Pregnancy." *Family Planning Perspectives* 9: 164–9.

Munnell, Alicia H. 1986. "Lessons From the Income Maintenance Experiments." Proceedings of a Conference Held in September 1986. Federal Reserve Bank of Boston, Conference Report 30.

National Employment Law Project. 1996. "Statement Submitted to the Council of the City of New York Oversight Hearing on the Human Resource Administration's Work Experience Program (WEP) and its Impact on the City Workforce." 26 March.

Nichols, Marion and Kathryn Porter. 1995. "General Assistance Program: Gaps in the Safety Net." Center on Budget and Policy Priorities, March 14.

O'Neil, June A., Douglas A. Wolf, Laurie J. Bassi, and Michael T. Hannan. 1984. "An analysis of time on welfare." Washington, DC: Urban Institute.

Parrott, Sharon and Robert Greenstein. 1995. "Welfare, Out-of-Wedlock Childbearing, and Poverty, What Is the Connection?" Center on Budget and Policy Priorities, Washington, DC, January.

Pawasarat, John *et al.* 1992. "Evaluation of the Impact of Wisconsin's Learnfare Experiment on the School Attendance of Teenagers Receiving AFDC," University of Wisconsin-Milwaukee Employment and Training Institute, February 5.

Pawasarat, John and Lois Quinn. 1995. "Integrating Milwaukee County AFDC Recipients Into the Local Labor Market." University of Wisconsin-Milwaukee Employment and Training Institute, November.

Piven, Frances Fox and Richard A. Cloward. 1987. "The Contemporary Relief Debate," in Fred Block, Richard A. Cloward, Barbara Ehrenreich, and Frances Fox Piven (eds.). *The Mean Season: The Attack on the Welfare State*. New York: Pantheon Books.

Quinn, Lois. 1995. "Using Threats of Poverty to Promote School Attendance: Implications of Wisconsin's Learnfare Experiment for Families." *Journal of Children and Poverty* 1(2)(Summer).

Quinn, Lois and Magill, Robert S. 1994. "Politics versus Research in Social Policy." *Social Service Review*, December.

Rank, Mark Robert. 1994. *Living on the Edge: The Realities of Welfare in America*. New York: Columbia University Press.

Rector, Robert. 2002. "The Baucus 'Work' Act of 2002: Repealing Welfare Reform." *The Heritage Foundation Backgrounder, 1580*. September 3.

Rein, Martin and Lee Rainwater. 1978. "Patterns of Welfare Use." *Social Service Review* 52: 511–34.

Ross, H. L. and Isabel Sawhill. 1975. *Time of Transition: The Growth of Families Headed by Women*. Washington, DC: Urban Institute Press.

Rueschemeyer, Dietrich and Theda Skocpol. 1995. *States, Social Knowledge, and the Origins of Modern Social Policies*. Princeton: Princeton University Press.

Schattschneider, E. E. 1960. *The Semi-Sovereign People*. New York: Holt, Rinehart, and Winston.
Schram, Sanford. 1995. *Words of Welfare: The Poverty of Social Science and the Social Science of Poverty*. Minneapolis: University of Minnesota Press.
"'Poor' Statistical Accounting: Rewriting the Boundaries of Social Science and Politics in the Public Sphere of Cyberspace." 1996. Unpublished manuscript. June.
Scott, James C. 1998. *Seeing Like A State*. New Haven: Yale University Press.
Spalter-Roth, Roberta M. and Hartmann, Heidi I. 1994. "Dependence on Men, the Market or the State: The Rhetoric and Reality of Welfare Reform." *Journal of Applied Social Sciences* 18.
Toulmin, Michael. 2002. *Return to Reason*. Berkeley: University of California Press.
Wilson, William Julius and Kathryn M. Neckerman. 1986. "Poverty and Family Structure: the Widening Gap Between Evidence and Public Policy Issues," in Sheldon H. Danziger and Daniel H. Weinberg (eds.). *Fighting Poverty: What Works and What Doesn't*. Cambridge, MA: Harvard University Press.
Winegarden, C. R. 1974. "The Fertility of AFDC Women: An Econometric Analysis." *Journal of Economics and Business* 26: 159–66.
Wolock, I. *et al.* 1985–86. "Forced Exit From Welfare: The Impact of Federal Cutbacks on Public Assistance Families." *Journal of Social Service Research* 9.
Zedlweski, Sheila R. 2002. "Are Shrinking Caseloads Always a Good Thing?" No. 6 in series *Short Takes on Welfare Policy*. Washington, DC: The Urban Institute.

6 The study of black politics and the practice of black politics: their historical relation and evolution

Adolph Reed, Jr.

The study of black politics as an academic enterprise evolved within the terms of the segregation era regime, which was congealing around a principle of elite-brokerage as the common sense form of black political activity roughly by the 1920s. (Alain Locke's 1925 anthology, *The New Negro*, reflected the sensibility of this evolving politics and trumpeted it, and its ideological underpinnings, as a new consensus within the black opinion-leading strata.) I argue here that the academic subfield's formation in the context of that political common sense has been consequential because the discourses shaping academic inquiry have naturalized the premises and worldview of the Jim Crow era black political regime. To that extent, the study of black politics has been an agent in legitimizing and reproducing the regime's presumptions and the politics flowing from them, even into the present, when whatever efficacy it may have had, as explanation or as practice, is exhausted.

Making this argument, first of all, requires establishing the contours of the form of black politics that emerged as hegemonic with the consolidation of white supremacy in the South. I argue that that politics was in part an artifact of the white supremacist regime as well as an adaptation to it. That politics was not the only form of political action that enjoyed a significant constituency among black people during the three decades or so after Hayes-Tilden. Its undergirding political rationality did not necessarily delimit the only form of political activity that could plausibly have become dominant or efficacious. From this perspective, the segregation era black regime appears as the historically contingent product of the dynamic interaction of institutional and ideological forces operating within and upon the black population.

I then propose an admittedly schematic account of the genesis of the academic study of black politics, chiefly within political science, that traces the field's embeddedness within the common sense of black political discourse and practice in the segregation era – most crucially, the presumption that black political activity reduces to a generic politics of racial advancement that dissolves or transcends more particularistic interests and programs. This account focuses first on the pioneering work of Harold Gosnell and the prewar Ralph Bunche. I then discuss the field's central concern with leadership studies during the two decades after World War II and the shift to studies of urban politics and group empowerment in the 1970s and 1980s. I conclude this account with an examination of three contemporary studies that proceed explicitly from a concern to thematize the issue of interest differentiation in the postsegregation era: Katherine Tate's *From Protest to Politics*, Carol Swain's *Black Faces and Black Interests*, and Michael Dawson's *Behind the Mule*.

Black enfranchisement after the Civil War opened an exhilarating period of popular black political and civic participation in the South. Contrary to proslavery ideologues' characterizations and subsequent white supremacist mythology, freed slaves approached their new opportunities for exercise of citizenship rights with a zest. And this flood of participation was truly popular; the extent of political engagement among the freedpeople during the two decades after Emancipation may rank as one of the most exciting periods of popular political participation in American history, certainly in the history of the South.[1] By the 1880s a clearly articulated node that presumed elite priority in racial agenda-setting was visible, though by no means hegemonic, in black political debate. This node's prominence grew as the national Republican Party's retreat from support of black interests and the corollary ascension of white Redeemers to political power in the South after the mid-1870s altered the context of black political strategizing and threatened to curtail or nullify the gains made since Emancipation. This tendency, which converged around an ideology of racial uplift and *noblesse oblige*, also was steeped in views – commonly articulated among upper-status people on both sides of the color line – that the better classes should lead the society. The black variant was propelled in particular by theories that the race's genteel and cultivated strata were best equipped to organize and guide the development of the remainder of the population (Meier 1963: 3–82; Stein 1974–75; Toll 1979; Moses 1988; Reed 1997; Gaines 1996; Mitchell 1998).

This tendency was never uncontested in black politics (Stein 1974–75; Anderson 1988; McMillen 1989; Bacote 1955; Harlan 1972, 224–7).

[1] As an illustration of the extent of popular participation during the Reconstruction era, an inventory of black office-holding reveals that the nearly 1,500 black men who held office between 1865 and 1877 included scores of blacksmiths, farmers, barbers, coopers, carpenters, butchers, laborers, shoemakers, tailors, a baker, a basketmaker, cabinet makers, harness makers, a machinist, an upholsterer, a teamster, servants, wheelwrights, millers, millwrights, mechanics, an oysterman, and a tannner, in all well outnumbering the healthy complement of planters, teachers, ministers, and lawyers (see Foner 1993: 253–61).

However, the black population's disfranchisement and expulsion from civic life in the South, and the concomitant consolidation of a white supremacist order, in the two decades bracketing the turn of the twentieth century altered the context of black public debate in four ways that distorted the scope and participatory foundations of black political discourse sharply to favor the custodial and tutorial pretensions emanating from the race's elite strata. First, disfranchisement raised the cost of popular participation by eliminating the most accessible forms of political speech – voting and other aspects of electoral action. The effect, as in any polity, was to inflect black political discourse toward the perspectives and programs of those elements of the population with access to resources that could enable other kinds of expression – for example, newspaper publishing, business or other property ownership, and affiliation with institutions that provided visibility. Second, the danger that any public expression of opposition to the white supremacist regime would provoke state-sanctioned retaliatory violence imposed a yet greater potential cost and reinforced the voices of those who counseled accommodation to white supremacy, at least by silencing some of those who might have disagreed.

Third, the nature of the challenge posed by disfranchisement and the consolidation of the Jim Crow order in general exerted an understandable pressure toward a defensive and group conscious orientation that also buttressed elite interpretations and programs. In a civic discourse otherwise biased toward elite views, the imperative of articulating a general group interest necessarily would reproduce that bias in defining collective objectives and shaping political strategies. Post-Emancipation black political activity had always tended strongly to be group conscious. The terms of discourse routinely focused, reasonably, on pursuing the good of the race, with debate – shaped most significantly through public forums such as Union Leagues, local Republican Party organizations, Colored Farmers Alliance locals, school boards and other government bodies, lodge or church groups, as well as the ballot box – centering precisely on the programmatic content of that good for specific communities and the individuals constituting them in specific circumstances. The totalistic nature of the white supremacist threat, which in principle affected all black people equally, buttressed the impetus to craft singular racial agendas.

Civic exclusion and the centripetal reaction to the tidal wave of white supremacist counterrevolution combined to install as unreflected-upon common sense a political rhetoric that accepted synecdochic projection of the outlook of the race's articulate elite strata as the collective mentality of the whole. The result was a default mode of politics in which individual "leaders" could determine and pursue agendas purportedly on the race's behalf without constraint by either prior processes of popular deliberation or subsequent accountability.

Its impact on issues of accountability and constituency is central to the segregation era's fourth significant effect on the character of black political life. Not only did the segregationist regime hinder normal – i.e., electoral and similarly public – mechanisms of accountability, but to the extent that a strategic vision stressing the elite's custodial and tutorial mission became dominant in black politics, concern with accountability to a popular constituency easily was rendered a non-issue. A leadership of the best men need not trouble itself with deliberatively based ratification and input from a population in need of tutelage and custodianship. This disposition, in fact, was one of the reasons that advocates of accommodationism frequently were willing almost blithely to accept disfranchisement. Many of them shared the view that the uncultivated should not vote and were willing to endorse disfranchisement on class grounds so long as it applied to whites as well. White Democrats often exploited black elites' belief in the possibility of interracial elite collaboration to gain acquiescence in disfranchisement schemes, disingenuously encouraging the blacks' hopes that electoral restriction was not simply a ruse for a principally racist agenda.

Whether accommodationists' acquiescence in supporting the white supremacist juggernaut stemmed from an honest, if miscalculated, desire to make the best of a bad situation for blacks or from pursuit of narrower self-interest are the axes on which debate about black accommodationism characteristically has pivoted in both popular and scholarly discussion of black politics. These issues are crucial for judgment in a discourse that seeks to establish the place of particular leaders because they inform assessment of the quality of individuals' leadership. However, concern with the quality of black leadership as the basis for primary interpretation and judgment is the artifact of precisely the elite-centered way of thinking about black politics that became hegemonic as a result of disfranchisement and the white supremacist consolidation.

Of course, the accommodationist impulse was not the only, or even the most popular form of black response to the political climate of the late nineteenth century. In addition to the elite-brokerage model characteristic of accommodationism, other black people pursued very different kinds of strategies during the period between Hayes-Tilden and white supremacist consolidation. The Colored Farmers Alliance, linked with the Populist movement, claimed at its crest over 1,250,000 members (Woodward 1951: 192; Bloom 1987: 41). The Knights of Labor had 3,000 members in the late 1880s even in Mississippi, where white supremacist victory came earliest and was most thorough, and a majority of those Mississippi Knights were black, even with some local units organized and operated on a racially integrated basis (McMillen: 164–5).[2] The Knights were active and had significant black membership through much of the South during the entirety of the organization's active existence, and Sterling D. Spero and Abram L. Harris consider an estimate that the Knights' black membership at one point totalled approximately 60,000 as reasonable (Woodward 1951: 229–30; Fink 1983: 169–72; Letwin 1998: 75–80; Spero and Harris 1968: 42). In addition to the Knights, black people were involved during that period in labor and trade union activism all through the South (see Spero and Harris 1968; Letwin 1998; Arnesen 1991; Arnesen 1994: 7–12; Rachleff 1989; Stein 1974–75).[3]

Elite figures such as Frederick Douglass, prominent journalist and publisher T. Thomas Fortune and former Mississippi Congressman and Republican official John R. Lynch endorsed this sort of political activity during the 1880s, even as the distinct node of conservative, self-help uplift ideology – the strain from which accommodationism emerged – was forming in petit bourgeois black political discourse. However, this support was tempered by their other, pragmatically, and materially, based commitments to Republican Party politics (Meier 1963: 46–7; Meier and Rudwick 1968; McMillen 1989: 165, 373n; Stein 1974–75: 434–5).[4] Douglass counseled against independent political action; even as he

[2] It is important to note that whether local organizations were racially integrated was not likely to be as salient an issue during that period as it would become later, when commitment to organizational interracialism assumed greater symbolic meaning as an indicator of whites' reliability as allies.

[3] The case of New Orleans dock workers is particularly interesting as an instance of black workers enacting a relatively successful race-conscious strategy of carefully negotiated class alliance with white counterparts. Clarence A. Bacote reports that the Populists enjoyed considerable black support in Georgia through the early 1890s, noting that twenty-four black delegates, from eleven different counties, participated in the party's 1894 state convention and that blacks voted in substantial numbers – perhaps even a majority – for the Populist gubernatorial candidate in that year's election (Bacote 1955; Edmonds 1951; Gilmore 1996).

[4] Fortune apparently continued to support labor union activity, even after he retreated from the radicalism he displayed in the late 1880s around the Knights and became an ally of Washington, and presumably through his late life association with Garvey, though, as Meier and Rudwick note, the main focus of his politics after the 1890s was more conventionally petit bourgeois (Meier and Rudwick: 37). Fortune is an especially interesting figure in this regard in that his simultaneous commitments to labor organization and the ideal of a racial group economy may help to map the evolution of those two strains in black political ideology from the republican, free labor discourse that seemed the dominant intellectual frame of black politics before Hayes-Tilden. Lynch also held to his support of labor union activity, though he was never tempted by labor radicalism (see Thornburgh 1972: 81–2, 158; Lynch 1970: 476–8). Lynch actually invoked republican principles to oppose labor's organization of class-based political parties.

acknowledged the importance of organizing and agitating on a class and economic basis, as early as the 1870s he discouraged the National Negro Labor Union from supporting an independent Labor Party (Stein 1974–75: 431).

For Douglass and others, reluctance to deviate from a politics based in the Republican Party rested on a pragmatic argument that the GOP was the only dependable ally of black interests in national politics; they expressed fear that desertion of the party would lead to unmitigated Democratic power in the South and perhaps nationally, an outcome that could only be dangerous for blacks. Yet their interpretation of the paramountcy of Republican partisanship also coincided with the articulation of a petit bourgeois racial politics that reflected the diminished expectations of the post-Reconstruction era. What came to be known as Black and Tan Republicanism pivoted on delivering a shrinking black vote to support the national ticket every four years in exchange for control of a limited set of patronage appointments (Stein 1974–75: 432; McMillen 1989: 57–65).

Particularly in the context of the narrowed black public forum, it is impossible to disentangle honestly pragmatic assessment and dependence on patronage politics as causes of elite aversion to independent political action. Either way, the ideology of racial uplift reinforced the group custodianship that was Black and Tan Republicanism's foundation. Uplift, after all, presumed a mass population that was by definition not capable of steering its own programmatic course or mobilizing on its own behalf. Because its main resource was the claim to connection to a bloc of black voters, patronage politics was most compatible with goals and organizational strategies that emphasized racial collectivity. Uplift ideology's foundations in the era's theories of racial development reinforced that practical, material bias by giving it an ontological rationality. Securing patronage appointments for elite blacks appeared as generic gains for the race partly because of the premise that elevating the best men advanced the group as a whole, as well as because the power of scientific racism in broader American political rhetoric raised the significance of any black achievement, by any individual, as a challenge to the ideological foundations of white supremacy. The potential for felicitous pursuit of a politics that took class interest as synecdochic for race interest was overwhelming.

I have taken this excursion to the dawn of the segregation era to underscore the extent to which black political activity was differentiated prior to the white supremacist consolidation. Black people engaged, as individuals and groups, in many different kinds of political action. While black Americans' racial status always at least partly shaped the character of their political involvements, the goals of those involvements were by no means always exclusively, or even primarily, racial. Nor was their politics always enacted on behalf of a generically racial population. In addition to pursuit of more broadly racial objectives, black farmers organized and acted as black farmers; planters acted on their interests as planters; dockworkers acted as dockworkers – even as they linked those activities to the larger goal of improving the race's condition.

The victory of the custodial approach to politics warranted by the ideology of racial uplift narrowed the scope of what was understood to be appropriate black political action to endeavors undertaken on behalf, or in the name, of the race as an undifferentiated, corporate entity. This focus is exemplified in the common reference to "the Negro" as a collective subject.[5] Uplift ideology became hegemonic in black politics because its adherents' social position enabled them to establish that outlook's interpretive and strategic imperatives as the boundaries of legitimacy under prevailing political conditions, not because of inevitable historical forces or popular racial consensus. An intellectual consequence of that petit bourgeois hegemony was a forgetting; as the premises and practices of uplift ideology came to monopolize the effective substance of publicly visible black political discourse and action, they became naturalized, and held to exhaust black American politics as a thinkable category.

The emergence of the study of black politics as an academic subfield reinforced this process of naturalization by formalizing its outcomes through a discourse that presumes black Americans to be a corporate

[5] Even attempts to reflect strategic and programmatic variety have tended to presume the possibility of – and to seek – the racial least common denominator implied in this usage, as is indicated in the titles of key anthologies of black thought during the segregation era: *The New Negro: An Interpretation* (Locke 1968); *What the Negro Wants* (Logan 1944); *Who Speaks for the Negro?* (Warren 1966). The genesis of the Logan volume highlights an aspect of the contradictory character of this project. The idea for the volume originated not with Logan but with two white racial liberals at the University of North Carolina, W. T. Couch and Guy B. Johnson, who approached Logan to compile and edit the collection. According to Jonathan Scott Holloway, who describes the controversy, Couch at least was motivated by a desire to respond to Gunnar Myrdal's *An American Dilemma: The Negro Problem and Modern Democracy* that appeared the same year. Serious disagreements developed between Logan and Couch and Johnson over ideological balance in the volume, with the latter two hopeful that a leavening with conservatives would temper the volume's anti-segregationist tilt. (Ironically, Logan initially considered excluding radicals such as Doxey Wilkerson, but was persuaded by Couch and Johnson to adhere to a format of one-third radicals, one-third liberals, and one-third conservatives.) When the desired effect turned out not to be forthcoming, Couch balked and threatened not to publish the volume (see Holloway 2001: 6–9).

This episode captures the plausible foundation of the impulse to imagine a corporate black politics; after all, it is reasonable to expect that blacks would share almost universally an opposition to white supremacy and Jim Crow. It also illustrates the problem that the impulse itself is implicated in a style of politics that centered less on mobilizing collective black action than on communicating with white elites.

racial body politically. This presumption requires anchoring conceptions of the black political universe to a strategic least common denominator of pursuit of racial advancement in general, thus rendering observable political differences as tactical or superficial (e.g., owing to quirks of personality and individual character), or – if too great to be minimized in that way credibly – as reflecting deviant or inauthentic political commitments.

To be sure, a scholarly focus on activities consistent with the norm of advancing collective interests of the racial group has seemed appropriate in part because institutionalized discrimination and explicitly racial exclusion have been such major factors in black American life, particularly during the field's formative period. Concern with countering racism and its effects obviously has figured centrally in black political practice and rhetoric; that it should be a central element in scholarly accounts of that practice and rhetoric is in one sense, therefore, only the expression of a faithful representation of empirical realities. However, the field's foundation in a scholarly discourse that takes the racial least common denominator unreflectively as the conceptual orienting point for inquiry both artificially limits the content of black politics and distorts examination of its internal dynamics. Reduction of black political differences to disagreements about tactics, for instance, supports construction of Procrustean, ahistorical typologies such as the variations on the theme of a dichotomy of tendencies advocating protest and accommodation (or Harold Cruse's recasting of the dichotomy as between nationalism and integrationism) as the animating tension in black politics. As Judith Stein argues, with specific reference to August Meier's typology of black political thought as an oscillation between self-help and protest or accommodation and militance, this presents an impoverished view of black political activity "by excluding actions which encompassed but transcended racial goals, and at the same time misrepresents the racial movements themselves because of the abstractions which lie at the base of the theory" (Stein 1974–75: 424). Reliance on such an abstract and formalistic construction of black politics washes out significant substantive distinctions among kinds of political action. Again referring to Meier, Stein points out the conceptual inadequacy and potentially distorting effects of treating "legislators' attempts to pass civil rights bills, black laborers' entrance into unions, and social workers' negotiations to find jobs [as] equivalent acts" (Stein 1974–75: 424).[6]

[6] "Although these three initiatives, perhaps, formally sought integration, the basis of the actions and their relationships with other forces in society were very different. Similarly, the formation of the Colored Farmers Alliance and the creation of black professional organizations shared formal attributes of nationalism. But defining them solely in this way distorts the dynmics of the organizations – their class origins, the future course of their actions, and their impact on the social order" (Stein 1974–75: 424).

Early studies of black politics, such as Gosnell's *Negro Politicians* and Ralph Bunche's more engaged critical analyses during the 1930s and 1940s, demonstrate both the reasonableness of the field's thematization of the corporate racial subject and its ultimate inadequacy (Gosnell 1967; Bunche 1973; Bunche 1995a, 1995b, 1995c; Holloway 2002). Gosnell provided a subtle and textured account of the intricacies of black political development in Chicago during the first three decades of the twentieth century. To Gosnell's left, Bunche laid out a comparably sophisticated view of the tendencies operating within black politics at the time. Both described the variety of organizational forms and patterns of alliance that shaped black politics.

Gosnell examined the relations between the structures of black electoral mobilization – chiefly churches, press, economic and professional groups, and fraternal and uplift/racial interest groups – and specific interest configurations in the larger local political system, arguing that those relations were partly constitutive of black politics insofar as they defined the substantive payoffs, material and symbolic, of political action and shaped issue positions among black elites and opinion leaders (Gosnell 1967: 93–114).[7] He indicated, for instance, that the general prevalence of anti-black racism in the local polity created a very narrow context of options from which an instrumental logic drew black politicians toward commitment to the dominant machine and business interests. He also noted the potentially contradictory implications of that logical alliance, as when Oscar DePriest, the first black elected to Congress in the twentieth century, was forced in the second year of the Great Depression to retreat from his initial opposition to federal relief provision by his black constituents' extensive need for aid (Gosnell 1967: 371–2).[8] In Gosnell's view, mainstream black politicians "felt that it was necessary to go along with the white leaders in power in order that they might protect their own people against attacks on the Civil Rights Act and against intolerant movements" (Gosnell 1967: 372).

Gosnell recognized that a politics largely centered on obtaining the kinds of employment and appointments he described could not begin to be adequate as a material inducement for popular participation. He observed, though, that these gains had broader symbolic significance (Gosnell 1967: 368).

This assessment was no doubt accurate. Even black people who did not themselves gain materially from the regime's distributive arrangements

[7] Among the material payoffs of black electoral participation were first of all jobs and appointments, which Gosnell details (1967: 196–318).

[8] Gosnell also discusses black legislators' opposition to the trade union agenda, including the eight-hour day for women (Gosnell 1967).

certainly viewed its partial and incremental benefits as symbolic accomplishments for the race at large. Such sentiments had a rational foundation in the objective limitations imposed on black aspirations by the environment of racial subordination. In a context defined by limited possibility for generalized material benefits, advances by individuals understandably came to represent successes for the group as a whole, and could be regarded as steps along the way to more general benefits. Moreover, to the extent that racial subordination rested on ideological justifications alleging black incapacity, examples of black accomplishment took on significance as points of empirical refutation.

Nevertheless, the fact remains that the material benefits realized through the regime of racial politics Gosnell describes were sharply skewed toward the stratum of upper-status blacks who also propagated the outlook that imbued their material advances with broader symbolic meaning. Gosnell did not recognize that circumstance's potentially complicating effect on his account of black politics because his *a priori* acceptance of the premise of a corporate racial subject provided no conceptual space for him to perceive fundamental distinctions between individual and group impacts. Seeing black politics as reducible to the generic "struggle of the Negro to get ahead," the least common denominator of racial advancement, encourages a naive view of the relation between the differential benefits of the prevailing regime of racial politics and black elites' arguments for its optimality (Gosnell 1967: 11).

The hegemonic power of the presumption of the collective racial subject appears even more strikingly in Bunche's critical analysis of black politics. Bunche grounded his analysis of black politics during the 1930s and early 1940s on a rather doctrinaire Marxism. He was not blind to, or sanguine about, class stratification among blacks in general, and he was incisively critical in his estimation of the petit bourgeois class character of the programs of racial advocacy organizations. He also recognized the narrowness and inadequacy of a politics focused solely on eliminating discrimination and argued that its limitations reflected a petit bourgeois acquiescence to ruling class interests (Bunche 1995a: 52; Bunche 1995b: 74–80; 1995d).[9] Yet Bunche's analysis proceeded no less than Gosnell's from an undergirding conception of a corporate racial standpoint in politics. Indeed, the acuity of his judgment of racial advocacy organizations derived largely from his perception that their strategies poorly served the general group interest.

Bunche's Marxism accommodated this notion in an argument that the caste aspect of blacks' social position warranted a political strategy that

[9] A flavor of Bunche's Marxism can be gleaned from his 1935 essay, "Marxism and the 'Negro Question'" (Bunche 1995d).

would "defend and represent the basic interests of the Negro workers, of the Negro peasants, and of the Negro petty bourgeoisie, to the degree that the latter constitutes a progressive historical force" (Bunche 1995d: 45).[10] His notion of the collective racial interest was purposively linked to a specific teleology, the march to socialist revolution, without which, he contended, the struggle for racial equality could not ultimately succeed. By the lights of that teleology, in principle (though not yet in fact), "the Negro proletariat constitutes historically the natural leadership of the Negro people in its social struggle in American society" (Bunche 1995d: 42). Therefore, even when Bunche characterized black Americans as a corporate entity – e.g., the Negro people as a singular possessive – he did not construe that entity as an organic unit devoid of problematic differentiation or intrinsic tensions. His "Negro people" more closely approximated a coherent, but internally stratified population made up of different strata with different interests and agendas. Certain of those interests and agendas were consonant with the direction of the overarching teleology, and the strata identified with those interests constituted the "natural" or correct interests of the group.

For instance, a decade after his especially doctrinaire formulations in "Marxism and the 'Negro Question'," Bunche expressed frustration with the National Negro Congress for operating on a "united front" principle. He objected that "the Congress has proceeded on the assumption that the common denominator of race is enough to weld together, in thought and action, such divergent segments of the Negro society as preachers and labor organizers, lodge officials and black workers, Negro businessmen, Negro radicals, professional politicians, professional men, domestic servants, black butchers, bakers and candlestick makers." Judging that approach to be immobilizing, he argued instead that a "Negro Congress with a strong labor bias and with its representation less diffuse and more homogeneous in its thinking could conceivably

[10] Bunche (1995d) contended that because of "the characteristic caste status of the Negro people in American society, the Negro petty bourgeoisie is destined to play a far more significant and progressive social role in the struggle of the Negro people for emancipation and in the general social struggle than is the white bourgeoisie in the analogous situation. As a significant factor in the life and development of the Negro race, the petty bourgeoisie is second only to the Negro proletariat" (1995d: 40). Consistent with an orthodoxy of the time, Bunche maintained that the racial caste system was "the only form of an uncompleted bourgeois revolution in the United States today. In that respect it is similar to the liberation of subject nations for colonies" (43). This sort of interpretation, particularly the colonial analogy, would become a conceptual linchpin of post-black-power radicals' efforts to craft variants of a Third Worldist Marxism, or Marxist/nationalist hybrids. Bunche, however, insisted that black Americans were not a colonial people or a true "national minority" and emphatically opposed demands for black self-determination (36–7).

work out a clearer, more consistent and realistic program" (Bunche 1995b: 80).

Bunche was almost apologetic in his support of a politics of racial identity, noting:

The white voter ballots according to his individual, sectional, and group interests. The Negro votes on the basis of identical interests, but the social system of America dictates that he must give prior consideration to his racial group interests. So long as the dual social system persists in the United States, just so long must the Negro justifiably expect that political parties desiring his political support will devote specific attention to ways and means of Negro betterment in framing their platforms. The Negro finds himself in the uncomfortable position of decrying racial differentiation, while being compelled to demand it when important political policies are being formulated, in order to hold his ground in an uncongenial social milieu. (Bunche 1973: 104–5)

Bunche considered black racial solidarity to be entirely a function of the "dual social system" and was emphatic and rigorous in his rejection of race-based social theory as well as of what we now would call racial essentialism.[11] Nevertheless, he constructed his critiques of black politics around the conventional notion of a collective racial subject. As his own statement indicates, his doing so was an expression of the reality of black life at the time, the centrality of explicitly and generically racial subordination as a fetter on black Americans across the board.

Bunche's attempt to articulate a popular, class-based critique within the political discourse oriented around the idea of a corporate black interest throws into relief its interpretive shortcomings, most consequentially its reliance on a naturalistic rhetoric of racial authenticity as the standard of political legitimacy. His invocations of the collective subject sometimes hovered ambiguously between a formulation resting on the teleological assertion that the interests of the mass of Negro toilers, to borrow the parlance of the day, represented the true interests of the group and a formulation that seemed less mediated, and thus more akin to the naturalized notion of corporate racial interest inherited from uplift ideology. However, even Bunche's more complex construction of group interest also fitted uncontroversially within the discourse that cohered around the latter notion, which – in treating the racial interest as given and not the result of contestation or other contingent processes – inscribes elite race spokesmen as arbiters of group strategy.

Bunche surely would have objected to such a characterization. He certainly wanted no part of the belief on which the standard of racial

[11] His most extensive critique of racial thinking, including the idea of race itself, is in *A World View of Race* (Bunche 1936).

authenticity ultimately rests – that racial identity confers a single, common sensibility. His notion of authenticity was concretely grounded and based on an analysis of Afro-American social structure; unlike others, his was tied strategically to a long-term goal of broad social reorganization along very specific lines and did not claim to represent a universal racial consensus, and his notion of proper race leadership insisted on a major role for labor organizations (see Bunche 1995b: 82–3). Bunche was astute in identifying and challenging the limited class character of the main racial advocacy organizations' interpretations and programs; he drew attention to the programmatic and ideological significance of social differentiation and hierarchy among black Americans, which was an important intervention both analytically and politically. However, in positing a general racial interest that can be determined *a priori* – apart from processes of popular deliberation – his critique engaged strategic debate on terms that reproduced, most likely without intention, the prevailing rhetoric of racial authenticity and thus the fundamentally custodial political discourse within which that rhetoric was, and remains, embedded.

Bunche's reading of the practical entailments of the "caste" system that he described combined with his particular teleological commitments to lead him to accept, without hesitation, a frame for critical inquiry and debate that pivoted on determining authentic spokesmanship for the racial group. The working class articulated the group's true interests because it was the segment of the population whose social position and perspective advanced the larger struggle for social transformation that was necessary to win racial equality. That interpretation was consistent with the exigencies of race-solidaristic politics at he time. Systematic racial exclusion and discrimination required broadly based responses that united as much of the black population as possible, and crafting such responses was understandably the default mode for black political debate. However, in unskeptically proposing the black proletariat as agent of the entire group's true interests, Bunche tacitly accepted the most insidious foundations of the class-skewed ideology that he criticized.

As a result of his acquiescence in the rhetoric of authenticity, Bunche's critique did not confront the custodial model of political action that derived from and rationalized commitment to elite primacy for steering the affairs of the race. Nor was he attentive to the anti-democratic entailments of the presumption of corporate racial interest, which evaded potentially complicating questions concerning participation, legitimacy, and accountability by positing an organic unity as black Americans' natural state. And, of course, by operating within it he could not address the rhetoric of authenticity of race spokesmanship as a mechanism facilitating those evasions. To that extent, Bunche's radicalism left intact the core of the ideology of racial uplift on which the black petit bourgeoisie rested its claims to political and cultural authority. Bunche did not explicitly ratify this discourse, but his historically reasonable acceptance of it at least instantiated, and arguably helped to establish, a pattern of radical social science discourse about black politics that strained against but did not fully transcend the petit bourgeois ideological hegemony.

During the post-World War II period, largely as an expression of the changing political climate, the study of black politics began to congeal as a conventional academic specialty area. Robert E. Martin provided an early institutional account of the sources and effects of increased black electoral participation and officeholding. He described how the decades of black migration from the disfranchised South had culminated in the growth of black influence in national politics that buttressed the challenge to civic exclusion in Dixie (Martin 1953).[12] Thereafter, however, reflecting heightened awareness of black presence in national and big city politics, and the explosion of activist protest in the South, the subfield clustered initially around questions concerning the nature, dynamics, and composition of black political leadership (see Ladd 1966; Burgess 1962; Wilson 1960; Thompson 1963; Matthews and Prothro 1966; Cox 1950; Walker 1963; Cothran and Phillips 1961). Instructively, despite often careful and nuanced case descriptions of leadership styles and much attention to the construction of empirically based typologies, this scholarship generally bypassed or touched only superficially on questions of accountability and actual processes of leadership selection.[13]

In part this lacuna resulted from the fact that the intensifying struggle against Jim Crow had strengthened programmatic solidarity within black politics, as was attested by the anti-segregationist unanimity expressed in the 1944 anthology, *What the Negro Wants*. In that context, unless scholars were predisposed to pursue questions of accountability and legitimacy, there would be little reason for them to arise. In addition, the reputational method through which the case studies typically ascertained leadership was useful for identifying characteristics of those recognized as leaders, and thus for describing how the politics of race advancement was enacted pragmatically at the level of local polities. This method by itself, however, did not prompt inquiry into issues bearing on the

[12] Henry Lee Moon also examined the growth and significance of the black vote in national politics in particular (see Moon 1948).

[13] Ladd, in distinguishing between black leaders selected by whites and those selected by blacks, was among the few to approach the issue of leadership selection as a problematic issue. However, he did not follow up the implications of that distinction by considering whether and how both "in-group" and "out-group" selectees can be seen as black political leaders (Ladd 1966: 113–16).

shaping of strategic consensus and the processes of legitimation within black politics because its focus was on describing the settled outcomes of those processes; nor did it facilitate critical examination of how different interests were weighted in defining the concrete substance of the racial agenda in particular circumstances. In effect, this approach detached the idea of political leadership from participation and other group processes, and it presumed a simplistic identity of elite and non-elite interests and objectives.

These limitations on intellectual perception derived more from ideology than methodology, however. As Ladd's reduction of black leadership to the promotion of generic, unproblematized group interests indicates, the scholarly discourse that formed around inquiry into black political leadership reproduced without question the presumption of the black corporate political entity. It is telling that none of the main characterizations of black leadership considered either the possible existence – much less significance – of interest differentiation among black Americans or tension and ambiguity in the relation between leaders and led as factors shaping the dynamics of black leadership.[14] Perhaps most revealingly, this was true even of Oliver C. Cox, whose work and outlook were at least sympathetic to Marxism.

Cox's typology of black leadership began with the postulation of unitary racial interest: "Negroes, probably more than any other group of Americans, have an abiding common cause." He then defined the black leadership as "those who, through their energy and insight, have become advocates of means and methods of dealing with this common cause and whose advocacy has been significantly accepted by the group" (Cox 1950: 228). Like Bunche, Cox viewed that cause as partly transcending race;

[14] "Ethnic group leadership is, by definition, concerned with promoting the interests of the ethnic group . . . Negro leadership in the United States has been and remains issue leadership, and the one issue that matters is race advancement" (Ladd 1966: 114–15). Thompson defined the black leader as "one who for some period of time identifies overtly with the Negro's effort to achieve stated social goals" (Thompson 1963: 5). Burgess, a sociologist, was purely reputational in her formulation of leadership, defining a leader as "an individual whose behavior affects the patterning of behavior within the community at a given time" (Burgess 1962: 77). Matthews and Prothro, while noting that it is less reliable than office-holding as a barometer of political leadership, were forced by the realities of black exclusion also to opt for a simply reputational criterion: "In view of these special difficulties, *we shall regard those people most frequently thought of as leaders by Negro citizens as being their leaders*" (Matthews and Prothro 1966: 178 emphasis in original). Wilson as well relied on purely reputational means for ascribing leadership and explicitly eschewed attempting any more substantive definition (Wilson 1960: 10). He constructed his list of leaders, in contrast to Burgess and others, more from culling names from the public record than from snowball interviews; however, he seemed not to notice that that approach begs questions concerning the interests participating in determining which individuals would be recognized and projected.

he saw it instead as "an aspect of the wider phenomenon of political-class antagonism inseparably associated with capitalist culture. A principle involved in the process of democratic development is at the basis of the Negro's cause" (Cox 1950: 229). Again like Bunche, this larger purpose provided Cox's foundation for making judgments regarding the efficacy of types of black leadership. He proposed that the truest leadership was that which in charting a course for black people also advanced the historic "march toward democracy" (Cox 1980: 269). This led him to characterize protest leadership as "genuine," though "negative" in that its focus stopped short of challenging the basic social system, and radical leadership as both genuine and positive because "it is not only conscious of the source of the limitation of [blacks'] citizenship rights but also recognizes that the limitation itself is a function of the social system" (Cox 1980: 271).

Like Bunche and others before and since, Cox was motivated in part by a concern to theorize a coherent relation between black Americans' struggle against racial oppression and a Marxian strategy for more general social transformation. But his perception of class as a source of political tension did not extend to his analysis of black leadership. His account presumed the black population to be an inert entity, a singular mass to be mobilized by one or another strain of leaders.

The bias in Cox's critique toward presumption of corporate racial identity, as in the leadership studies generally, was a rational expression of the manifest realities of black politics at the time. On the one hand, regular mechanisms for popular selection of leadership were rare, and, on the other, the primacy and immediacy of the struggle against segregation imposed a practical coherence and commonality of purpose. The latter circumstance no doubt made the former all the less likely to arise as a matter for concern. A need for formalized systems of participation is only likely to be perceived where substantial differences are apparent.[15] The

[15] Following Drake and Cayton, Susan Herbst examines the contests from the 1930s through the early 1960s for honorary Mayor of Bronzeville in Chicago. These elections, sponsored by local black newspapers, fell somewhere between pure popularity contests and selections of men (one woman was elected, in 1959) held to represent community aspirations and values. Herbst notes that some holders of the title sought to use it to pursue ends linked to the politics of racial advancement, though the title was understood to be non-ideological and devoid of politics (see Herbst 1994: 71–95; Drake and Cayton 1944: 383). In any event, the enthusiastic response to the Bronzeville mayoralty contests suggests at least a willingness to take advantage of opportunities for participation in leadership selection. At the same time, though, the contest itself attests to the hegemony of the elite politics of racial advancement. The field of candidates was limited in principle to non-ideological "race men"; elections were candidate-centered and heavily favored prominent businessmen and professionals, and the prohibition on political programs actually reinforced the naturalization of conventional petit bourgeois

hegemonic status of the outlook that defined the sphere of black political action narrowly, as the elite-driven politics of racial advancement, also conduced to inattentiveness toward issues of participation, accountability, and legitimation. Social science reinforced this overdetermination because, in establishing the terms of inquiry around an unquestioning acceptance of the prevailing model as the quintessential facts of black political life, scholarly discourse inscribed it as the natural truth of Afro-American politics. To that extent, the study *of* black politics has been a moment in the dialectical reproduction of a class-skewed hegemony *in* black politics.

With the rise of black officialdom in the 1970s, Afro-Americanist scholarship in political science focused centrally on assessing the significance of the new, institutionally based black politics. This scholarship primarily examined such issues as the ways in which black public officials represent black political interests, the pressures operating on them to balance advancement of black concerns with other imperatives flowing from their formal positions and the alliances outside the race on which they depend, and the impact of systemic incorporation on black political styles. In particular, Robert Smith provided an institutional account of the development of the new black officialdom from protest politics in the 1960s. Matthew Holden and Michael Preston very early explored the constraints that the autonomous imperatives of political institutions and entrenched patterns of political behavior in local polities placed on the capacities of the newly elected black mayors. Lenneal Henderson examined the effects of similar constraints on the growing cohort of non-elected public administrators. This current of scholarship began, albeit most often tentatively, to examine structural characteristics and tensions endemic to black politics. However, Martin Kilson was virtually unique in extending the analysis to take for granted political differentiation among black Americans – along class, ideological, and other lines – as a normal and predictable feature of systemic incorporation (see Holden 1971; Preston 1976; Cole 1976; "Symposium . . ." 1974; Henderson 1979; Keller 1978; Nelson and Meranto 1977; Karnig and Welch 1980; Woody 1982; Eisinger 1984; Willingham 1986; Smith, R. 1981; Preston, Henderson, and Puryear 1987; Jones; 1987; Picard 1984; Kilson 1971, 1974, 1980; Reed 1986).

ideas of racial uplift as a non-controversial black agenda. In addition, the fact that the office carried no institutional power or resources militated against linking candidacies to durable poitical alliances or membership-based organizations. According to Herbst, *The Defender* newspaper's leadership saw the contest as a way to create backchannel leverage to City Hall, but the overarching context for that objective remained the elite politics of race advancement.

As black officeholding and governance became more established, scholars naturally enough began to consider its efficacy. Albert Karnig and Susan Welch, Peter Eisinger and others sought to determine the policy impact of the new black officialdom. The overlap between the black politics and urban politics specialty areas grew, as scholars like William E. Nelson, Jr., Browning, Marshall and Tabb, Clarence Stone, Larry Bennett, and James Button probed at the outcomes of the new minority politics (Nelson 1987; Browning, Marshall, and Tabb 1986; Stone 1989; Bennett 1993; Button 1989; Reed 1987, 1988; Judd 1986; Picard 1984). Nelson, Stone, and Bennett were exceptional, however, in problematizing the differential distribution of the benefits of black officialdom within the black constituency. More typically, this issue did not arise. Instead, the main line of scholarly discourse assumed, as had been the norm in the field since Gosnell, that racial benefits delivered by black officials were indivisible, that even benefits to individuals – such as public procurement contracts or executive appointments – accrued equally, if differently, to the black population as a whole.

That assumption was not without empirical basis. Black citizens by and large no doubt did regard gains made by black individuals, especially in the early years of postsegregation era black governance, as having larger racial significance. As in Gosnell's Chicago of the 1920s and 1930s, to many people those individual benefits symbolized generic racial accomplishment and a promise of more broadly disseminated concrete gains to come. The rhetoric of racial advancement that undergirded the rise of the new black political regime encouraged and reinforced such views. But scholars' acceptance of that ultimately elite-driven interpretation simply as transparent fact also contributed to its persuasiveness as ideology and rationalization of class privilege. Even criticisms of the inadequacies and unrealized potential of black governance were couched in terms that presumed the existence of a coherent black community agenda and that most often absolved black officials of agency by stressing their powerlessness and dependence.[16] The presumption of corporate racial interest continued to shape the field's conceptual foundation.

Perhaps the most striking contemporary indication of that presumption's persisting force is that it frames recent scholarship explicitly concerned with examining black political differentiation. Studies by Katherine Tate and Michael Dawson inquire directly whether increased

[16] Jones' "Black Political Empowerment" (1978) is an instance of the former tendency, though not the latter. Ironically, leftists may have tended more than others toward implicitly absolving black officials of responsibility for their actions by emphasizing their powerlessness. See, for instance, Rod Bush's *The New Black Vote: Politics and Power in Four American Cities* (1984).

systemic integration and economic mobility since the 1970s have had a centrifugal impact on black political behavior and attitudes. Carol Swain, in a study centered on the quality of Congressional representation of black interests, declares at the outset that black Americans' political interests are "varied and complex." However, each author proceeds by assuming the primacy of corporate racial interests (Tate 1994; Dawson 1994; Swain 1995).

Tate anchors her inquiry conceptually to an uninterrogated notion of "the black vote," or the "new black voter," as the baseline expression of black political activity (Tate 1994: 9). This formulation, which, as I have noted above, has its origins in the elite brokerage regime of Black and Tan Republicanism at the dawn of the Jim Crow era, naturalizes the idea of a unitary racial voice. It biases interpretation toward assigning both logical and normative priority to high racial cohesion as the standard from which differentiation is the aberration. A politics centered on the idea of the black vote must be geared strategically toward defining the scope of political action in ways that facilitate mobilization around broad racial appeals. This in turn requires grounding the analytical universe of black politics on those programmatic and issue agendas that seem to have very broad or consensual black support – the program of generic racial advancement. To that extent, Tate's project verges on tautology; if the sphere of black politics is limited to areas of broad racial agreement, then interest differentiation within black politics will appear to be slight almost by definition. Although the range and number of practices and issues that compose that universe may vary at any given moment, the universe itself will always exhibit little differentiation on pertinent issues because its boundaries are stipulated so that group homogeneity is one of its essential features. Moreover, insofar as high racial cohesion is treated as a norm and at least an implicit goal, differentiation will appear as a problem to be explained and, ultimately, overcome. This logical problem's embeddedness in hegemonic ideology renders it invisible, but it is the source of conceptual difficulties in Tate's account.

Tate asks: "Are Blacks still race-conscious? Do Blacks have group interests, or have their policy views become more individualistic" (Tate 1994: 18)? These questions set up false opposition between race-consciousness and political differentiation, as well as between group and individual interests. Race-consciousness can support many different, conflicting political programs, ranging across the ideological spectrum; therefore, the category is practically meaningless other than for demonstrating what the idea of "the black vote" already presupposes, the existence of a politics nominally centered on advancing the interests of the group in general. It would be very difficult, moreover, to imagine the disappearance of black race-consciousness in the face of persisting racial stratification and resurgent anti-black rhetoric and political initiatives.

And the crucial question regarding group interests is not whether they exist; their existence is also already presumed in the notion of the black vote. The more important issues for making sense of black politics are how concrete group interests are determined, who determines them and how those determinations, including the strategies and objectives they warrant, affect different elements of the black American population. Reducing black politics to the black vote hinders addressing these questions because its presumption of cohesiveness as a natural state overlooks the processes of interest formation to take their outcomes as given. This mindset also bypasses questions pertaining to the distribution of costs and benefits resulting from those outcomes, as well as consideration of possible alternative processes, strategies, and objectives.

Because it leaves no space for conceptualizing structural or institutional processes *within* black politics, that presumption of cohesiveness – the corporate racial interest – also leads analysis of differentiation to focus on aggregates of individual attitudes and opinions and the extent of their approximations to or divergence from putative group norms. This disposition may stem most immediately from the bias toward that sort of approach within the conventions of political science's larger American politics subfield. However, that line of inquiry also comports well with and reinforces the dominant tendency within Afro-Americanist discourse to naturalize formulations of group interests and agendas; the effect is to reify the group by screening out consideration of questions bearing on the internal and external dynamics of its constitution as a political entity. Here again the relation between ideological and intellectual commitments is dialectical, and that dialectic shapes the choice of approach and method for inquiry by informing the definition of problems and questions to be pursued. By vesting it with the appearance of a settled finding of social science, the interaction of unexamined ideology and approach to inquiry in this case buttresses the perception of black interests as given and unproblematic. At the same time, this convergence of ideological common sense and political science convention hampers Tate's ability to produce a coherent or textured account of the new black politics.

Finding continuities or incremental shifts on issue positions over time within a universe of black respondents does not tell us anything about the agenda-setting and opinion-leading dynamics operating within the new black politics; nor does comparing black and white attitudes, which substitutes conveying a sense of generic racial difference for describing politics among black Americans (Tate 1994: 20–49). The limitations of her approach and its encompassing mindset show through clearly in the way that Tate seeks to address the relation of race and class in black politics.

She begins with the currently common move of juxtaposing race and class as alternative bases of political identification. Consonant with the underlying presumption of racial cohesion, that formulation implicitly denies the possibility that class dynamics may arise from and operate autonomously within black politics and the stratification systems of black communities. For example, Tate finds that among black Americans racial consciousness seems to vary by class and that "it is not the poor and working-class Blacks but the higher-status Blacks who possess the stronger racial consciousness and identities" (Tate 1994: 25).[17] She observes that these findings contradict conventional social theoretical wisdom that "poor Blacks, particularly the urban poor, possess the strongest racial identities within the Black community" (Tate 1994: 27). Her attempts to locate possible explanations for this apparent anomaly focus on factors related to class status that could strengthen or weaken individuals' racial identity, which she takes as, in effect, monadic and prior to those other identities.

She does not consider that the problem that makes these findings seem anomalous could lie in the presumption of a singular, fixed racial identity that exists in greater or lesser degrees (Tate 1994: 90–2). That view preempts the possibility that class-linked variations in expressed race consciousness indicate differences of kind as well as degree of racial consciousness and that those differences are linked to, among other factors, class-based ideological, institutional, and programmatic dynamics at least partly internal to black political life. The pattern that she finds could represent different ways of understanding racial identity and its political warrants, and those differences could reflect the different material stakes attaching to the idea of racial identity for different social positions. A black housing authority director's, an inner-city public school principal's, a real estate developer's, or corporate functionary's experience of racial

[17] Tate (1994) notes that one study found that "More than 60 percent of self-identified middle-class Blacks . . . stated that they felt closer to their racial group than their class group. In contrast, a mere 5 percent of self-identified poor and working class Blacks preferred their racial group over their class group." She also found a positive relation between education and race-consciousness, noting that "College-educated Blacks, in particular, possessed the strongest racial common-fate identities." Yet she finds as well that those who "identify with the middle, upper-middle and upper classes were less likely to identify strongly with the race" (28). This contradictory finding could result from limitations of the 1984 National Black Election Study, from which it and the education finding are derived. Dawson notes that existing surveys of black Americans are not equipped to capture complexities of class structure and relations (Dawson 1994: 75).

consciousness and the substantive political and economic entailments of racial identity will quite likely differ from those of a resident of low-income public housing, a person underemployed without health care benefits in the low-wage consumer service sector, an industrial worker stalked by job insecurity and the threat of downsizing, a public sector line worker, a Deep South poultry plant worker, a college student besieged by aid and budget cuts, or a workfare assignee.

Tate's perspective does not provide for differences such as those. Instead, it yields only a standardized racial consciousness – measured largely in terms of aggregated issue positions and abstract attitudes – and deviations from it, also measured in terms of recorded attitudes and positions on issues already given as being of general racial significance. And she does not consider the possibility that those issue positions and attitudes take shape within a discourse defined by agenda-setting elites and researchers. This is a construction of black politics devoid of politics among black people. To that extent, it cannot recognize the possibility that the articulation of black political interests may result from the operation of discursive and organizational practices – interest-group activity – within black communities. Therefore, her account does not provide the conceptual wherewithal to adjudicate competing claims to represent racial interests in given contexts. It cannot, for example, inform interpretation and judgment of instances – such as gentrification and redevelopment controversies featuring black gentrifiers, policymakers, developers, and displaced – in which competing, perhaps incompatible interests purport to represent racial interest.

Carol Swain explicitly engages the issue of determining black interests and the reality of black differentiation. However, she acknowledges differentiation only to subordinate it to a notion of collective group interests. She argues that:

> the interests of blacks must vary in important ways; still, it would be a mistake to place more emphasis on the variations than on the commonalities. Broad patterns of objective circumstances and subjective orientations characterize American blacks, and striking differences continue to exist between black and white Americans. (Swain 1995: 7)

Swain relies largely on surveys of political attitudes and comparison of aggregate black and white socioeconomic conditions to construct an account that, notwithstanding her initial declarations, reproduces the premise of the corporate political entity. Her narrative distinguishes "objective" and "subjective" black interests. The former can be deduced from the extent of aggregate black socioeconomic disadvantage relative to whites. The latter are discernible through identifying differences in black–white distribution of opinions on significant issues and a simple majoritarian approach to black opinions as expressed in the findings of survey research (Swain 1995: 10–13).

Swain's characterization of objective interests seems reasonable enough. However, it implies a condition of racial parity as the goal of black aspiration and standard of equity, and that is a questionably modest ideal of egalitarianism or social justice. If black and white unemployment rates were equally high, for example, it is not clear that there would be no objective black interest in reducing unemployment. In addition, linking determination of black interests to cross-racial comparison in that way perpetuates the practice of defining black interests in terms of an exclusively racial agenda – although many of the most pressing socioeconomic concerns of a great many black Americans are not purely or most immediately racial. Therefore, while she does not naturalize the category of racial interests, Swain's formulation presumes the restriction of black politics to the program of generic racial advancement.

Swain's reduction of what she calls subjective black interests to the findings of survey research also in at least two ways reinforces the premise of a black corporate political entity. First, as with Tate, her reliance on a simplistic notion of "the black public," defined by those survey findings, skirts fundamental questions concerning the processes of agenda-setting and the opinion-leading role of black political elites, as well as the context-defining roles of the mass media and the interventions of researchers who themselves proceed unquestioningly from hegemonic assumptions about the scope and content of black politics. The attitudes and opinions recorded in those surveys most likely do express a black public opinion, especially in the circular sense that public opinion generally is understood to mean what is expressed in such surveys. They indicate how groups of black voters most likely would line up on a range of given positions within given terms of debate in a particular political conjuncture. Treating those expressions as autonomous preferences, however, is problematic intellectually because doing so obscures their character as the products of political and ideological processes. Reifying those responses in that way also asserts a conservative force in political discourse by denying the fluidity and contingency of the terms of debate and positions held at a given moment. Obscuring the ideological and political processes that shape those terms of debate – the dimension of power and hegemony – works to similar effect. Swain notes the role of interest groups in black politics, but she does not extend her purview to take account of their role in shaping the discursive field.

Second, Swain's reliance on a majoritarian standard for determining subjective black interests may avoid the pitfall of naturalizing black interests, but it does so only technically. Why should 50 percent +1 be the threshold of racial authenticity? What does that standard say of the views of the very large remainder? If the standard were even a supermajority – say, two-thirds or even three-fourths – how could the one-third or one-fourth minority not be seen as also expressive of black interests? Swain finesses these questions by resort to the device of black–white comparison. But noting apparent racial differences in clustering of opinions on a range of issues does not indicate a corporate "black" point of view. In fact, Swain does not escape the presumptive rhetoric of racial authenticity, which underlies her discussion of black interests. She deploys it, apparently in service to her predilections in current politics, to suggest tension between rank-and-file black citizens, whom she casts as embracing a range of conservative social views, and the officially relatively liberal black political establishment (see Swain 1995: 10–14).[18]

Michael Dawson's examination of the relative significance of race and class in black politics also proceeds from a formulation of corporate racial interest, though he attempts to ground it in a theory of pragmatic action. Because race is such a powerful determinant of the political and economic circumstances of black Americans of all strata, he argues, it has been practical for black citizens to employ a "black utility heuristic" in deciding their political views. What this means is that "as long as African-Americans' life chances are powerfully shaped by race, it is efficient for individual African Americans to use their perceptions of the interests of African Americans as a group as a proxy for their own interests" (Dawson 1994: 61).

This construct, however, fails to ground the idea of corporate interest. The "black utility heuristic" is a theoretical evasion; at bottom, it only stipulates the idea of group perspective at a one-step remove and thereby does not escape reifying and naturalizing it. Dawson's formulation begs critical questions as to how individuals form their perceptions of group interest: How do individuals determine that certain initiatives or conditions generate racial common effects? And, more significantly, how do they ascertain which interpretations, issues, and strategies actually represent the interest of the collectivity and not some more narrowly partisan or idiosyncratic agenda? He does not address such questions; in effect, he treats group interests as given and transparent to members

[18] Swain's claim that 60 percent of black voters supported Clarence Thomas's appointment to the US Supreme Court illustrates the limitation of relying on such data as evidence of firmly held views. Claims of polls reporting majority black support for Thomas often were exaggerated; that support evaporated when respondents were informed of Thomas's views. And, no matter what polls at the time might have shown, Thomas hardly has enjoyed substantial black support since his elevation to the Court (Swain 1995).

of the group. In so doing he partly answers the question concerning the relation of racial and class interests by stipulating the former to be entirely independent of the latter.

Dawson's avowed commitment to methodological individualism may obscure this problem because – as it does with Tate and Swain – it leaves him with no clearly articulated unit of analysis between individual and race (Dawson 1994: 11–12, 45, 75–9).[19] Insofar as he attends to the region between individual and corporate racial body, his discussion both presumes group interest to be a coherent, out-there thing, rather than itself the product of political processes and contestation, and at least evokes the rhetoric of racial authenticity. He maintains that using group status as a proxy for individual status is efficient,

> not only because a piece of legislation or a public policy could be analyzed relatively easily for its effect on the race but also because the information sources available in the black community – the media outlets, kinship networks, community and civil rights organizations, and especially the preeminent institution in the black community, the black church – would all reinforce the political salience of racial interests and would provide information about racial group status. (Dawson 1994: 11)

Dawson bases this description on retrospective characterization of the segregation era, when the politics of generic racial advancement had its

[19] On the other hand, other contemporary scholars who employ more institutional approaches are no less likely to presume and defend the principle of corporate racial interest (see Smith 1996; Keiser 1997). Keiser actively argues for seeing black Americans as a corporate political entity even while acknowledging the existence of significant intraracial stratification and differential policy impacts; he rests this argument on a notion of group political empowerment, which he defines as "the incremental growth of group political power" (1997: 9). The main problem with this defense, however, is that Keiser proposes no account of how power is exercised by the group as a collectivity and no convincing argument that we should accept on face value politicians' claims that the agendas they pursue represent the most that can be won for the race; he simply assumes that racially descriptive representation and membership in a governing coalition suffice as evidence of empowerment of the racial group as a whole within the pertinent polity. Though he is not a political scientist and not immersed in the disciplinary study of black politics, Manning Marable's *Black Leadership* (1998) also illustrates the pervasiveness of the tendency to naturalize the premise of generic racial interest. Even as an ostensible leftist, Marable embraces a notion of black leadership that requires no systematic consideration of questions of representativeness or accountability to specific constituencies.

Curiously, Dawson himself has noted the "shortage of sophisticated treatments of the internal political dynamics of African-American communities and populations," though it may be revealing that he and his coauthor characterize this as "the internal poiltical experience of *the* black community" (my emphasis) (Dawson and Wilson 1991: 194). Moreover, opting for an approach centered on attitude and opinion surveys does not necessarily preclude attention to historical, structural, and processual dimensions of politics; see, for an especially intelligent use of such an approach, Edward Carmines and James Stimson, *Issue Evolution: Race and the Transformation of American Politics* (1989).

clearest and solidest material foundation within the black American population. The straightforwardness that he imputes to the mediation of individual and group certainly is more plausible as an account of a context dominated by issues bearing on explicitly racial exclusion and subordination than in application to contemporary politics. Even then, however, such cut-and-dried racial questions did not exhaust the universe of black political concern and action; conflicting views existed with respect to specifics of strategic and tactical response, as well as larger, though still intraparadigmatic, interpretive and normative differences within the dominant group politics. And, as we have seen, the hegemonic status of that politics did not spring from nature or result from open, deliberative processes, and its program was not the pristine expression of a collective will. Most important, assuming a context that, if only superficially, presents a best case circumstance for defense of Dawson's black utility heuristic is ultimately conceptual sleight-of-hand because doing so also sidesteps precisely the most crucial questions that his defense must address.

Because the black utility heuristic is based on an idealized model of unproblematic racial *Gemeinschaft* in the segregation era, it is at least vulnerable to accommodating a rhetoric of racial authenticity. Construing the relation of race and class as a tension between group and individual identities reinforces this tendency. From that perspective, class interests can operate among blacks only through departure from, if not rejection of, the interests of the racial group. Because he presumes racial identity to be homogeneous or monadic, he can approach class differentiation only as a deviation from, not a constitutive element of, racial identity and consciousness.

In line with his formulation, Dawson suggests that increasing upward mobility (and, curiously, the "underclass's" supposed isolation as well) could weaken race consciousness, partly by "weakening links between individual African Americans and the race as a whole" (Dawson 1994: 80). He elaborates a view that is hardly distinguishable from a theory of racial authenticity:

> Consider two middle-class individuals, one with strong ties to family and other black community institutions, the other *with the same household or individual utility function* but with very weak ties to the black community. The individual with strong ties to family and community will be slower to deemphasize racial group interests than the individual with weak ties because of the greater impact of information from the black community for that person. (Dawson 1994: 67)

He finds that, instead, upward mobility appears to this point to increase race consciousness and racial "common-fate" identity, and this finding supports what is in effect a validation of the politics of racial advancement, which remains the most legitimate and efficacious focus for black political activity (Dawson 1994: 67, 87–8, 211–12).[20] He takes this finding as evidence that, because of linked-fate identification, the black middle class by and large remains connected to the racial interest and political agendas. This is tantamount to reducing assessment of the significance of class in black politics to a determination of whether or not individuals have "forgotten where they came from." Because of Dawson's reified notion of black interest, the possibility that prevailing forms of racial consciousness and identity might reflect class or other forces within black life cannot arise in his analysis. His approach to evaluating the significance of class among blacks never turns toward the dynamics of group interest formation.

Like Swain, Dawson adduces evidence of persisting racial disparities, in particular between black and white middle classes, to support his claim that racial identity trumps class identity by creating the material basis for a primary community of interest among blacks (Dawson 1994: 15–34).[21] But, although it arguably would induce toward race consciousness and race-solidaristic perception, the existence of disparity does not automatically yield any particular strategic interpretation or response; it does not decree, that is, that race consciousness would take any particular form or content. This once again underscores questions concerning the political processes through which concrete notions of collective interest are shaped – who determines the boundaries of group activity, who participates in those processes, and who is advantaged and disadvantaged by them – questions that have no place in Dawson's analysis.

Dawson reports that, despite the power of race consciousness, "individuals' economic status plays a large role in shaping African-American public opinion in several issue domains" and that "affluent African-Americans are much less likely to support economic redistribution than those with fewer resources [and] the most affluent African Americans hold views more consistent with those of the conservative white mainstream" (Dawson 1994: 205; see 181–99). And he notes that upper-status

[20] "The future of African-American politics may well depend on how the racial and economic environment of twenty-first century America dictates which African Americans perceive that their fates remain linked" (Dawson 1994: 212).

[21] Black–white comparison is the standard in contemporary academic discussion of the black middle class, which hardly ever addresses issues such as the systemic features and entailments of social stratification and hierarchy among blacks. Discussion of the role of upper strata within black social life typically centers on questions of problems of racial identity, as with Dawson, or on role modeling and other forms of *noblesse oblige* (see Landry 1987; Willie 1979: 1989).

blacks are more likely than others to see themselves as having benefited from the politics of racial advancement (Dawson 1994: 83). Yet, he does not consider the possibility that the "linked-fate" phenomenon that he observes may reflect not a mitigation of class consciousness but its expression as an ideology through which those very strata enact the dominance of a particular definition of the scope and content of black political activity. His findings are consistent with the view that the generic politics of racial advancement is skewed to programmatic agendas that confer concrete benefits disproportionately on those petit bourgeois strata. From this perspective, the perception of a common fate – without regard to the sincerity with which individuals adhere to it – could be an ideological mechanism that preserves and buttresses individual or stratum advantages by harmonizing them with a notion of larger group interest. His "group interest perspective" forecloses that possibility because it continues the pattern of reading politics *among* black people out of inquiry into black politics.

Dawson, Swain, and Tate stand out among students of Afro-American politics in that they attempt, in distinct but related ways, to problematize the issue of interest differentiation among blacks. Their concern with this issue is a response to a more broadly, if tentatively and inchoately, articulated suggestion in contemporary public discourse – usually propounded by conservatives in support of their own counter-narratives of racial authenticity[22] – that systemic incorporation has destabilized conventional understandings of black political life. However, all three short-circuit inquiry from the outset, by positing the notion of corporate racial interest as an unexamined background assumption that organizes their approaches conceptually and procedurally. In accepting that premise they perpetuate the restriction of black politics thematically to the domain of collective struggle against generic racial inequality and subordination. This in turn makes it unnecessary to acknowledge or examine, perhaps even impossible to notice, the autonomous political processes and structural, ideological, and institutional tensions that constitute the matrix of concrete black political action.

In thus insisting from the first that black politics be seen only as a corporate, group endeavor, it is as if these scholars raise the question of black

[22] Preston Smith discusses at length this rhetoric and the ideological agenda in which it is embedded (see Smith 1999). Robert Woodson, Clarence Thomas and the next-most-recent incarnation of the recently reborn Glenn Loury are among the most prominent, and tendentious, exemplars of this rightist appropriation of the rhetoric of racial authenticity to challenge the legitimacy of the black liberal political establishment (see Legette 1999).

differentiation only in order to refute it. The political sea change that has prompted questions about the presumption of corporate racial interest thereby calls forth active defense of it.[23] Ironically, as it purports to a skeptical attitude toward the new black politics, this scholarship may be more explicitly and willfully engaged than its antecedents in reproducing ideological legitimations for the politics of generic racial advancement and in obscuring its class basis.

This characterization of the trajectory of scholarship on black politics is schematic and has no pretense to exhaustiveness. It nonetheless specifies a conceptual organizing principle that has underlain the subfield's discursive conventions throughout its history. Reconstruction of that history also highlights the dialectical character of the relation between the study and practice of black American politics. That relation is hardly unique to black politics, of course, though the precariousness of the black situation in the United States and the leading role of mainstream academic discourses in rationalizing racial injustice and inequality for most of the last century probably have led Afro-Americanists to be more self-conscious and less diffident than others about it. The problem with the main lines of scholarly inquiry into black politics is not, therefore, that they have been partisan or engaged. It is that their partisanship has been unselfconsciously fastened onto the racial vision of the black petite bourgeoisie – a singular class vision projected as the organic and transparent sensibility of the group as a whole.

This bias is problematic not only because it has produced accounts of black politics that are incomplete and static, devoid of political dynamics among black people; its reductivist postulation of the corporate racial subject also reinscribes essentialist presumptions regarding the black American population. Such constructions as "the Negro (or the black community) believes/wants, etc." originate in nineteenth-century race theory's notions of racial temperament and ideals and other formulations asserting in effect that blacks think with one mind. Both recent scholarship's insistent minimizing of the political significance of racial differentiation and the more general sanctification of racial unity as the substance of a political program in its own right are anchored in such

[23] Indeed, it may be that an understandable desire to meet the right-wing attack with which that questioning is usually associated partly motivates scholars like Tate and Dawson to vindicate the premises of the politics of racial advancement and thus contributes to the short-circuiting of their critiques. In any case, the fact that the intellectual challenge to the rhetoric of racial authenticity and presumptive corporate group interest originates most conspicuously from the sophistries and political disingenuousness of conservative ideologues has not helped the quality of the debate.

formulations. Both also are active interventions that seek to impose those categories on an increasingly recalcitrant world of manifest black experience.

While this scholarly orientation around an unacknowledged class vision of racial politics may have been adequate pragmatically to its intellectual and political objectives during the segregation era in which it emerged and evolved, it does not equip us well to make sense of the imperatives and possibilities that define current political conditions. The subfield's main discourse revolves around the dual objectives of defending a coherent and homogenized racial image and articulating putatively collective mentalities and aspirations. These objectives, and the deeply sedimented presuppositions on which they rest, direct the lens of critical investigation away from careful, skeptical scrutiny of interior features of black political life.[24] The field is therefore left without well-developed critical standards for judgment in black political debate, aside from a norm of racial authenticity. It has, moreover, produced a scholarship on black politics that characteristically is blind to the operation of political processes *among* black people.

However powerful it may have been rhetorically when universal group consensus seemed plausible in opposition to Jim Crow, the proposition that racial authenticity is a basis for intellectual and political judgment was always extremely problematic. Ideologically, as I have noted, it is an artifact of racial essentialism and presumes a given and objectively discernible mode of correct thought for the race. Practically, it begs questions regarding how authenticity is determined and who chooses its guardians. Because it focuses critical vision around ascriptive rather than substantive criteria, this standard does not make for clear, open political debate. Racial authenticity is a hollow notion that can be appropriated by nearly anyone to support or oppose any position, and it can be absurd and self-defeating.

The inadequacy of the notion of racial authenticity as a basis for political judgment is a thread that unravels the entire outlook that continues to shape both the study and practice of politics among black Americans. Without that notion the assumption of an *a priori* corporate group interest

[24] This orientation is not idiosyncratic to the black politics field. The institutionalization of black studies in niches within mainline academic disiplines – itself a phenomenon of the postsegregation political context – has shown that this mindset is pandemic in black intellectual life. Afro-Americanist scholarship in literary studies, psychology, and anthropology is dominated by it, and it is a prominent node threatening dominance in the Afro-American history field. Black studies discourse in sociology appears grounded on an ambivalent amalgam of this racial vindicationism and the discipline's fetishized discourse of inner-city social pathology.

and program cannot be sustained. Without the premise of a norm of authenticity, prevailing forms of political debate – which pivot on claims to represent the genuine interests of the racial collectivity – lose their fulcrum, and the corollary notions of generic race leadership also collapse. Elimination of the notion of racial authenticity undermines the politics of generic racial advancement as a conceptual boundary for black politics and the crucial presumption of black Americans' default position as a corporate political entity. Without the standard of authenticity, appeals to unity as a means to resolve conflicts or as an explanation of failure appear all the more clearly as the wish fulfillment and platitudes that they are.

Stripping away that complex of mystifications also removes layers of ideological evasion that have obscured the partial, class-skewed program that has dominated black politics for a century. What remains is a clear view that the dominant politics of generic racial advancement – though it has won genuinely popular victories and has had many beneficial consequences, especially during the anti-segregation period – has always rested on a non-participatory, undemocratic foundation of elite custodianship and brokerage.

We need a different way of thinking about and doing politics among black Americans. I hesitate to proclaim a need for a *new* approach to black politics, partly because calls for new ideas are an academic marketing cliché, but also because I think it preferable to understand what I believe we need as more of a return to the path that black political development was on before the distortions of the Jim Crow era.

We need an approach to black political activity that most of all recognizes the diversity of black life and concerns and that therefore presumes norms of popular, democratic participation and broad, open debate. Rejection of claims to racial authenticity and other forms of organicist mythology means that democratic and participatory values must be the cornerstone of credibility for the notion of black politics; group consensus must be constructed through active participation.

We need to reconceptualize black political activity as a dynamic set of social relations and interests that converge around some issues as consequential for broad sectors of the black population – and that diverge around others based on other identities and interest aggregations. This is a black politics that does not pretend to exhaust the totality, or even necessarily the most important, aspects of all black people's political concerns and activity. It is a notion of black politics in which black people, as individuals and groups, organize, form alliances and enter coalitions freely on the basis of mutually constituted interests, crisscrossing racial boundaries as they might find pragmatically appropriate.

Anchoring inquiry on normative commitment to a democratic and popular politics of this sort also cannot tolerate models of "natural" or organic leadership. Acknowledging the autonomy of black individuals as the basis of group solidarity and the variety of black experience means that, as scholars no less than citizens, we should, as a predisposition, be wary of claims to legitimacy that do not arise from concrete models of volitional political association. There can be no legitimate leadership that is not accountable to specific constituencies in clearly delineated ways. Claims to speak on behalf of some constituency should be received with skepticism unless the spoken-for have direct and accessible mechanisms at their disposal for ratifying or repudiating those claims. A credible, demystified black political leadership must be based on actual membership organizations and electorally, or otherwise demonstrably, accountable political institutions, not self-proclamation or recognition by third parties.

Trivialization of the mechanisms of popular participation and political accountability has been a central element of legitimation and reproduction of what has been understood as black politics by scholars and activists, and across the ideological spectrum, since the initial articulation of the ideology of racial uplift in the decades after Hayes-Tilden. By contrast, the two moments of greatest political gains for black Americans were also the moments of most extensive popular participation and mobilization: the thirty-year period before consolidation of the hegemonic form of black politics and the high period of popular activism from the 1950s to early 1970s when it broke down temporarily in the face of popular insurgency.

Presuming a norm of liberal democratic, popular politics is also, ironic though it may seem to many on the left, the clearest and likeliest way to redirect the focus of black political debate along more substantive lines by insisting on making explicit its class and other asymmetries. Only by acknowledging differentiation is it possible to discuss the issue of unequal impact of political strategies and unequal distribution of the costs and benefits of political action. Thus, naturalistic or corporate racial justifications of inequality should warrant skepticism – including claims that an unequal distribution of material benefits to a few is compensated by symbolic benefits allegedly accruing to the many as a byproduct.

Ian Shapiro's argument for a dispositive skepticism toward all hierarchies as a foundation of practical democracy is noteworthy in this regard. Shapiro observes that hierarchies,

> will benefit some, harm others, and sometimes benefit some at the price of harming others. Hierarchies also tend to become hostage to the imperatives for their own maintenance, and these imperatives frequently involve the creation and

propagation of fictions about either the nature or the arbitrariness of the hierarchy in question. Those who benefit from the existence of a hierarchy can be expected to try – more or less consciously – to obscure its hierarchical character through argument or to insist that it is rational or normal. (Shapiro 1996: 124)[25]

In his view, it should be incumbent on those who benefit from hierarchies to shoulder the burden of proving them justified. The context for fulfilling this requirement must be a sphere of open political debate based on substantive, not ascriptive, criteria. There has rarely been such a sphere in black American political life since the end of the nineteenth century. From this perspective, we need to generate an approach to analyzing and enacting black politics that pivots on the fulcrum of Marx's famous heuristic: *Cui bono*? Who benefits? Who doesn't? Who loses? As a civic matter, this requires perceiving politics from the bottom up, as shaped around the specific interests and concerns of specific groups of black individuals in specific places and specific social circumstances.

REFERENCES

Anderson, James. 1988. *The Education of Blacks in the South, 1860–1935*. Chapel Hill: University of North Carolina Press.

Arnesen, Eric. 1991. *Waterfront Workers of New Orleans: Race, Class and Politics, 1863–1923*. New York and Oxford: Oxford University Press.

"'What's on the Black Worker's Mind?': African-American Workers and the Union Tradition." *Gulf Coast Historical Review* 10(1): 7–30.

Bacote, Clarence A. 1955. "The Negro in Georgia Politics, 1880–1908." Unpublished doctoral dissertation. University of Chicago.

Bennett, Larry. 1993. "Harold Washington and the Black Urban Regime." *Urban Affairs Quarterly* 28: 423–40.

Bloom, Jack H. 1987. *Class, Race, and the Civil Rights Movement*. Bloomington: University of Indiana.

Browning, Rufus, Dale Rogers Marshall, and David H. Tabb. 1986. *Protest Is Not Enough: The Struggle of Blacks and Hispanics for Equality in Urban Politiccs* (reprint edn). Berkeley: University of California Press.

Bunche, Ralph J. 1936. *A World View of Race*. Washington, DC: Associates in Negro Folk Education.

1973. *The Political Status of the Negro in the Age of FDR*. Chicago: University of Chicago Press.

[25] Shapiro points out that he does not claim "that all social hierarchies should be eliminated but rather that there are good reasons to be nervous about them. Escapable hierarchies can be alleged to be inescapable, oppression can be shrouded in the language of agreement, unnecessary hierarchies can be declared essential to the pursuit of common goals, and fixed hierarchies can be shrouded in myths about their alterability. Accordingly, although many particular hierarchies might in the end be conceded to be justified, this concession should truly *be* in the end and *after* the defender of hierarchy has shouldered a substantial burden of persuasion" (Shapiro 1996: 125–6).

1995a. "A Critical Analysis of the Tactics and Programs of Minority Groups," in Charles P. Henry (ed.). *Ralph J. Bunche: Selected Speeches & Writings*. Ann Arbor: University of Michigan Press. 49–62.

1995b. "The Problems of Organizations Devoted to the Improvement of the Status of the American Negro," in Charles P. Henry (ed.). *Ralph J. Bunche: Selected Speeches & Writings*. Ann Arbor: University of Michigan Press. 71–84.

1995c. "The Negro in the Political Life of the U.S.," in Charles P. Henry (ed.). *Ralph J. Bunche: Selected Speeches & Writings*. Ann Arbor: University of Michigan Press. 93–112.

1995d. "Marxism and the 'Negro Question,'" in Charles P. Henry (ed.). *Ralph J. Bunche: Selected Speeches & Writings*. Ann Arbor: University of Michigan Press. 35–45.

Burgess, M. Elaine. 1962. *Negro Leadership in a Southern City*. Chapel Hill: University of North Carolina Press.

Bush, Rod (ed.). 1984. *The New Black Vote: Politics and Power in Four American Cities*. San Francisco: Synthesis.

Button, James. 1989. *Blacks and Social Change: Impact of the Civil Rights Movement in Southern Communities*. Princeton: Princeton University Press.

Carmines, Edward and James Stimson. 1989. *Issue Evolution: Race and the Transformation of American Politics*. Princeton: Princeton University Press.

Cole, Leonard. 1976. *Blacks in Power: A Comparative Study of Black and White Elected Officials*. Princeton: Princeton University Press.

Cothran, Tillman C. and William Phillips, Jr. 1961. "Negro Leadership in a Crisis Situation." *Phylon* 22 (Winter): 107–18.

Cox, Oliver C. 1950. "Leadership Among Negroes in the United States," in Alvin W. Gouldner (ed.). *Studies in Leadership*. New York: Harper and Brothers. 228–71.

Dawson, Michael C. 1994. *Behind the Mule: Race and Class in African-American Politics*. Princeton: Princeton University Press.

Dawson, Michael C. and Ernest J. Wilson, III. 1991. "Paradigms and Paradoxes: Political Science and African-American Politics," in William Crotty (ed.). *Political Science: Looking to the Future, The Theory and Practice of Political Science*. Vol. I. Evanston, IL: Northwestern University Press. 189–234.

Drake, St. Clair and Horace R. Cayton. 1944. *Black Metropolis: A Study of Negro Life in a Northern City*. New York: Harcourt, Brace.

Edmonds, Helen. 1951. *The Negro and Fusion Politics in North Carolina, 1894–1901*. Chapel Hill: University of North Carolina Press.

Eisinger, Peter. 1984. "Black Mayors and the Politics of Racial Economic Advancement," in Harlan Hahn and Charles Levine (eds.). *Readings in Urban Politics: Past, Present and Future*. New York: Longman. 249–60.

Fink, Leon. 1983. *Workingmen's Democracy: The Knights of Labor and American Politics*. Urbana and Chicago: University of Illinois Press.

Foner, Eric (ed.). 1993. *Freedom's Lawmakers: A Directory of Black Officeholders During Reconstruction*. New York & Oxford: Oxford University Press.

Fortune, T. Thomas. 1968. *Black and White: Land, Labor and Politics in the South* (reprint edn). New York: Arno.

Gaines, Kevin. 1996. *Uplifting the Race: Black Leadership, Politics, and Culture in the Twentieth Century*. Chapel Hill: University of North Carolina Press.

Gilmore, Glenda Elizabeth. 1996. *Gender & Jim Crow: Women and the Politics of White Supremacy in North Carolina, 1896–1920*. Chapel Hill: University of North Carolina Press.

Gosnell, Harold F. 1967. *Negro Politicians: The Rise of Negro Politics in Chicago* (reprint edn). Chicago: University of Chicago Press.

Harlan, Louis R. 1972. *Booker T. Washington: The Making of a Black Leader, 1856–1901*. New York and London: Oxford University Press.

Henderson, Lenneal J. 1979. *Administrative Advocacy: Black Administrators in Urban Bureaucracy*. Palo Alto, CA: R & E Associates.

Herbst, Susan. 1994. *Politics at the Margin: Historical Studies of Public Expression Outside the Mainstream*. New York and Cambridge: Cambridge University Press.

Holden, Matthew. 1971. "Black Politicians in the Time of the 'New' Urban Politics." *Review of Black Political Economy* 2 (Fall): 56–71.

Holloway, Jonathan Scott. 2001. "The Black Intellectual and the 'Crisis Canon' in the Twentieth Century." *The Black Scholar* 31(1): 2–13.

2002. *Confronting the Veil: Abram Harris, E. Franklin Frazier, and Ralph Bunche, 1919–1941*. Chapel Hill: University of North Carolina Press.

Jones, Charles E. 1987. "An Overview of the Congressional Black Caucus: 1970–1985," in Franklin D. Jones and Michael O. Adams (eds.), *Readings in American Political Issues*. Dubuque, IA: Kendall/Hunt. 219–34.

Jones, Mack H. 1978. "Black Political Empowerment in Atlanta: Myth and Reality." *Annals of the American Academy of Political and Social Science* 439 (September): 90–117.

Judd, Dennis R. 1986. "Electoral Coalitions, Minority Mayors, and the Contradictions of the Municipal Policy Agenda," in Marc Gottdiener (ed.). *Cities in Stress*. Beverly Hills, CA: Sage Publications. 145–70.

Karnig, Albert and Susan Welch. 1980. *Black Representation and Urban Policy*. Chicago: University of Chicago Press.

Keiser, Richard A. 1997. *Subordination or Empowerment? African-American Leadership and the Struggle for Urban Political Power*. New York & Oxford: Oxford University Press.

Keller, E. J. 1978. "The Impact of Black Mayors on Urban Policy." *Annals of the American Academy of Political and Social Science* 439 (September): 40–52.

Kilson, Martin L. 1971. "Political Change in the Negro Ghetto, 1900–1940s," in Nathan Huggins, Martin L. Kilson, and Daniel Fox (eds.), *Key Issues in the Afro-American Experience*. Vol. II. New York: Harcourt Brace Jovanovich. 167–92.

1974. "From Civil Rights to Party Politics: The Black Political Transition." *Current History* 67 (November): 193–9.

1980. "The New Black Political Class," in Joseph Washington (ed.), *Dilemmas of the Black Middle Class*. Philadelphia: Joseph R. Washington. 81–100.

Ladd, Everett Carl, Jr. 1966. *Negro Political Leadership in the South*. Ithaca, NY: Cornell University Press.
Landry, Bart. 1987. *The New Black Middle Class*. Berkeley: University of California Press.
Legette, Willie. 1999. "The Crisis of the Black Male: A New Ideology in Black Politics," in Adolph Reed Jr. (ed.), *Without Justice for All*. Boulder: Westview Press. 291–324.
Letwin, Daniel. 1998. *The Challenge of Interracial Unionism: Alabama Coal Miners, 1878–1921*. Chapel Hill: University of North Carolina Press.
Locke, Alain (ed.). 1968. *The New Negro: An Interpretation* (Reprint). New York: Atheneum.
Logan, Rayford W. (ed.). 1944. *What the Negro Wants*. Chapel Hill: University of North Carolina Press.
Lynch, John R. 1970. *Reminiscences of an Active Life: The Autobiography of John Roy Lynch*. Chicago: University of Chicago Press.
Marable, Manning. 1998. *Black Leadership*. New York: Columbia University Press.
Martin, Robert E. 1953. "The Relative Political Status of the Negro in the United States." *Journal of Negro Education* 22 (Summer): 363–79.
Matthews, Donald R. and James W. Prothro. 1966. *Negroes and the New Southern Politics*. New York: Harcourt Brace Jovanovich.
McMillen, Neil. 1989. *Dark Journey: Black Mississippians in the Age of Jim Crow*. Urbana and Chicago: University of Illinois Press.
Meier, August. 1963. *Negro Thought in America, 1880–1915: Racial Ideologies in the Age of Booker T. Washington*. Ann Arbor: University of Michigan Press.
Meier, August and Elliott M. Rudwick. 1968. "Attitudes of Negro Leaders Toward the Labor Movement from the Civil War to World War I," in Julius Jacobson (ed.), *The Negro and the American Labor Movement*. Garden City, NY: Doubleday. 27–48.
Mitchell, Michele. 1998. "Adjusting the Race: Gender, Sexuality and the Question of African-American Destiny, 1870–1930." Unpublished doctoral dissertation. Northwestern University, Evanston, Illinois.
Moon, Henry Lee. 1948. *Balance of Power: The Negro Vote*. Garden City, NY: Doubleday.
Moses, Wilson J. 1988. *The Golden Age of Black Nationalism, 1850–1925*. New York & Oxford: Oxford University Press.
Nelson, William E. 1987. "Cleveland: The Evolution of Black Political Power," in Michael B. Preston, Lenneal J. Henderson, and Paul L. Puryear (eds.), *The New Black Politics: The Search for Political Power*. New York: Longman Publishers. 172–99.
Nelson, William E. and Philip J. Meranto. 1977. *Electing Black Mayors: Political Action in the Black Community*. Columbus: Ohio State University Press.
Picard, Earl. 1984 "New Black Economic Development Strategy", *Telos* (Summer): 53–64.
Preston, Michael B. 1976. "Limitations of Black Urban Power: The Case of Black Mayors," in Robert Lineberry and Lousi Masotti (eds.), *The New Urban Politics*. Boston: Ballinger. 111–32.

Preston, Michael B., Lenneal J. Henderson, and Paul L. Puryear (eds.). 1987. *The New Black Politics: The Search for Political Power*. New York: Longman.

Rachleff, Peter. 1989. *Black Labor in Richmond, 1865–1890*. Urbana and Chicago: University of Illinois Press.

Reed, Adolph L. Jr. 1986. *The Jesse Jackson Phenomenon: The Crisis of Purpose in Afro-American Politics*. New Haven and London: Yale University Press.

1987. "A Critique of Neo-Progressivism in Theorizing about Local Development Policy: A Case from Atlanta," in Clarence N. Stone and Heywood T. Sanders (eds.), *The Politics of Urban Development*. Lawrence: University of Kansas Press. 199–215.

1988. "The Black Urban Regime: Structural Origins and Constraints." *Comparative Urban and Community Research* 1: 139–59.

1997. *W. E. B. Du Bois and American Political Thought: Fabianism and the Color Line*. New York and Oxford: Oxford University Press.

Shapiro, Ian. 1996. *Democracy's Place*. Ithaca and London: Cornell University Press.

Smith, Preston. 1999. "'Self-Help', Black Conservatives, and the Reemergence of Black Privatism," in Adolph Reed Jr. (ed.), *Without Justice For All*. Boulder: Westview Press. 257–90.

Smith, Robert C. 1981. "Black Power and the Transformation from Protest to Politics." *Political Science Quarterly* 96 (3): 431–43.

1996. *We Have No Leaders*. Albany: State University of New York Press.

Spero, Sterling D. and Abram L. Harris. 1968. *The Black Worker* (reprint edn). New York: Atheneum Press.

Stein, Judith. 1974–75. "'Of Mr. Booker T. Washington and Others': The Political Economy of Racism in the United States." *Science and Society* 38: 422–63.

Stone, Clarence N. 1989. *Regime Politics: Governing Atlanta, 1946–1988*. Lawrence: University of Kansas Press.

Swain, Carol M. 1995. *Black Faces, Black Interests: The Representation of African Americans in Congress*. Cambridge, MA and London: Harvard University Press.

"Symposium: Minorities in Public Administration." 1974. *Public Administration Review* 34 (November/December): 519–63.

Tate, Katherine. 1994. *From Protest to Politics: The New Black Voters in American Elections*. Cambridge, MA and London: Harvard University Press.

Thompson, Daniel C. 1963. *The Negro Leadership Class*. Englewood Cliffs, NJ: Prentice-Hall.

Thornburgh, Emma Lou. 1972. *T. Thomas Fortune: Militant Journalist*. Chicago: University of Chicago Press.

Toll, William S. 1979. *The Resurgence of Race: Black Social Theory from Reconstruction to the Pan-African Conferences*. Philadelphia: Temple University Press.

Walker, Jack L. 1963. "Protest and Negotiation: A Case Study of Negro Leadership in Atlanta, Georgia." *Midwest Journal of Political Science* 7: 99–124.

Warren, Robert Penn (ed.). 1966. *Who Speaks for the Negro?* New York: Random House.

Willie, Charles V. 1989. *Caste and Class Controversy on Race and Poverty: Round Two of the Willie/Wilson Debate* (2nd edn). Dix Hills, NY: General Hall, 1989.

Willie, Charles V. (ed.). 1979. *The Caste and Class Controversy*. Bayside, NY: General Hall.

Willingham, Alex. 1986. "Ideology and Politics: Their Status in Afro-American Social Theory," in Adolph Reed Jr. (ed.), *Race, Politics and Culture: Critical Essays on the Radicalism of the 1960s*. Westport, CT: Greenwood Press. 13–28.

Wilson, James Q. 1960. *Negro Politics: The Search for Leadership*. Glencoe, IL: Free Press.

Woodward, C. Vann. 1951. *Origins of the New South, 1877–1913*. Baton Rouge, LA: Louisiana State University Press.

Woody, Bette. 1982. *Managing Urban Crises: The New Black Leadership and the Politics of Resource Allocation*. Westport, CT: Greenwood Press.

7 External and internal explanation

John Ferejohn

I. Introduction

Should the social sciences focus more than they now do on solving real (explanatory) problems and less on developing methodologies or pursuing methodological programs? Two distinct worries animate this question. One is that too many resources may be devoted to the development and refinement of methodologies and theories, while too little attention is paid to the actual things needing explanation. In this sense there may be a misallocation of social scientific resources. The other worry is that when proponents of some methodology turn to explaining a particular event or phenomenon, they tend to produce distorted accounts; they are deflected by their inordinate attention to and sympathy for their favorite method. Method-driven social science comes up with defective explanations. Proper attempts to explain things, one might think, ought to be open ended and responsive to the phenomenon to be explained and not be committed in advance to any particular explanatory methodology. Such a commitment smacks of dogmatism or *a priori*-ism. These complaints are often illustrated by the familiar metaphors of drunks searching under street-lamps and the law of the hammer.

My inclination is to resist the question as not quite usefully posed. The development of systematic methodologies and theories is what permits the social sciences – or particular approaches to social science – to make distinctive and sometimes valuable contributions to understanding the events that interest us.[1] There are several reasons why this is the case. A methodological focus can throw new light on old issues in various ways; things that might be taken for granted from one perspective look problematic and in need of explanation from another. It can show how new kinds of evidence can bear on the explanation of an event, and how evidence – old as well as new – ought be interpreted. Even if commitment to a particular method tends to produce uneven or partial explanations

[1] For purposes of this chapter, I shall use "theory" and "methodology" interchangeably.

in some cases, such a commitment can enhance our understanding of phenomena by providing a new perspective on events that had previously been thought to be adequately understood.

Indeed, one might ask whether there is any sense to the idea of "an" explanation of any particular phenomenon. If one accepts the notion that explanations are answers to questions and that questions themselves are dependent on methodology, the notion of a unique or best or privileged explanation seems hard to defend. The best we can hope for is to pose method-driven questions about an event of interest (for whatever reason) and then try to make explanatory headway from the viewpoint of that question. By examining an event from more methodological perspectives, we can hope to increase our "understanding" or grasp of it. This is not to deny progress in understanding but it is to say that understanding is a multifaceted idea and that improving the understanding of an event does not imply the existence either of a unique explanatory viewpoint (or methodology) or of a best explanation.[2] In this sense, we cannot avoid methodology any more than Molière's gentleman could avoid speaking prose.

So, as a general matter, methodological eclecticism seems not only defensible, but unavoidable. But perhaps we cannot approach all interesting or important events eclectically. It is often thought that certain kinds of historically important events – "singular" events – are inaccessible to method-driven social science. To understand such occurrences we must resort to other forms of understanding, especially techniques of narrative or interpretive reconstruction. There may be some truth to this claim if we identify methods-driven social science with the pursuit of empirical regularity and causal explanation, of the kind sought by practitioners of the natural sciences. But I think that is much too narrow a conception of social science. Of course we are interested in causal explanation but also

[2] In fact, methodological programs generate events needing explanation. This idea is a commonplace among those social scientists who generate their data in a theory-based manner – as did, for example the sociologists who defined and operationalized the concept of alienation or the economists who formulated the idea of consumer confidence with survey research methodology. While both of these ideas may have originated as explanatory variables to account for other phenomena, they have long since become free-standing dependent variables needing explanation.

Besides, what is an "event" anyhow? We often understand an event requiring explanation as a kind of primitive pretheoretic idea but that is, at best, just a rough starting point. Event descriptions are themselves dependent on method or theory. Events are theoretically unstable in that as accepted theory changes, the events needing explaining change too, as the voting example suggests. And, if any doubt remains one can point to the ubiquitous use of prisoners' dilemmas and collective action problems to turn Aristotle's axiomatic conception of man as inherently social on its head. Group formation and social organization become things in need of explanation rather than givens.

in other questions as well. Perhaps considering the example of singular events will help sharpen some of these other explanatory aims, and show that methods-driven social science can help in their pursuit.

There are so many different methodological programs within the social sciences that it would be impossible and distracting to try to discuss even a fraction of them. But much of what I have to say can be illustrated by considering two broad approaches to explanation found within the social sciences. Externalists explain action by pointing to its causes; internalists explain action by showing it as justified or best from an agent's perspective. Externalist explanations are positivist and predictive; internalist explanations are normative or hermeneutic. Externalists tend to call themselves political scientists; internalists, political theorists. And, both externalists and internalists agree, if they agree on little else, that they are engaged in different enterprises.

I shall argue against this widely shared belief. I believe that what distinguishes social science from humanistic accounts of action is a recognition of the importance and validity of externalist explanation – seeking to understand events by finding causal information about them. And, what distinguishes social and non-social scientific explanatory practice is a recognition of the importance and validity of internalist accounts: seeing actions as more or less justified from the perspective of normatively guided actors. I do not think either of these commitments can be surrendered without giving up much of the distinctiveness and vitality of the social sciences. This is not to deny that there are differences between these broad methodological approaches, nor that these differences can run deep. But, it is to insist that the two approaches have in common a shared concern to explain roughly the same range of social phenomena. And I think that this shared concern often produces significant pressures for convergence in practice if not in theory.

II. Singular events: an example

Singular events are sometimes understood to mark the beginnings or ends of eras, to punctuate longer periods of slow or evolutionary development, or to carry special meaning or significance as events. Certain especially important wars, rebellions, assassinations, historic compromises, or elections supply typical examples. Within American political history, for example, the elections of 1800 (sometimes called the "revolution" of 1800), 1828, 1896, or 1932 might each qualify on grounds of their significance at the time or later. Truman's decision to use an atomic weapon against the Japanese, or his Marshall plan would probably count too, at least from some perspectives. And, of course, the outbreak of certain wars, the American Civil War especially but also the American Revolution, are often thought of as singular and calling for a special kind of explanatory strategy.

Singular events are distinctively heterogeneous – each is supposed to be singular after all – and so one is entitled to doubt that much can be said about them as such as there may be nothing, other than their rarity and their *ex-post* significance, they have in common. We may well be forced to think of such events as bearing no more than a relationship of family resemblance to one another. Moreover, because such events are thought to be freighted with meaning – indeed having some kind of special meaning is part of why the event is given this status – the designation of an event as "singular" is bound to be controversial and "theory dependent."

If this is so, singularity is not really definable in an extensional sense. Events are typically taken to be singular partly because of what they mean or signify, or because of some consequences that follow from them. The concessions to Hitler at Munich in 1938 are thought to be singular, for example, partly because of Hitler's subsequent invasions of Poland and Western Europe, and partly because of the horrific character of his regime. Had he been run over by a car late in 1938, or had he turned out to be a small-time bully who had no further territorial ambitions, Munich itself would probably have faded from view and lost any claim to special importance. Munich came to be characterized as an act of appeasement of a tyrant – indeed it has come to stand for any such act – because of how Hitler acted afterwards. The significance of Munich then rests on the beliefs that Hitler had an insatiable lust for domination, that he was there given extra time and opportunity to develop his forces for future aggressions, and that the results of these aggressions turned out to be momentous for the rest of Europe. In this respect, the singularity of Munich depends both on subsequent events and shared meanings.

The explanation of such occurrences seems especially difficult from a scientific viewpoint for several reasons. The first issue is, as already mentioned, that such events however identified have nothing in common.[3] I am happy to grant this because not much turns on it; there is no reason to think that good explanations of singular events would necessarily have much in common either. The designation "singular" only conveys rarity and special significance for a people. Nothing more than that. But the most important practical explanatory difficulty seems to be that singularity itself implies that there is a relatively thin base of information to draw

[3] This is so in two senses: first, their meanings and subsequent consequences are heterogeneous. And second, they are identified by reference to shared beliefs or meanings and this implies that singularity is not robust to changing beliefs or understandings about the event.

on – at least thin in some important sense. In another sense, we might have quite a lot of narrative information about such an event precisely because of its importance to contemporaries or successors. Participants will tend to save their papers, write memoirs or produce apologies; historians and others will try to understand and explain it; politicians will invoke it as justifying some further course of action.

So, we will tend to have at once both too little and too much information about a singular event. Moreover, our capacity to explain or understand such events sometimes is important from a policy-oriented standpoint. Witness, for example, the role of "Munich" in debates about how the United States or the United Nations should deal with Saddam Hussein. Those who cite Munich argue that a particular "causal" process was in motion in contemporary Iraq – roughly the same kind of process that was in motion in Germany in 1938 – that would have led to the development or acquisition of weapons of mass destruction unless there was an outside intervention to put a stop to it. This argument rested on the claim that we have an explanation or understanding of developments in Germany in the late 1930s that instructs us as to how to behave toward (relevantly similar) new tyrants. But how could this understanding be relevant to Iraq, given the assumed singularity of Munich?

One strategy of explanation is to embed Munich (and Saddamite Iraq) in a class of relevantly similar cases and then attempt to find generalizations that seek to connect, in some standard sense, appeasement and its consequences. This is, in effect, to deny that Munich was actually a singular event – or at any rate to make its singularity explanatorily invisible. Large-N studies of the outbreak of wars and rebellions – many of which might count as singular when viewed from another perspective – are examples of this strategy. Another approach is to break up the larger event into complex sequences and ensembles of actions – subevents – where these sequences and subevents are more mundane and perhaps explicable by resorting to ordinary explanatory strategies.[4] For example, in explaining the American response to the placement of Russian missiles in Cuba, Graham Allison (1971) draws heavily on the organization theories of the time. From these more or less ordinary behavioral predictions

[4] There are a lot of ways this might be done. One is to break the event into its sequential happenings, horizontally across time. Another is to layer the event vertically into simpler subprocesses. One lesson of Allison's (1971) analyses of the Cuban missile crisis is that there are many ways in which this may be done. Another issue is what kinds of external background information, theoretical or empirical, may be usefully called upon to analyze the patterns of subevents. Allison himself draws upon a complex mixture of theoretical knowledge from various areas of social science in order to cast light on the complex events associated with the crisis. The fertility of this approach is perhaps best seen from the illuminating responses of critics.

he crafts or aggregates a story of the crisis and its resolution. This strategy involves assuming that the singularity of the event does not infect its subevents in any strong way. Allison assumed that more or less ordinary regularities of organizational behavior would be exhibited throughout the missile crisis. It was in the concatenation of these events that the singularity of the crisis arose.

Both of these explanatory strategies seem completely compatible with ordinary social science practice and I have nothing to say against their pursuit. There is nothing in the nature of the explanations offered that makes them different, in principle, from any other social scientific explanation. The thinness of the data and plausibility of the comparisons, and indeed the cleverness of the explanatory strategies, may tend to make the particular accounts incompletely convincing and controversial but the resulting disagreements are not in any way mysterious. But both approaches involve, in one way or another, denying the assumption that the event in question is in some way singular.

There is another familiar strategy of explanation of singular events that seems more unusual from the standpoint of ordinary social science explanation and which is predicated directly on singularity itself. This strategy does not depend on constructing larger classes of similar events, or on breaking down the event into smaller sequences. It focuses instead on explaining the event from the point of view of the actor or actors involved. The availability of this approach, everyone will quickly notice, does not necessarily turn on the singularity of the event at all but rather on our access to the perspectives of involved agents. Any action, however ordinary, could be examined from the viewpoint of the agent or agents involved if we had some way of getting access to those viewpoints. In the case of singular events, however, it seems to me that this strategy – of agent-centered explanation – seems to recommend itself especially strongly to historians and others who are convinced of the event's significance. Who better to consult about Munich than Chamberlain or Hitler or perhaps those of their agents who were present for the discussions? Those involved seemed have a special access to what was said and thought and, perhaps, to whatever it is that made Munich singular (though this last claim seems disputable if singularity arose partially from later events that could not be foreseen).

Moreover, when an event is seen to be of special importance, participants and close observers are probably more likely to remember what they did, to save their records, to speculate and explain, justify, and criticize what happened. Such recollections will need to be treated with caution because participants may have scores to settle, people to protect, and further agendas to pursue. Or it may be that memory processes are particularly likely to distort recollections in particular and invidious ways. Participants may, like the rest of us, have a tendency to reconstruct events that they think they are remembering. Still, such information, if sensitively handled, might seem especially privileged material for explanatory purposes.

Narrative access facilitates the resort to a deliberative perspective on the event. From an agent's perspective it is often quite natural to invoke some idea of what would have happened (or what the agent believed or should have believed would have happened) in counterfactual circumstances to explain (or justify) her choices. Of course, we can have little evidence about the accuracy of counterfactual conjectures. But from an agent-centered viewpoint, that does not matter very much. To explain her actions, we need to invoke what she thought would have followed from alternative courses of action. Her beliefs can be criticized of course, and if we think that they were badly formed or incoherent, that might affect our assessment of the overall course of events. One might think about Munich specifically, that there was, in fact, a lot of motivated belief formation occurring – that participants came to believe what they wanted to believe – and that such phenomena are characteristic of events like Munich. Of course, insofar as beliefs at the time evolved under these pressures, we might think that those beliefs came under quite distinct pressures as subsequent events unfolded.

In any case, the worry is that narrative or agent-centered explanation might be overly seductive and therefore tend to crowd out other kinds of explanations. This is not worrying if there is reason to think that each of these explanatory strategies will converge in some sense. But I think there are some reasons to doubt such a convergence and to think that agent-centered explanations will tend to diverge systematically from the comparative and counterfactual approaches outlined above. Are these reasons good? Or do they turn on misapprehensions about either social scientific or agent-centered explanation?

III. Two approaches to explanation

I will distinguish two kinds of explanation – external or more or less causal explanation, and internal or deliberative explanation. External explanations represent agents as doing things because of some configuration of causal influences, broadly speaking. I mean to include not only ordinary causal explanations but also functional or structural explanations as well. Functional or structural accounts imply the existence of some vindicating causal processes; functional explanations in biology, if true, are vindicated by the existence of causal processes described under the rubric of natural selection. Structural explanations, to be complete, require some underlying causal processes standing ready to police structural irregularities. So, for present purposes I shall speak only of causal explanation where it is to be understood that I use this expression to stand for a broader class of social scientific accounts.

Obviously, the notion of "cause" needs to be spelled out and how it is spelled out will turn out to depend on the theory of behavior that is supposed to be at work in the agents. But, whichever theory is in play could work in one of two general ways: one, causal factors might determine behavior without "running through" the agent's deliberative or reasoning processes. When a doctor hits your knee with a hammer, it moves automatically. Or, causal factors may drive or shape the agent's deliberations by presenting her with reasons for action so that she "chooses" certain actions rather than others based on those reasons (this is developed in more detail in Pettit (1993)). Either way, we could understand her action as caused, but in the latter case she is also acting deliberatively (for reasons) as well.

We may consider some more or less classical examples of external explanations: Durkheim's account of societal suicide rates. Some fraction of people in certain kinds of societies will, Durkheim thought, tend to kill themselves at certain rates and these rates are said to be predictably related to certain attributes of the society. So, one might want to say that a society's suicide rate is explained causally. It is not clear which kind of causal explanation this is supposed to be, but it seems that it could be causal in the second (deliberative) sense. In any case, for this to be a causal explanation of any kind seems to require that there are mechanisms that tend to lead the particular individuals who commit suicide to do this. It is possible that a mechanism that produces this result might work by presenting individuals with reasons and that the individuals themselves all act deliberatively.[5] That the rate turns out to be predictable, while the individual event is not, might reflect the fact (if it is a fact) that the aggregate-level causal factors relate statistically to the circumstances of the individuals.

Drug addiction offers a different kind of example. Here, the agent does something compulsively, in spite of her judgment as to what she thinks she should do or in spite of an intention not to do it. There are various ways to understand this phenomenon. There may be deliberation and choice involved even though the outcome of the deliberative process is a

[5] Persons in certain social locations, for example, might find themselves faced with particular conflicts among reasons and there may be more such locations in some societies than in others.

predictable consequence of the working of a chemical addiction. The drug might work, for example, by altering the preferences of the individual so that she has a strong reason to take the drug when in certain chemical induced circumstances. Or, the drug might work by suspending some or all of the agent's deliberative capacities so that she takes the drug in spite of knowing that she has sufficient reason not to do so. In the first case, the drug works through agent deliberation; in the second, it works around agent deliberation. But either way, there is a causal relation between having the addiction and consuming the drug.

Internal or deliberative explanations work differently. An internal explanation for an action identifies reasons that an actor had that would rationally lead to or produce the action. An action is explained internally as an outcome of a deliberative process in which the agent is assumed to act for reasons; to take actions which are best on the reasons available to the agent. To "explain" in this sense is to "justify," in that an actor with the goals and beliefs held by (or attributed to) the actor would have, or could rationally have adopted that action. Internal explanation is, in this respect, a normative account of an agent's behavior in that an action is explained only if there are sufficient reasons, from the agent's viewpoint, for doing it. From the agent's perspective, the action was the best thing to do.

There are some qualifications and ambiguities that need to be addressed from the internal perspective that may be important later on. One is that an actor can have reason to do *X*, and do *X*, but do it for some other reason. I might have reason to go to Chicago tomorrow to complete some transaction, but I might actually go to Chicago because I was abducted by kidnappers at gunpoint. In this case I have two sufficient reasons to do *X* and so, in either case, *X* is justified and therefore internally explained. But there are, however, two different explanations and they are not compatible. In this case, I suppose it is clear that the reasons emanating from the kidnapper "trump" the other reasons and so the kidnapper-centered explanation is the correct internal account.[6]

Internal explanations do not involve any comparative test, at least not directly or obviously. We explain an agent's behavior by considering

[6] Some will want to say that the kidnapper-centered explanation, in which I comply with the kidnappers' requests because they are pointing guns at me, is causal. I am with Hobbes on this issue: the agent might have chosen to die or to risk death even if these were unattractive choices. Going along with kidnappers is a choice which, no doubt, has an awful lot to recommend it, and the agent's complying with the kidnappers' request is explained by its being the best course of action on the reasons presented. So, I resist the idea that this is a causal explanation. If one takes sufficiently compelling internal accounts to be causal, it is difficult to separate this example from one in which the kidnappers bop the person on the head, stuff them in a bag and dump them in Chicago, which does strike me as a causal story.

whether she had sufficient reason to take the action and, in fact, took the action because of that reason. This may involve appeal to the agent's beliefs about counterfactual circumstances, of course, and perhaps this would include her beliefs about what would happen in those circumstances. She may decide to act in order to bring about some desired state of affairs or to prevent the occurrence of some event. Still, the explanation itself does not appeal to anything other than the agent's own reasons for action, and not to what other agents would do in similar circumstances.

Examples: When we seek to understand or explain everyday events – why you were late to dinner last night – we normally seek an internal explanation. You say that your daughter needed help with her homework and you needed to delay your departure from your house for a few minutes in order to assist her. Your belief that your daughter was in need of assistance (together with your belief in the unimportance of being a bit late for dinner) provides a reason for delay and is the reason you acted on in leaving your house as late as you did. This account provides a justification for your leaving your house late, and an explanation to me for your tardiness. Obviously, the explanation in this case is also meant to excuse your lateness too. To say that we expect an internal explanation in this case does not mean that external accounts are unavailable. Presumably we would be reluctant to accept as true an internal explanation unless we believed that agents like you would generally find it reasonable to respond to a request for homework help. But it would be odd for you to offer the external account as an account of your action.

Ethnographic explanations are typically internal. For example, Richard Fenno (1978) explains how each of his Congressmen tried to present themselves, in their districts, by appealing to reasons they gave or appeared (to the analyst) to have and to have acted on. More elaborate ethnography seeks to explain the system of reasons that are available for motivating action in a community or culture. In his book, Fenno hoped also to provide an external account: he argued that one might be able to explain which kinds of Congressmen and/or which kind of congressional districts would tend to produce which kind of "home style." If he is right, the internal explanations that he offered might converge on a unified external or causal account of homestyles.

But there are reasons to be skeptical of such a convergence, at least when put forward as a general claim. The main reason is that there seems to be a fundamental distinction between causal accounts and justificatory accounts of action. Causal accounts of how things go seem to be, in their nature, positivistic in that they are offered as true accounts of the phenomena to be explained. Internal explanations or justifications seem to be inherently normative and so not to be anchored in the way the world is. They certainly don't offer causal explanations for the action and would not be expected if a causal explanation was sufficient. If you could not help doing *X*, there is no need to give reasons or justifications for doing *X*. This is rough worry, and one that I will argue against, but I think it captures part of the skepticism about convergence. Moreover, there are a lot of people – social scientists, lawyers, philosophers – who seem to think that one or another of these two kinds of explanations is either impossible or unimportant. One area where this battle has raged is in law.

For example, consider the problem of explaining the structure of contract law within the United States. There are roughly two approaches to this problem within the legal academy: one account, law and economics, claims that most of the structure of this law can be explained as the structure of legal rules that would produce the most total wealth in the society. The law and economics of contracts explains how contracts are made, which contracts can be breached, what damages should be paid, which kinds of evidence count in ascertaining the existence of a contract and so on. The alternative approach explains these doctrinal elements as the consequences of shared and deeply held values that are at stake in contracting. An explanation for a rule, in this latter case, is a justification of the rule in terms of values embodied in the legal system. Unsurprisingly, each approach is better at explaining some legal structures than others and, for this reason, proponents of each approach sometimes try to fuzz the distinctions between them. Internalists sometimes recognize that wealth maximization may itself be a value and externalists are sometimes forced to endorse broader sets of values than wealth maximization as the driving causal motor for their theories. But the important point here is that proponents of each side tend to think of arguments from the other viewpoint as irrelevant, irreverent, and unimportant.

Part of the reason for this disagreement lies in different attitudes about causation in human affairs. Internalists, at least since Aristotle, see actions as taking place for reasons and tend to admit the possibility that an agent may fail to do what she has reason to do. Indeed, the concerns of internalists are at least partly driven by normative or moral concerns and a causal account of human action tends to undermine the kind of human agency needed to make normative appraisal relevant. Externalists are concerned to explain regularities they see in the social world and tend to believe that the only or the best way to account for such regularities is by appealing to more or less general causal mechanisms. From an externalist viewpoint, normative concerns are wholly separate from explanatory ones.

Some have argued in favor of a practical preference for seeking external rather than internal explanations when both kinds of explanation might be available. Gary Becker (1976), for example, believes that actions are in fact brought about by agents seeking preference satisfaction given their beliefs in circumstances of constraint or scarcity and, in this sense, would be expected to believe that deliberative or internal accounts of economic activity are freely available. But he also thinks that because 'we' (economists) cannot really know much about the preferences or beliefs held by individual agents – we have no special access to these mental attitudes – good economic explanations will focus on showing how it is that the constraints do most of the causal work in producing actions, and making little appeal to subjective or deliberative factors. As far as any particular agent is concerned, the constraints themselves are likely be produced causally (through variations in prices or technologies). Thus, as I understand him, Becker has a practical preference for external or causal explanation even when deliberative or reason-based explanations are, in principle, available.

IV. Are internal and external accounts necessarily in conflict?

Superficially, external explanations claim to be more or less causal accounts whereas internal explanations explain actions by making them intelligible – showing how they are justified from the agent's perspective – which doesn't seem to be a causal story. One might try to convert internal accounts into causal accounts by taking reasons as causes. This would involve saying that if an agent has most reason to do *X* in circumstance *C*, that *X* is caused by facts about the agent and her circumstances. On the face of it this seems to involve denying that the agent has a capacity to choose not to do *X*. Or, it is to deny agency. Is this objectionable?

Maybe not. We are, after all physical/biological creatures embedded in the material world and, in that sense, whatever it is that our bodies do must be compatible with physical laws. So, if an agent does *X* (where *X* is described physically), there seems to be little objection to saying that *X* is caused. But, there are reasons to think that *X*, taken now as an action, cannot be completely described physically. The description of an act also includes the intentions with which it is done – waving my hand (a bit of behavior) might be a signal (an act) if I have a certain intent, or it might be part of an effort to dry my hands (a different act) if that is what I intend, or it might simply be a twitch (a mere bit of behavior) – the physical description doesn't tell us whether *X* is an act or what act it is. So, while there is no problem with saying that the physical aspect of *X* is caused, it is a stronger claim to say that *X* as an act is caused. Whether that stronger claim is sustainable depends on whether we are willing to attribute causal force to intentions.

There two issues here: first, what kind of linkage is there between intentions and actions? One might be willing to say that once an intention to do *X* is formed, the act of doing *X* follows automatically. This might make sense for certain kinds of simple intentions and actions – intending to pick up a fork might be so closely connected to picking it up that there is little room for deliberation or choosing between the intention and the action. But most acts of interest to us as social scientists are much more complex. If I form an intention to vote on election day, for example, there are many things I need to do to actually vote: I need to see to it that I get registered to vote, I need to find time during the day to get free from competing obligations, I have to find out where the polling place is and get there. Michael Bratman (1999) argues that intentions are best seen as something like partially specified plans that serve to regulate our further deliberations and choices as to what actions to take. If this is right, there is often a big gap between intentions and actions and a causal story seems implausible.

Second, we need to ask how intentions are formed. Some kinds of intentions might arise causally: if you haven't had any liquid in couple of days and a glass of water appears in front of you, you might "automatically" form an intention to drink it. But most intentions do not arise automatically in that way. On the deliberative account, the intention to do *X* arises from *X*'s being the best thing for you to do in circumstance *C*. But being the best thing to do is a normative notion and so seeing *X* (including its intentional aspect) as caused seems to involve saying something like that norms have causal force. I agree that they have force, but of a different kind.

I think it is more natural to work the other way around and to try to assimilate external to internal explanation. That is, perhaps we should bite the bullet and admit that human action is deliberative and that we cannot escape the necessity of recognizing the priority of an internal perspective if we want to understand action. In particular, I want to argue that the notion of causation in human affairs needs to be removed from the physicalist model and given an understanding that fits with internalism. This will involve restricting the role that causal explanation can play in explaining human action. This may seem like a strange thing to do but I think there is a familiar model for making exactly this move: rational choice theory, or at least what I propose as the best way to understand the relation of that theory to actual human behavior.[7]

[7] For purposes of this chapter I am not going to keep reminding the reader that there are many variants of rational choice theory, indexed largely by the preferences that agents are assumed to have. Some variants of the theory posit that people are motivated to maximize

V. A social ontology of rational choice theory

Rational choice theory can be understood to offer both internal and external accounts of action. Internally, rationality explanations work through a mental model in which beliefs, desires, and actions are supposed normally to be related to each other. This is an internal idea if we think of this relation being established deliberatively, in reasoning about what to do, or think, or believe. But externally, rational choice accounts seem to rest on some presumed causal regularities as well. Having beliefs B and preferences P can be understood to cause act A where A is a best action given B and P. There are some problems for the external interpretation of course: what if A is not the unique best act? How do we understand the failure of actors to take best actions, or to exhibit variability in their choices? From an externalist viewpoint we can solve these problems pragmatically (by representing decision problems in such a way that there are always unique best responses, or by taking beliefs, preferences, and actions to be random variables whose values are determined according to stochastic assumptions, or by recognizing measurement error, etc.).

Even with these assumptions in place, however, I don't think that the external view of rational choice theory is compelling. It makes causal claims that simply seem to deny that actual agents make choices and could refuse to make them as well. At best, the external account offers a causal story about perfectly rational machines that are in fact not deliberating at all but are infallibly picking best actions given their "preferences" and "beliefs," together with a claim that real human actors will behave in approximately the same way. I shall argue that there is a plausible way to put this story. It entails seeing the rational machines of the previous sentence as machines that are designed to implement a norm of rationality. The machines will behave rationally according to some causal processes – which depend on how they are built internally, whether out of gears and levers, digital computing devices, or whatever – because they are designed to implement a rationality norm. From the standpoint of the norm, the connection among B, P, and A is of course not causal

their wealth, or perhaps their social status, or in some other more or less private-regarding way. Other versions, more common in political science, posit ideological motivation. And still others make no substantive assumption as to what ends agents desire – only that the desires, beliefs, and actions are connected in the way the rationality hypothesis requires. Sometimes this distinction is summarized by saying that the latter agents are "thin rational" and the others "thickly rational." This is an unfortunate usage in that it suggests a continuum or dimension of thickness, whereas like Tolstoy's unhappy families there are many different kinds of thickly rational agents, and therefore many different thick rational choice theories.

but is conceptual or, if you prefer, constitutive. Choosing best acts given beliefs and preferences is what it means to be rational; we would not call our machine rational unless it actually did implement the rationality norm.

Putting the matter this way also illustrates the connection between internal and external explanations: real human agents regard rationality as normative – they take it as a defect of their action choices that they fail to fit with their beliefs and desires – and that is why internal accounts are illuminating.[8] And, the fact that real human beings are effective enough in actually behaving as rationality requires gives us reason to think that we can explain some of their behavior in the external way. This metaphorical talk of machines needs now to be grounded in human behavior. Indeed, if humans were not successful in this way, it is hard to see how social life would even be possible. So at least as a rough statistical matter, we can understand much of what others do, by taking an external explanatory "stance" with respect to their behavior. But, it must be remembered that a stance or perspective – even a highly successful one – is not the same thing as an ontological claim about how humans act.

Pettit (1993) has argued that human beings are not merely intentional systems (entities that seek ends), though they are that, but are also thinking beings.[9] So, satisfactory social explanations – and here he means internal explanations – need to take account of both aspects of humanity. The notion of an intentional subject is familiar enough not to need discussion; what is new in Pettit's idea is that humans have a distinctive capacity (thinking) that distinguishes them from merely intentional systems (animals, thermostats, etc.). He argues that this capacity – perhaps the most distinctive such capacity – is the capacity to identify and follow rules. By this he means the capacity to identify and follow rules – of the kind Wittgenstein, Kripke, and others have discussed – that are potentially applicable in indefinitely many circumstances. Philosophers writing on this topic typically choose their examples from language or mathematics, and are usually concerned to show that rule identification is impossible. But as Pettit argues, these examples support the idea that

[8] This statement needs more defense and qualification than I can offer here. I suppose actually that agents do not really take formal rationality as a norm directly, but are instead disposed to act in ways that further their aims. Effective pursuit of aims, however, entails that the agent is generally acting rationally in a formal sense. So, I would argue that the norms that agents accept are concretely related to the aims they actually have, and the alternative actions they may actually take.

[9] My account differs somewhat from Pettit's in that I argue that rationality is a norm and not a property of human beings as intentional systems. Moreover, I think that the rationality norm has the characteristics of a rule that actually requires thinking to follow.

the capacity to follow rules, if there is one, must rely in some way on background assumptions and cannot be infallible. Norms may be taken to be special examples of rules in that they are supposed to apply in indefinitely many situations and to direct action in those cases. Therefore, on this account, the capacity to be guided by norms is an instance of the capacity to identify and follow rules more generally. Moreover, the rationality hypothesis, or any variant of it, may also be seen as a rule in Pettit's sense, in that it directs the choice of action in indefinitely many circumstances and conforming to rationality hypothesis is an example of rule following. Thus, if Pettit's conception of rule following is right, we might expect human beings to exhibit (at most) imperfect rationality, in the same way that they would be capable of complying (imperfectly) with other norms and rules.

On this account we can understand that human beings are physical entities, subject to ordinary causal laws, and that these causal regularities produce in them certain (more or less hard-wired) capacities to act as particular norms direct (however imperfectly). One could well ask why physical human beings would be impelled to conform to a norm even if they could do so. This is a question of motivation and is a familiar problem for moral theories. Philosophers since Aristotle have insisted that seeing what one ought to do is one thing, and being motivated to do it is something else altogether. Indeed, a common criticism of a moral theory is that it fails a test of motivation: that it makes demands on people that they cannot ordinarily be motivated to respond to. Various versions of utilitarianism are often criticized in this way.

Conversely, some versions of rational choice theory are supposed to offer particularly compelling reasons to people. Material self-interest, for example, is thought to be a strong motivation in many circumstances (one can imagine evolutionary causal mechanisms that might explain why this would be so). If this is right, one would expect the behavior of physical humans to conform roughly to the operation of that version of the rationality hypothesis, at least where there are not other competing and compelling normative reasons to refrain from such action. One can also imagine causal mechanisms that would support status motivation and many biologists think that one could similarly explain the attraction of certain forms of altruism as well.

So, there is a role for ordinary causal processes in this story; ordinary causal mechanisms will no doubt play a part in shaping human capacities. And, the attraction of a norm to an agent may also be explained causally. But there is no asserted causal relation among the mental attitudes of the human beings and their actions. The relation among these entities is tautological (with beliefs *B* and preferences *P*, a rational entity would, definitionally do *A*),[10] and normative (doing *A* is attractive to the agent on the grounds that this is what rationality recommends). In this respect, the rationality hypothesis offers an internal account of action: rational agents choose best actions, based on the reasons they have, and so their choices are justifiable. But it offers an external account of action insofar as capacities and motivations play a causal role in setting up the parameters for rational deliberation.

This view is, it seems to me, quite different from the standard interpretation of rational choice theory, one that seems accepted by both proponents and critics. On that account, rationality can be understood as a property of the human mental or neural apparatus that generates behavior causally. Put a (fully described) rational creature in a certain circumstance and it automatically does a specific thing. As I suggested before, this view seems (needlessly) to deny that people make choices. The construction I offered above leaves room for deliberative rationality and choice while, at the same time, recognizing some room for causation (in forming capacities and motivations). But the causal aspects of an explanation have little to do with rationality itself. In that sense the main work of rational choice *theory* is internal. Whether there are, additionally, external/causal implications turns on other facts about humans (capacities and motivations).

One implication is that the propositions from rational choice theory cannot be expected to represent or approximate causal regularities exhibited in human action. They are statements of how rational creatures would act – they are statements about how normatively directed creatures would act and statements about how people trying to follow that norm should act – and whether actual people come near to attaining that standard is going to depend on how closely they either can or want to comply with the norm itself. I have little to say about the capacity to follow rules (including rationality) in general or how that capacity might vary over contexts. But the attractiveness of various rationality norms seems likely to vary significantly. Consider, for example, the wealth-maximizing

[10] I won't go into the matter here but there is a deceptive simplification in this formulaic expression of the rationality hypothesis. Beliefs and preferences are both normative objects too. There are things, based on what I have experienced, that I ought to believe. No doubt my actual beliefs only approximate to the beliefs I should have. Perhaps more controversially, I think the same is true of preferences. There are things I should want and perhaps I do not want them at present. There are various ways to make sense of this; perhaps I don't fully appreciate the need for a gas-driven generator because I don't assign a high enough probability to a power failure, or because I don't fully appreciate what it would be like not be able to run my expresso machine in the event of a power outage. For now, these issues are off the table.

variant of rationality. Such a norm is likely to conflict with other attractive norms in various contexts. An agent might, for example, find wealth maximization an attractive guide where she is deciding on a retirement portfolio. But, she may resist following that norm if it recommends purchasing shares of tobacco firms. So we might think that any particular version of the rationality hypothesis will work better in domains without attractive competing norms.

VI. Limitations of rationality: a digression

I should add here that I am taking rationality explanations as examples of social scientific explanations in that they purport to offer causal or positivistic explanations. I do not deny that other "paradigms" may offer other positivist accounts of behavior. But, a purely causal explanation of human action seems defective on grounds I argued above; a theory that fails to recognize and represent how things seem from the agent's viewpoint, deliberatively, is missing something important about human action.

As an example, bounded rationality theories purport to offer alternative more or less causal or external accounts of human behavior, perhaps sometimes better accounts in some contexts than explanations offered from a rational choice perspective. Indeed, the rule following account offered here might seem, on the surface, to offer a reason to believe that as causal explanation, bounded rationality would likely be superior to rational choice theory. After all, bounded rationality stories seem to rely on what humans, as they are hard wired, are capable of doing. Such stories emphasize the incapacities of people to be rational even when they want to. And, on my account, these incapacities are likely to be the kind of thing that can be explained causally. So we might very well expect such theories to lead eventually to superior causal explanations of human behavior. But will such theories offer better accounts of human action – of what people do intentionally or deliberatively? Here I have doubts.

Can there be norms of bounded rationality? On the surface there seems to be no problem. Bounded rationality has the outward appearance of a norm: it directs an agent to do or refrain from doing things in indefinitely many circumstances. But, there are some internal problems too. Assuming, as I do, that we are cognitively constrained, we can nevertheless imagine, or build, systems capable of more rationality than we have. That is, at least in specific domains, we can build systems that transcend our limitations or capacities to behave rationally. We buy spreadsheet programs and programs to compute our taxes, presumably because, within their task domains, they can be counted on to do a better job at complying with rationality requirements than we would do ourselves. Expert chess computers can by now beat all human players, and are presumably less cognitively constrained on chess problems than we are. Indeed, long ago it was proved that fully rational strategies exist in chess (or at least in the near relatives of chess that have finite game trees; i.e., the versions of chess that are played in professional chess tournaments).

So, at least in these cases, we could imagine a norm of bounded rationality but probably few of us would find such a norm compelling. We want to find out the best or most powerful chess program, not one that best tracks our limited capacities. We remain dissatisfied if we are told that there are better answers available even if we ourselves cannot reliably find them. So, in this sense, a boundedly rational "norm" has some real defects from an internal perspective; we know it would be giving us wrong or inadequate advice. It may represent the best that we can do on our own behalf, being wired as we are. And, indeed insofar as bounded rationality theories are inductively derived from empirical experience, their appeal is pretty much purely external.

From an explanatory standpoint, the attraction of bounded rational theories is that they provide a wedge by which recognizably causal factors can come into play directly in explaining human behavior. We can say that an actor took the action she did in part because she was cognitively unable to recognize the existence of better alternatives, or even to see what she did as an occasion to "choose" at all. Perhaps those statements are parts of good explanations. But then, have we really made her action intelligible? We may have explained why she failed to see a choice that was really there but, at the same time, we have made the action she took less than a full-blooded deliberatively chosen action. That may be the best route to a more or less causal account – perhaps what we initially thought was a deliberate action was actually something less than that. But, in some respects, adopting this explanatory strategy involves abandoning the idea of explaining human action rather than mere behavior. Explaining action – deliberatively chosen behaviors – requires something more than this. It requires showing that what the agent did was a best or most effective way of pursuing her purposes. And this entails establishing the embedded normative assertion.

VII. Conclusions

Rational choice theory provides one way in which the internal and external perspectives can be bridged. And the hermeneutical perspectives taken by ethnographers exemplify another way of understanding these perspectives. Obviously, there are very important ways that practices of descriptive ethnography and theoretical economics differ, but they share one important feature. Both are, at bottom, normative enterprises aimed at showing how the agents – real or imagined – can act and have reason to act in ways that comply with norms they accept. Both also make more or less implicit causal claims that agents will generally tend to act as norms they accept direct them to. But this causal claim is not really a part of either theoretical apparatus. That the causal assertions tend, in many circumstances, to be empirically true makes these approaches useful to understanding social action.[11]

Naturally enough, these two bridging paradigms have generated distinct methodological programs and are fitted for addressing distinct explanatory problems. It is too easy, however, to overstate these differences and to lose the sense in which the common subject matter – understanding human activity – yokes these perspectives. Most often, the real differences between these approaches is driven by practical considerations that arise from the explanatory questions that are asked rather than any deep division between them.

Methodology, on this account, is fundamental to the social sciences. Methodological awareness forces us to recognize the pluralism of social reality: the fact that description and explanation are relative to perspective; that human beings are embedded in causal processes but are also responsive to and guided by norms. So a self-conscious focus on methods cannot be abandoned without giving up on social science, and indeed history, in favor of mere uncritical chronicle.

REFERENCES

Allison, Graham. 1971. *The Essence of Decision*. Boston: Little, Brown, and Company.

Becker, Gary. 1976. *The Economic Approach to Human Behavior*. Chicago: University of Chicago Press.

Bratman, Michael. 1999. *Faces of Intention: Selected Essays on Intention and Agency*. Cambridge University Press.

Durkheim, Emile. 1966. *Suicide: A Study in Sociology*. John A. Spaulding and George Simpson (trans.). New York: Free Press.

[11] Debra Satz and I have argued elsewhere that different causal processes might explain why and when agents respond to normative recommendations that arise from rationality. Some of these might be internal to the agent, some might be learned, and some might arise from the environment of interaction. But whatever it is that explains when rational choice theory provides a successful account of human action is itself part of rational choice theory. See Satz and Ferejohn (1994).

Fenno, Richard. 1978. *Homestyle: House Members in their Districts.* Boston: Little, Brown and Company.

Krasner, Stephen D. 1972. "Are Bureaucracies Important? (Or Allison Wonderland)." *Foreign Policy* 7: 159–79.

Pettit, X. 1993. *Common Mind.* Oxford University Press.

Satz, Debra and John Ferejohn. 1994. "Rational Choice and Social Theory." *Journal of Philosophy* 91: 71–87.

Part II

Redeeming rational choice theory?

8 Lies, damned lies, and rational choice analyses

Gary W. Cox

What is rational choice? Green and Shapiro (1994) treat it as an empirically worthless but nonetheless highly pretentious *theory*. Shepsle (1995) treats it as the worst social scientific *paradigm* we have, except for all the rest. And I have elsewhere (Cox 1999) reduced it to game theory, called it a *methodology*, and analogized it to statistics – hence my title.[1] In this essay, I consider rational choice both as a paradigm, identifying what I view as its "hard core," and as a methodology, again reducing it to game theory and comparing it to econometrics.

Rational choice as a paradigm: behaviorism, rational choice, and cognitive choice

One way to locate rational choice theory is in terms of what it assumes about human psychology. Social science does not necessarily need to incorporate a full-blown theory of psychology, just as cell biology does not necessarily need to incorporate a full-blown theory of molecular interaction. However, if one does not need *all* the complexities that psychologists or cognitive scientists discern, what is the minimum one does need?

Radical behaviorism in social science would take the position that one can get by without any theory of psychology at all – and should do so since mental phenomena are inherently (or at least practically) unobservable, hence outside the purview of an empirically based science.[2] Rational choice theory can be viewed as refining what philosophers call "folk psychology" (Ferejohn and Satz 1994b), the root idea of which is that actions are taken in light of beliefs to best satisfy desires. Belief-desire

[1] The original reference, Mark Twain's, was to "lies, damned lies, and statistics."

[2] A radical behaviorist would claim that whether or not mental processes intervene between antecedent observables and predicted behaviors, nothing scientifically useful can be said about them. Moreover, a radical behaviorist might argue, via Hempel's "theoretician's dilemma," that theories are superfluous and one is better off relying on empirical regularities. On the dilemma, see Kaufman 1967 (272). What I call "radical behaviorism" Ferejohn and Satz (1994a) call "radical externalism."

psychology is the implicit model of a vast horde of novels, narrative histories, and journalistic accounts; and the explicit model of many treatises in economic, sociological, and political science.[3] What one might call *cognitive choice theory* is an emerging body of social scientific inquiry in which something more complicated than folk psychology is (formally) assumed about individuals. A good deal of novels, histories, and journalistic accounts might stray (informally) into this camp, too.

By this accounting, rational choice can be distinguished from what it is not on two fronts. On the one hand, it is not as squeamish about referring to unobservable mental states as was radical behaviorism. On the other hand, it confines itself to just two categories of mental phenomena – beliefs and desires – where in principle one might have indefinitely more complicated psychologies than that.

Distinctions within the paradigm

Let us call the definition of rational choice just given – whereby actions are chosen in light of beliefs to best satisfy desires – *broad*.[4] There are distinctions within this broad category that others have offered. Sometimes one hears a contrast between hard and soft rational choice. Sometimes one hears that rational choice is limited to the study of instrumental rather than consummative utility and, moreover, that this distinction separates economics from sociology (Barry 1970). I shall consider each of these distinctions in turn.

Hard rational choice

One can tell a story about another person's actions and impute to them (a) inchoate or emerging preferences and (b) beliefs that are not fully responsive to information – and still remain pretty squarely within belief-desire psychology. The principle behind hard rational choice analysis is to focus on the purest sort of belief-desire stories, in which preferences have crystallized, beliefs are fully responsive to all information, and pursuit of preferences is maximal (within stipulated constraints). In this way, the main analytical engine – namely, "pursuit of goals in light of beliefs" – is as clear as can be.

[3] If one is going to attribute any psychological underpinnings to rational choice theory at all, folk psychology seems the natural attribution (Ferejohn and Satz 1994b). However, one can reasonably argue that rational choice is not about or much dependent on individual psychology; it is about and dependent on structural conditions (Ferejohn and Satz 1994a).

[4] This definition is similar to "a rational choice is one that is based on reasons . . . ," per Lupia, McCubbins, and Popkin (2000: 7).

But what are "crystallized" preferences and "fully responsive" beliefs? If agents view objects of choice as possessing various attributes and have not established stable cardinal tradeoffs between those attributes, then their preferences have not "crystallized." They may exhibit cyclic choices if they use only ordinal rankings by criterion as the inputs to their decision (Arrow and Reynaud 1986). They may exhibit temporally unstable choices if cognitive limitations mean that they only consider a sample of criteria at any one time (Zaller and Feldman 1992). Only if agents have established stable cardinal tradeoffs will their preferences crystallize and satisfy the usual technical assumptions (e.g., transitivity).[5]

To understand what "fully responsive" beliefs are, begin by noting two ways in which beliefs might *not* respond fully to information. First, limits on agents' memories might prevent them from accessing all the relevant information at the time of decision. Second, limits on agents' deductive capacity might prevent them from wringing all the implications out of a given amount of information. In order to deflect attention away from these very real limits, hard rational choice analysis (as defined here) assumes them away. What is left is perfect recall and unlimited information-processing capacity, typically represented by updating according to Bayes' rule.[6]

The principled reasons for ignoring limits on information processing are two: it is hard to identify those limits; and the limitless model provides an interesting baseline, one that eons of evolutionary competition between myopic agents or shorter periods of competition between clever agents will sometimes approximate. Note, however, that if we knew precisely how to model limits on memory or deductive capacity, there would be nothing in the broad conception of rational choice that would prevent us from incorporating those insights. Moreover, there is a growing body of work in economics and game theory that does seek a justified modeling of memory and deductive limits (e.g., behavioral economics; game theoretic modeling of learning).[7]

Possibly we should call some of these new approaches "cognitive choice" rather than "rational choice." But many of them retain as their essence the engine of optimizing behavior in light of beliefs and have simply allowed a fuller picture of the information-to-beliefs mapping. Thus, I consider them rational choice too, albeit with not quite the hard

[5] For a fuller discussion of the assumptions needed to establish various levels of "crystallization", see Schwartz 1986.

[6] Note that one implication of this definition is that early work in game theory on games of imperfect recall is not hard rational choice.

[7] On behavioral economics, see e.g. Mullainathan and Thaler 2000. On game-theoretic modeling of learning, see e.g., Fudenberg and Levine 1998.

edge of models that abstract away from deductive frailties and other limits.

Note that this particular definition of *hard* rational choice leaves as *soft* rational choice anything in which the information-to-beliefs mapping is a bit more complicated, or preferences a bit less consistent, but the motive force in the theorizing is still pursuit of goals in light of beliefs. This surely covers a very wide range of theorizing, some of which is highly mathematical, some of which is completely non-mathematical.

Instrumental rationality

One sometimes sees *instrumental* rationality touted as the defining characteristic of hard rational choice analysis and also as separating economics from sociology (e.g., Barry 1970). The definition of instrumental rationality that some critics seem to have in mind is this: instrumentally rational agents are those who have internalized no norms of any kind, so that they can take any action without feeling good or bad about it; only the distal consequences of their actions motivate them. For example, an instrumental view of voters would be that they turn out, not because of any internalized sense of duty, but rather because they think their one vote's chance of affecting the outcome is worth 15 minutes of driving (cf. Downs 1957).

The prominence of instrumental utilities in rational choice notwithstanding, I see no defensible way to confine hard rational choice to instrumental utilities. There are many act-contingent utilities that fit straightforwardly into even a view of humans as solely wealth-maximizing. In the case of the turnout decision, for example, payments to sit the election out are non-instrumental reasons not to vote (Cox and Kousser 1981), while free doughnuts at the polling place are non-instrumental reasons to vote.

Neither do I see any defensible way to exclude internalized norms from hard rational choice analysis.[8] If I really do feel good about saluting the flag, is it irrational for me to salute for that reason? Must I wait till there is a devious instrumental reason to do so? If hard rational choice just means that beliefs are maximally responsive to information, desires are clear, and pursuit of desires is maximal (within stipulated constraints) – then taking actions prescribed by internalized norms is certainly within the ambit of the paradigm.

[8] I say this even though rational choice analysis in the study of American politics arose in contradistinction to sociological approaches emphasizing norms. Rational choice can accommodate norms but only certain explanations of them comport well with the paradigm – and the view of norms prevalent in the study of American politics was not one of those.

In addition to this elementary point, one might add that the root idea of a norm – a shared expectation about appropriate actions in given situations – is a naturally occurring one within game theory. Moreover, one can find hard rational choice analyses of norms in Ullmann-Margalit (1977), Coleman (1990), Sugden (1986), and elsewhere. I consider one example more extensively below.

Put in disciplinary terms, much of sociology is within the ambit of rational choice as defined here. Much of psychology, however, goes beyond the paradigm or begins to stretch it – as we shall see in the next section.

Criticisms within the broad paradigm

Many critics of rational choice argue that people are psychologically different than the core model assumes. We might array the critics under two broad banners: those who are satisfied with belief-desire psychology but dissatisfied with game theory's treatment of it; and those who wish systematically to go beyond belief-desire psychology. In the first, less radical camp there is a great variety of positions of which I shall note only two.

More realistic belief-desire psychology: contra hyper-rationality

Some critics disapprove the hyper-rationality that agents in some games exhibit. In particular, the speed and accuracy with which information is translated into beliefs troubles many. Indeed, the agents in some game models would never miss a question on a math test and infallibly deduce subtle implications from scant information. If one thinks of brains as evolved computing machines, however, then they should be only so good at mapping information into beliefs – and then only in evolutionarily frequently occurring situations.

I would make two points about assuming perfect information processing. First, although it is clearly unrealistic, that fact alone should not lead us to reject theories based on such an assumption. Second, competitive pressures will, in certain contexts, lead to better information processing – with perhaps a sufficiently close approximation of the perfect benchmark so that models based on that assumption are useful.

To clarify the first point – about unrealistic assumptions – recall that planets in Newtonian celestial mechanics are reduced to point masses subject to gravitational attraction. Among other things, point masses do not have weather, oceans, or inhabitants. Should we discard Newtonian celestial mechanics because its assumption about planets is unrealistic? If all we want the theory to do is predict interactions between bodies in a system, it does a good job with a noticeably simplified view of planets. It would, however, do a bad job of predicting weather.

Much the same can be said about hard rational choice assumptions about persons. Persons are reduced to unitary actors with perfect information-processing capacities. Among other things, unitary actors do not have moods, teeth, or parasitic infections. If all we want the theory to do is predict (some features of) interactions between buyers and sellers in a market, it does a good job with a noticeably simplified view of persons.

A response to the analogy just posed between Newton's unrealistic planets and rational choice's unrealistic humans is to (1) point out that Newton's theory was considerably more successful than is rational choice; and (2) claim that this ruins the analogy. Implicitly, this line of response seems to adopt a version of Friedmanite instrumentalism: if a theory is successful enough, then we'll tolerate crazy assumptions; otherwise, we won't. If empirical success is the only reason we have for tolerating unrealistic assumptions, then the debate would of course return to how much empirical success rational choice has had (on which see Green and Shapiro 1993; Cox 1999).

There is another issue at stake regarding unrealistic assumptions, however. The planets of Newton's celestial mechanics and the molecules of the perfect gas laws both utterly lack internal structure. Planets and molecules are subsystems in the respective theories under discussion, while the main focus of concern is the larger system's behavior. The unrealistic assumptions serve the purpose of black boxing or bracketing the subsystems, denying them any explicit internal structure, for the purpose of getting on with analysis of the larger whole. It turns out that, once one opens up the subsystem, one does get an additional payoff in both cases: quantum mechanics supersedes both Newtonian mechanics and the gas laws (and does so by dealing with the internal structure of the planets and molecules).

Another way to think about rational choice in social science, then, is that it focuses on the system of human interaction and black boxes the constituent parts of the system (humans). The particular bracketing move is intermediate between radical behaviorism (which would be closer to Newton's planets in that it attempts to assume *nothing* about the internal structure of humans), and richer conceptions of humanity (e.g., cognitive science or traditional political theory). The argument is about how much internal structure, how much human nature, we need to bring in to our models of social interaction. However much you decide to bring in, you can presumably always be criticized for not appreciating even richer conceptions that tap into levels even lower in the architecture of complexity (Simon 1962). Moreover, however much you decide to bring in, the resulting social science is not comparable to Newtonian mechanics in precision.

Now consider the matter of competitive pressures pushing actual information-processing capabilities closer to the theoretical optima. We might *model* cheetahs as if they simply received a full and accurate three-dimensional picture in their brains of the antelopes in front of them. This would ignore the obvious reduction to two dimensions that physically occurs at the corneal surface and the tricky recreation of the fuller three-dimensional view in the visual cortex. Yet, to a pretty good approximation, cheetahs act "as if" they just got a full and accurate three-dimensional model.

Competitive pressures in social contexts may mean that the hard rational choice assumption about information processing is sometimes reasonably well approximated, though probably not as well as in the cheetah example. That is, there are choice contexts in which persons have relatively simple information-processing tasks to perform and have trained in them, so that their performance is nearly optimal.

The general point is this: information-processing systems that have evolved under competitive pressures may mimic first-best information processing, even if the system's actual processing of information cuts corners, makes not fully justified mathematical inferences, economizes on storage and so forth. We may not want to see how sausages (beliefs) are produced but in the end they are pretty good and might even be approximately Bayesian.

Of course, even if people process information better than a detailed look at how they process it would suggest, there is still ample room to conjecture that analysts can do better with more complex models than provided by the hard-rational benchmark. If our goal is to understand the details of why people take certain actions, or how they decide, then certainly we can do better. One can agree with Lupia, McCubbins, and Popkin (2000: 12) that "our treatment of how people reason should be informed by modern scholarship about how cognition and affect affect information processing." If our goal is to understand aggregate patterns of interaction between individuals, then the case for more realistic models of decisionmaking has to be made on different grounds. In particular, do more realistic agents interact differently than do hard-rational agents, in such a way that we can better predict human social interactions? As an example where one might answer "yes" to this question, consider recent work in behavioral macroeconomics, in which various kinds of cognitive costs lead to price stickiness, and price stickiness implies that monetary policy will have real effects (Akerlof 2002). Here, there is an important difference between fully rational agents who instantly adjust prices to reflect all new information and "near-rational" agents who are not quite so speedy.

Can we develop principles that allow us to select the most useful decisionmaking model (how complex?) for different purposes? Hopefully, we can agree on a class of more general information-processing assumptions which include the traditional hard position as a special case, and work from there.

Endogenizing belief and desire

A second criticism of game theory, from those basically satisfied with belief-desire psychology, is that game models stop just where things get interesting – at the origin of beliefs and desires. These critics do not wish beliefs or desires to be broken down to more primitive cognitive processes – that would take them into the more radical camp discussed below. Instead, they merely observe that economists take tastes (desires) as givens, leaving the investigation of where tastes come from to other disciplines. While it is true in a sense that economists observe this division of labor, there is nothing inherent in rational choice that would require such a limitation. Or, to put the point differently, some of the structural arguments that sociologists advance regarding the origin of tastes or beliefs are consistent with belief-desire psychology and with game-theoretic analysis.[9]

By way of example, consider the general idea that desires are "social constructions" and the specific case of foot binding. The physical facts of foot binding – a practice nearly universal in China from the Sung Dynasty (960–1279) to the nineteenth century – were gruesome. Female children's feet were tightly bound starting at an early age, sufficiently tightly to produce sustained and intense pain until the nerve cells in the feet were eventually killed, the bones disformed and the flesh putrefied. By adulthood, foot-bound women were all hobbled or crippled. Yet, knowing this final outcome, generation after generation of women in China inflicted years of excruciating pain on their daughters.

Rather than consider why this social practice arose and persisted (on which see Mackie 1996), I wish here simply to record the attitudes of Chinese themselves. Bound feet – malformed, crushed, putrefied – were desirable. Across the years of pain, girls were told incessantly that such feet were the essence of femininity, greatly to be desired. The boys were fed the same line. Tomes of poetry urged the erotic, esthetic, and other

[9] My point here is similar to that made by Ferejohn and Satz (1994a) regarding the compatibility of rational choice analysis with structuralism.

appeal of what would simply appear a grotesque injury to Persians or Parisians. It is hard to think of a better example of the social construction of an arbitrary, even unnatural, desire.

Was it easy to generate the desire for crushed feet? Was it easy to dissipate such desire? Should rational choice theorists hesitate to deal with such preferences, which some might characterize as "irrational"?

In this particular case, one can tell a very rational story about the whole melancholy mess (following Mackie 1996). It begins with a preference that is, arguably, hard to socially construct away – the desire by men to be sure that the children their wives and concubines bear are their own. The Southern T'ang emperor, possessing a large harem, felt the problem of paternity confidence acutely. Whether intentionally or by happenstance, the practice arose in the T'ang court of tightly binding the harem women's feet, to prevent them from running around – both literally and figuratively. The rich and powerful soon emulated imperial practice. Any parents seeking prestigious matches for their daughters entered a competition: be the first to bind, the first to bind more tightly, the first to crush. The competition for good marriages swept down the social scale and intensified the girls' agonies. No single family could opt out of the emerging practice; to do so was to condemn one's daughter to marrying well beneath her station or even not at all.

This storyline suggests an explanation for the practice of foot binding. But where does the poetry extolling the practice come from? Here, the answer presumably lies in the justifications that parents invented as they bound their daughters' feet. In order to inflict great pain with a manageable conscience, it was insufficient to explain the harsh practicalities of the marriage market, even if those were understood clearly. All manner of extraneous and more immediate benefits had to be constructed, and were. Eventually, the constructed desires percolated into high culture.

In a sense, the story just told is a rational choice story. The desires that are taken as exogenous in the story are men's desires to have paternity confidence and parents' desires to marry their daughters well. Actions taken in pursuit of these primary and universal goals, in light of the best available current information, led to a horrible coordination equilibrium. The "irrational" desires emerge endogenously as artifacts of many families struggling to make the best of a very bad socially constructed situation.

I believe one can similarly rationalize other examples of socially constructed values and desires. Whether one can so rationalize all such desires is another question but the work of Ullmann-Margalit (1977) and Coleman (1990) in some ways suggests such a project, as part of a larger project in which not just desires but beliefs and other social constructs are partly endogenized.

The social theory that results takes only certain desires as exogenous, the more universal and unreconstructed ones. It then seeks to endogenize many other desires at the level of social analysis, while largely retaining simple belief-desire psychology at the individual level. The analysis is straightforwardly rational choice at the individual level but can also be essentially rational choice at the level of social theory.

Beyond belief and desire

In this section, I consider a more radical camp, consisting of those dissatisfied with belief-desire psychology itself. The movement in economics toward integrating psychology with economics (Mullainathan and Thaler 2000); calls for integrating cognitive science with social science (Turner 2001); and sociobiology are examples. So also are some traditional understandings of human nature in the humanistic disciplines, which embrace any number of mental entities beyond belief and desire.

I have only two points to make about such radical departures from rational choice. First, some steps toward more complex psychologies are small enough that a good deal of the original rational choice flavor remains. Second, the question for social scientists is not whether there are better models of human psychology – surely there are – but which of these can be turned into a tractable workhorse to replace belief-desire psychology.

Mapping attributes into preferences

As an example of work that departs from the simplest belief-desire models, while still retaining a rational choice flavor, consider the work of Zaller and Feldman (1992) on public opinion and the work of Arrow and Reynaud (1986) on multicriterion decisionmaking. In Zaller and Feldman, the objects of desire are viewed as complex bundles of different attributes. Agents' preferences over attributes are taken as exogenous but their preferences over any given set of objects is then derived endogenously according to stipulated psychological rules. For example, agents do not necessarily remember all attributes when they decide; instead, the set of attributes they remember is determined by a statistical sampling process with temporal autocorrelation. Contingent on a given subset of remembered attributes (each with an attached salience weight which is also endogenously generated), preferences among the objects of desire are then determined.

A somewhat similar approach is taken by Arrow and Reynaud. The individual's preference over objects is built up by evaluating these objects according to various criteria. The criteria are representatives to a congress in the head, where they decide by a collective choice process. Each criterion is limited to stating its own rank ordering of the objects. The collective choice procedure then maps these ordinal rankings into a collective ranking or choice. Arrow and Reynaud do not allow forgetting (abstention by a criterion), as do Zaller and Feldman. But the analysis does dissect preference into considerations and then examine a wide range of ways in which these might be combined. An interesting conclusion in the Arrow and Reynaud set-up is that the individual's preferences over objects need not be transitive or even acyclic – a conclusion that resonates with Zaller and Feldman's interest in the lack of ideological consistency in the mass public.

The main point about work that begins with "attributes" and builds up to preferences is that some of it implicitly or explicitly brings other psychological processes into sight (e.g., memory), while retaining preference as a central concept.

Tractability

The question about more complex psychological models is whether they can provide a better foundation for social scientific models and theories. Does the fact that people are capable of anger systematically affect market-clearing prices, or can we study prices better by ignoring anger? (cf. Mellers 2000). Does our inability to deal well with small probabilities systematically affect candidates' electoral strategies (cf. Kahneman, Slovic, and Tversky 1982)? When social scientists ignore the computational model of the mind, do the savings in terms of simplicity justify the losses in terms of explanatory power (cf. Pinker 1997)?

The methodology of rational choice analysis

Probability theory and statistics emerged as attempts to improve thinking about correlation and causation. Game theory emerged as an attempt to improve thinking about belief-desire explanations. Statisticians focus on imperfections in causal inference – for example, those that arise from neglecting selection bias – and seek ways to detect and ameliorate each problem. Game theorists focus on imperfections in belief-desire explanations – for example, those that arise from relying on incredible threats – and seek ways to purge such reliance from our explanations.

As methodologies, statistics and game theory share many similarities. For example, both produce large cookbooks of standard models; both employ various principles for concocting new models; and both feature a division of labor between pure theorists, who study the logical properties of models, and applied theorists, who use models to explain real-world phenomena.

Both statistics and game theory also deal with *systems* of equations, representing multiple causal interactions. In game theoretic models, each actor's optimal choice is represented by a separate equation, with their beliefs about each other's decisions regulated by some equilibrium concept. The result is a system of interrelated choice equations, with sometimes complex interactions and feedbacks. Things are abstractly similar in econometrics. One has a system of equations, each pertaining to a given endogenous variable. These equations are interrelated via (various decisions the analyst makes concerning) cross-equation error correlations and the like.

I would note two similarities about how game theory and statistics deal with their respective systems of equations. First, both disciplines employ what I will call "parameterized black boxes" at various points in the causal chain. Second, both disciplines bring order to their potentially intractable equations via higher-order theoretical restrictions: equilibrium concepts in game theory; various structural-equation modeling tools in econometrics.

Parameterized black boxes

One way to think of structural equation models is that they introduce one-parameter black boxes to replace full models of certain intermediate causal chains that are not of immediate interest. Consider, for example, a standard plant experiment in which the growth-promoting properties of a new fertilizer are under study. In the statistical analysis of the experimental data, let us imagine that a bivariate linear regression is employed, with the dependent variable being the weight of each plant at the end of the experiment and the sole independent variable being the amount of fertilizer applied to each plant per week. Many other factors, such as water and sunshine, are of course held constant. In this regression, the fertilizer enters a black box representing the plant. The details of how the fertilizer is absorbed, how it affects internal plant chemistry, and how it ultimately affects growth are not modeled. In place of that intermediate causal mechanism, a single parameter – the regression coefficient – is substituted.

One way to think of rational choice models is that they introduce a two-parameter black box to replace a full model of psychology. Observable inputs of various kinds flow into the black box, such as price increases or changes in electoral rules. Observable outputs flow out, such as purchasing patterns or campaign tactics. Some inputs are thought to affect belief parameters only, others to affect preference parameters only, others to affect both. Various modeling decisions are guided by the maintained assumptions about these parameters (e.g., from expected utility maximization or from equilibrium concepts).

In our analysis of plants, the one-parameter black box representing internal plant chemistry can be viewed from an "as if" perspective. We conduct our statistical analysis *as if* there is a direct, homogeneous causal connection between fertilizer and growth. We believe in fact that the connection is mediated by numerous intermediate processes, not all of which are well understood. Yet, we introduce a single parameter to cover this multitude of excluded variables – not because we believe that the intermediate causal mechanism is in fact linear or homogeneous; rather, because we believe that a linear and homogeneous representation of that process will capture a good portion of the true relationship.

In our analysis of social, economic, and political interactions, the two-parameter black box representing human psychology can also be viewed from an "as if" perspective. We conduct our analysis as if there are only two internal free parameters (beliefs and desires) which interact in a particular way to produce observable behaviors from observable conditions. We believe in fact that the connection between incoming influences and outgoing behaviors is mediated by numerous intermediate processes, not all of which are well understood. Yet, we introduce a pair of parameters to cover this multitude of excluded variables – not because we believe that the intermediate causal mechanism is in fact literally one involving only beliefs and desires in some neatly defined neural sense; rather, because we believe that an expected utility representation of that process will capture a good portion of the true relationship.

Conceivably, we could improve our analysis of plants by increasing the number of unobservable parameters that stand in for internal plant chemistry and biology. Suppose, for example, that some biochemist tells us that growth in plants is a simple function of two chemicals getting together at the same spot at the same time. The fertilizer is a good source of one of the chemicals. Water promotes the supply of the other chemical. We could now vary the amount of water applied to each plant instead of holding it constant, include the relevant interaction term in our statistical analysis, and test the hypothesis. The parameters in the statistical model would still, however, be mere shorthand for the intermediate causal steps; and we would not have brought into the analysis actual observations on the intermediate steps, only some theoretically imposed structure.

Conceivably, we could improve our analysis of social, economic, and political interactions by increasing the number of unobservable parameters that stand in for human psychology. McKelvey and Palfrey's (1995) quantal response equilibrium, for example, introduces an additional parameter representing limits on rationality. The parameter is a very primitive model of cognitive limits and learning but no more primitive than is the one-parameter model of internal plant chemistry in the example above.

Equilibrium concepts and structural-equation modeling

If only the weakest assumptions are made in an econometric analysis, one ends up with all variables being endogenous and with perfectly general error terms and covariance structures. This does not much limit the possible lines of causal influence. There are a lot of possible causal mechanisms that can be stated as a (not necessarily estimable) structural equation system.

If the weakest equilibrium concept (rationalizability) is employed in a game-theoretic analysis, the possible storylines are not much limited. Everyone must act rationally in the root sense of pursuing desires in light of their beliefs. Moreover, everyone must realize that their opponents are also rational and will avoid choosing obviously non-optimal (dominated) strategies. But there are a lot of possible plots that satisfy these minimal requirements, even in detective novels.

Econometric theory features various principles that let one know what parameters in a given system of structural equations are estimable under certain conditions. If one is willing to stipulate that certain variables are both exogenous and excluded from one or more of the structural equations, one may be able to "identify" certain parameters of interest. If one is willing to stipulate that certain covariance restrictions are in place (certain causal lines of influence are nil), one may again be able to "identify" parameters of interest. Generally speaking, the more demanding the set of assumptions one makes, the more parameters can be identified.

Game theory features various stronger equilibrium concepts that sharpen or extend the root notion of rationality. If one is willing to stipulate that each actor will come to know the other's strategy, then one is on the road toward accepting the Nash equilibrium concept. If one is willing to stipulate that no actor will believe "incredible" threats, then one is on the road toward accepting the subgame perfect equilibrium concept. Generally speaking, the more demanding the set of assumptions one makes, the more precise are the equilibrium predictions.

Pathologies

In dealing with systems of equations, whether based on utility-maximizing choices (game theory) or causal relations (econometrics), there are many things that can go wrong. One perennial source of disgruntlement runs roughly as follows. If one starts with the root models – anything that is rationalizable in game theory, completely general structural equations in econometrics – almost anything goes: almost any pattern of behavior is rationalizable (in many games); no parameter can be identified. Thus, to get anything useful out of the models, one has to impose theoretical restrictions.

It is at this point that a variety of potential abuses arise, due to the variety of possible theoretical restrictions that can be employed, relative to the amount of data available to tame theorists' fancies. What earned statistics the barbs of Auden ("Thou shalt not sit / With statisticians nor commit / A social science")[10] and Twain ("lies, damned lies, and statistics"), was the potential ability of debaters to furnish a specious statistic to support any desired inference. What earned rational choice analyses the barbs of Green and Shapiro was (partly) the potential ability of scholars to furnish a specious model to rationalize any explanandum.

However, the potential to misuse statistics and game theory arises precisely from the flexibility and abstractness of their core concepts. It is the same with other successful methodologies. Consider, for example, Lakatos' (1970: 107) discussion of Newtonian celestial mechanics.

> A physicist . . . takes Newton's mechanics and his law of gravitation, (N), the accepted initial conditions, I, and calculates, with their help, the path of a newly discovered small planet, p. But the planet deviates from the calculated path. Does our Newtonian physicist consider that the deviation . . . refutes the theory N? No. He suggests that there must be a hitherto unknown planet p' which perturbs the path of p. He calculates the mass, orbit, etc., of this hypothetical planet and then asks an experimental astronomer to test his hypothesis. . . . But [the new planet is not where it should be]. Does our scientist abandon Newton's theory . . .? No. He suggests that a cloud of cosmic dust hides the planet from us. He calculates the location and properties of this cloud [and sends up a satellite to investigate]. But the cloud is not found. Does our scientist abandon Newton's theory . . .? No. He suggests that there is some magnetic field in that region of the universe which disturbed the instruments of the satellite. [And so forth.] (Lakatos 1970: 100)

Green and Shapiro have a similar discussion of various rational choice theories of this and that. The first go at modeling the phenomenon as rational behavior does not work. Models 2, 3, and 4 are duly trotted out. Green and Shapiro, and other critics, conclude that the whole enterprise is obviously degenerate. Lakatos, faced with a similar (hypothetical) example, wonders where to draw the line: "what kind of observation would refute to the satisfaction of the Newtonian not merely a particular version but Newtonian theory itself?" (Lakatos 1970: 101 n. 2).

10 See stanza 27 of *Under Which Lyre: A Reactionary Tract for the Times* (Phi Beta Kappa Poem, Harvard: 1946). www.crab.rutgers.edu/~goertzel/auden.htm.

I don't think we have much purchase on saying when a paradigm/methodology, with both free theoretical parameters and a ceteris paribus clause to protect its hard core, is degenerate. When dealing with paradigms, we are by definition dealing with a situation akin to one in which there are more parameters to be estimated than data. Thus, we ought to abandon paradigms primarily when better ones come along – where I would define "better" not simply in terms of empirical but also in terms of theoretical values (cf. Cox 1999). In other words, the question about both statistics and game theory is not whether they are immune from abuse or misuse. They aren't. The real question is whether there are any systematic ways of thinking through correlations and belief-desire stories, respectively, that are superior.

In answering this question for game theory, note that enriching our model of mental phenomena, by including some additional mental categories (emotion, memory, etc.), will only worsen the degrees of freedom problem that Green and Shapiro complain about. Any behavior that can be explained by reference to the pursuit of desire in light of belief can also be explained by beliefs and desires jostling around with a variably good memory or variably good powers of deduction.

Method-driven research

Sometimes one hears the claim that additional training in econometrics or game theory will push the discipline toward methods-driven, rather than problem-driven, research. If we invest more heavily in quantitative methodologies, so the argument goes, we will increasingly study problems that can be studied with those methods.

To some extent, these worries are no doubt well founded. Presumably those who invest in any given set of tools – say, area studies skills – will then look for problems that can be addressed using those skills. However, one must hope that each individual scholar chooses her research projects wisely, looking both to the problem's intrinsic importance and the probability that she will be able to contribute to its illumination, given her skills. Sheer importance may even induce some quantitatively skilled researchers to do archival research themselves, or bring area-studies specialists to "commit a social science."

But suppose this happy breadth of view does not arise in individual scholars. Suppose our methodologists remain relatively eclectic in substance and narrowly focused in method, while our non-methodologists remain relatively eclectic in method but narrowly focused in substance. In this case, there is still the possibility that collaboration will restore some breadth and balance of vision, both substantive and methodological.

Indeed, one might define modern science as simply the specialization of intellectual labor in order to solve practical and theoretical problems. From this perspective, the suggestion that we must choose between more scientific method-driven and less scientific problem-driven approaches to research is puzzling. Science applies methods to solve problems.

Conclusion

In this chapter, I have suggested that one way to locate rational choice theory is in terms of what it (appears to) assume about human psychology. From this perspective, rational choice can be distinguished from radical behaviorism on one side, and from cognitive choice theory, on the other. Radical behaviorism would rely as little as possible on statements about unobservable mental entities, such as preferences and beliefs. Rational choice theory regularly trots out these two entities. Cognitive choice theory would trot out emotions, moods, and other unobservable mental phenomena as well – and use them all in a consequential way in modeling social interactions.

From this perspective, any scholar who is basically satisfied with belief-desire psychology as a shorthand model in analyses that focus on social phenomena could be classified (broadly) as a rational choice theorist. The only ones who would escape this classification would be those either who insist on purging social science of references to unobservable mental phenomena, or who systematically and consequentially bring in an even richer psychology in their theories of social, economic, and political interaction.

By a narrower, harder definition of rational choice theory, only those scholars who make liberal use of (1) complex errorless deduction from information to beliefs and (2) common knowledge assumptions, would qualify. As a classification device, such a definition might have been useful twenty-five years ago but now it would hardly work to provide any useful sorting even of game theorists, much less any larger community of scholars. Nonetheless, the definition is useful in identifying the part of rational choice analysis that has attracted the most criticism: the theory's assumptions and modeling of beliefs.

Viewed as a methodology, rational choice analysis (in the guise of game theory) deals with systems of (choice) equations and uses equilibrium concepts (which include stipulations of what is common knowledge) and information-to-beliefs assumptions (e.g., Bayesian updating) in more or less the way that econometricians use cross-equation error assumptions (e.g., independence) or exclusion restrictions. Such assumptions help to pin down the causal pathway.

In both disciplines, assumptions used to tame multiequation systems need to be justified on substantive grounds, when employed in specific empirical investigations. Discussions of whether or not there is sufficient justification should, in principle, proceed on a model-by-model basis. Global discussions of pathologies afflicting all of econometrics or all of game theory seem unlikely to produce methodological or substantive advance, relative to specific diagnoses of what is wrong with the current models and what might be done to improve or replace them.

REFERENCES

Akerlof, George. 2002. "Behavioral Macroeconomics and Macroeconomic Behavior." *American Economic Review* 92(3): 411–33.

Arrow, Kenneth, and Hervé Reynaud. 1986. *Social Choice and Multicriterion Decision-making.* Cambridge, MA: MIT Press.

Barry, Brian. 1970. *Sociologists, Economists and Democracy.* Chicago: University of Chicago Press.

Coleman, James S. 1990. *Foundations of Social Theory.* Cambridge, MA: Harvard University Press.

Cox, Gary W. 1999. "The Empirical Content of Rational Choice Theory: A Reply to Green and Shapiro." *Journal of Theoretical Politics* 11(2): 147–69.

Cox, Gary W. and J. Morgan Kousser. 1981. "Turnout and Rural Corruption: New York as a Test Case." *American Journal of Political Science* 25: 646–3.

Downs, Anthony. 1957. *An Economic Theory of Democracy.* New York: Harper and Row.

Ferejohn, John and Debra Satz. 1994a. "Rational Choice and Social Theory." *Journal of Philosophy* 91(2): 71 –87.

Ferejohn, John and Debra Satz. 1994b. "Rational Choice Theory and Folk Psychology." Unpublished typescript, Stanford University.

Fudenberg, Drew and David K. Levine. 1998. *The Theory of Learning in Games.* Cambridge, MA: MIT Press.

Green, Donald, and Ian Shapiro. 1994. *Pathologies of Rational Choice Theory.*

Kahneman, Daniel, Paul Slovic, and Amos Tversky (eds.). 1982. *Judgment Under Uncertainty: Heuristics and Biases.* New York: Cambridge University Press.

Kaufman, Arnold S. 1967. "Behaviorism," in *Encyclopedia of Philosophy*, vol. 1, pp. 268–72. New York: Collier Macmillan.

Kreps, David. 1990. *Game Theory and Economic Modeling.* Oxford: Oxford University Press.

Lakatos, Imre. 1970. "Falsification and the Methodology of Scientific Research Programmes," in *Criticism and the Growth of Knowledge*, Imre Lakatos and Alan Musgrave (eds.). Cambridge: Cambridge University Press.

Lupia, Arthur, Mathew D. McCubbins, and Samuel Popkin (eds.). 2000. *Elements of Reason.* Cambridge: Cambridge University Press.

Mackie, Gerry. 1996. "Ending Footbinding and Infibulation: A Convention Account." *American Sociological Review* 61: 999–1017.

McKelvey, Richard and Thomas Palfrey. 1995. "Quantal Response Equilibria for Normal Form Games." *Games and Economic Behavior* 10: 6–38.
Mellers, Barbara. 2000. "Choice and the Relative Pleasure of Consequences." *Psychological Bulletin* 126: 910–24.
Mullainathan, Sendhil and Richard Thaler. 2000. "Behavioral economics." Working paper 7948 (National Bureau of Economic Research).
Pinker, Steven. 1997. *How the Mind Works*. New York: W. W. Norton.
Schwartz, Thomas. 1986. *The Logic of Collective Choice*. New York: Columbia University Press.
Shepsle, Kenneth. 1995. "Statistical Political Philosophy and Positive Political Theory." *Critical Review* 9: 213–22.
Simon, Herbert. 1962. "The Architecture of Complexity." *Proceedings of the American Philosophical Society* 106(6): 467–82.
Sugden, Robert. 1986. *The Economics of Rights, Cooperation and Welfare*. Oxford: Blackwell.
Turner, Mark. 2001. "Toward the Founding of Cognitive Social Science." *Chronicle of Higher Education*. October 5. Pp. B11–12.
Ullmann-Margalit, E. 1977. *The Emergence of Norms*. Oxford: Clarendon Press.
Zaller, John and Stanley Feldman. 1992. "A Simple Theory of the Survey Response: Answering Questions Versus Revealing Preferences." *American Journal of Political Science* 36(3): 579–616.

9 Problems and methods in political science: rational explanation and its limits

Alan Ryan

This chapter contains no novelties, no technicalities, no surprises, and almost no footnotes.[1] It gives an airing to thoughts I have developed only in lectures on the philosophy of social science to undergraduate audiences, and in conversations with the late Martin Hollis, which never quite reached consensus (Hollis 1994). I begin with some truisms about explanation generally, and the connections between problems and methods. I then argue that it is impossible to *start* explaining human behavior except on a "rationalizing" basis, and explain why this accords no priority to so-called rational choice theory. That rationalizing explanation is only the beginning of explanation I explain by showing that there is (i) room for argument about the difference between an actor's professed reasons and his "real" reasons, and (ii) further room for argument about the conditions under which reasons will be causally efficacious or the reverse. Very often the rational explanation of action is not wrong but simply uninteresting in comparison with questions about how actors came to adopt the goals they did, and how they came to perceive the situation in which they are acting.in one way rather than another. I then take up very briefly the suggestion that rational actor explanation is improperly individualistic, and discuss its relationship to "holistic," and "functionalist" explanation, as well as its relationship to historical narrative.

The platitude that we should adopt the methods best suited to our problems is just that – a platitude, but irresistible; so is the platitude that new methods may show up new problems or illuminate old ones in unexpected ways; and so is the platitude that we should distinguish between methods in the sense of explanatory paradigms such as rational choice theory, and methods in the sense of techniques of uncovering data, such as opinion polling, or laboratory simulations. It is obvious that new techniques, for

[1] This is largely because Chapter 1 of Sen 2002 says much of what I would wish to say if I had sixty pages in which to say it, and with a wealth of reference beyond anything I might mention.

instance the Galileo telescope, have brought to light new facts (in this case about the existence of satellites orbiting other planets), which required new astronomical theories for their explanation, much as the techniques of market research revealed in 1944 that most Americans knew nothing much about the functioning of the American political system and those who ran it, data that provoked all sorts of interesting ideas about the desirability of apathy and ignorance in a democratic electorate (Lazarsfeld, Berelson, and Gaudet 1948; de Luca 1996). Another situation is where the success of a method in one area provokes interesting questions about why it works there and not elsewhere. Sociology exists to explain why economics is not the only social science. The only general rule is not to behave like the drunk searching for his keys under a street light. Asked if he had dropped the keys there, he replied that he had not, but that there was no light where he had dropped them. We ought not to be unduly biased in our selection of topics for research by the fact that we have ways of finding answers to those but not to more interesting or more important questions. This chapter, however, is devoted not to providing a list of questions for my colleagues to answer, but to laying out some constraints on the demands of explanation in social science generally, and *a fortiori* in political science.

> *1. Explanation is* prima facie *rational actor explanation: complete explanations of political behavior are rooted in the desires and beliefs of individual actors.*

Explanation in the social sciences is *prima facie* rational actor explanation. This means nothing extravagant, and merely endorses Donald Davidson's long-ago insistence that the explanation of human conduct invokes the agent's beliefs and desires – what is often described as folk psychology – and that this is explanation by rationalization (Davidson 1980). Rationalization means that the action taken by the agent must be displayed as "the thing to do under the circumstances," that is, the *right* thing to do under the circumstances. "Right" gives rise to misunderstandings worth clearing up. The rightness involved is not *only* instrumental, but there is no priority to its being more than that. A deft assassin picks the right knife for a wrongful act. Its rightness includes both its lethal qualities and its ease of disposal. Failing to pick the right knife marks one out as an incompetent assassin. Although we all make innumerable mistakes almost all the time, the right action has a priority in explanation that wrong actions cannot have. If I lose my job by failing to keep proper financial records, it is the proper records that illuminate the mess I have made; there are innumerable messes one might make, but they do not (generally) illuminate what the proper conduct is like. They may provide different sorts of illumination, however, as when the frequency of some kinds of error leads us to institute ways of reducing their number. This claim is different from, and perhaps at odds with, Jon Elster's emphasis on the difference between explanations grounded in instrumental rationality and those grounded in normative attachments (Elster 1989a; 1989b). For my purposes, the second involve rational explanation as much as the first; they have features that are not shared with instrumentally rational explanation, but even those differences can be exaggerated. The idea that instrumental action is intrinsically flexible – that action will respond rapidly to changes of ends or of available means – while normative action is not, is one exaggeration; we commonly adjust the pursuit of one value in the light of what it costs in terms of another, and the person who pursued justice without ever thinking of *how* to achieve it would be a strange creature indeed.

So much for rationalizing; when I say that complete explanations must be rooted in the desires and beliefs of individual actors, I mean the following. The proposition that members of British SES groups IV and V vote Labour rather than Conservative by a margin of two to one (which was once true with a smallish margin of error), is not self-explanatory; indeed, it is not even a candidate explanation. It is a summative account of what members of these socioeconomic groups do at an election in Britain. Suppose we find a member of those groups who votes Labour, and we say "Jones voted Labour *because* he was a member of SES group V," what are we saying? It was antecedently half as likely that he would have voted Conservative, and yet we would not say "Jones voted Conservative *because* he was a member of SES group V." On the assumption that something about being a member of SES groups IV and V is causally connected to voting behavior, we must think that whatever it is bears on the reasoning implicit in the voting behavior. Here are three possibilities of how it is causally effective – the point of offering just these three will emerge in due course. The first is a simple payoff argument; Jones reckons that a Labour government will do better for people in SES groups IV and V, and votes on simple self-interest. (I omit questions about whether self-interest would get us to the polling booth in the first place; we may imagine that Britain has acquired the Australian rule of compulsory voting, and Jones would be fined for non-appearance.) Choosing between Labour and Conservative, and abstracting from his own circumstances, he reckons he will do better under Labour and votes accordingly. I stress two points that will come up again. The first is that these reasons are objective; they are *his* reasons, but they are also reasons that anyone in his position has; and even if a person in that position has other reasons for acting that outweigh these reasons, they *outweigh* but do not *cancel* reasons that he still has. The reasons do not go away when they are outweighed. The second is that appealing to these reasons is appealing to the causes of his behavior. Contrary to the conventional wisdom of several decades ago, there is no contrast between causal explanation and rational explanation; if reasons explain, they explain by being causes, and if they are not causes they do not explain (contrast Lesnoff 1974 and Davidson 1980). Consider the Conservative-voting member of SES groups IV or V; he *has* the reasons to vote Labour that his Labour voting peer has, but he does not act on them. To cite them cannot explain his behavior, since they do not make him vote Labour; if they are cited, it must be in the context of showing how they pushed him one way but not powerfully enough to offset the countervailing reasons.

Within the "what's in it for me?" paradigm, it is not difficult to construct a chain of reasoning that might counter the fact that Labour does better for members of SES groups IV and V. Although Labour might be better for members of SES groups IV and V on average, our hypothetical voter may not think of himself as an average member of the group. Perhaps he is in an occupation, low-paid at present, that will be given a great leg-up by the Conservatives. So the party to vote for is the Conservatives. This is why we might well say, "he voted Tory although he was a member of SES group V." In the case of the Labour voter, we assume that the reasons that he shares with his Conservative peer meet with no countervailing reasons, and so he votes Labour, while in the other case to say "although" simply draws attention to the fact that the countervailing reasons override the reasons on the other side. But, I want to offer two more thoughts that will stretch the argument a bit.

First, let us reintroduce normative reasoning into the argument. Norms provide reasons for action, just as self-interest does. Sometimes, it is true that norms do not influence action in a "more or less" sense, but in a "yes or no" sense. That is, it makes sense for me to buy cheaper apples if I can, but not if the walk to the shop is so far that it outweighs the advantage of their cheapness; the distance increases their cost as a function of the distance, so there is a flexible tradeoff between money saved and distance walked. The shopkeeper who knew my aversion to walking with sufficient exactness would know how much more he could afford to charge than his rival, and his rival could know by how much he had to undercut his competitor. If, on the other hand, I thought that one greengrocer was simply wicked and that my custom was helping in funding his wicked behavior, I ought not to buy from him, full stop. This is a formulation that I have already said was too extreme, but it will get us started. It is familiar in political science that voters who have (objective, class-based, self-interested) reasons for voting one way will do so in smaller numbers if they are surrounded by large numbers of voters of the opposite persuasion. One explanation is that they see "normality" as embracing voting for the "other" party, and they are led toward thinking that it's "what people do," in a normative as well as a statistical sense. If a large local majority votes for the "wrong" party, people will have friends who vote for it, and will feel they should stick with their mates; or parents will vote that way and their children will defer to their parents' views. And so on.

I said it was too extreme to say that normative considerations were "all or nothing" and that is right. Although the housemaid who said that she was only a little bit pregnant was making a crucial mistake, the person who thinks that his offence against morality is a small one is not always doing so. The point here, however, is simply that a person may vote the way he does out of a conviction that he *ought* to vote that way. This provides a perfectly good reason, and in the sense that it rationalizes the action and picks out "the thing to do" it provides the characteristic bottom line explanation of human action that we employ in everyday life. Neither our Labour voter's appeal to the obvious economic benefits to himself if "his" party wins, nor the Conservative voter's appeal to the norm of voting with his friends and relations is more rational than the other; both rationalize the agent's behavior in the sense explained. There are those, Jon Elster most notably, who insist that the two sorts of explanation operate in different universes and cannot be reduced to one another. And there are others, Robert Frank most notably, who insist that a rational person in the utility-maximizing sense of that term would choose to have a temperament that made himself susceptible to the pull of norms such as justice and honesty – on the grounds that the benefits will on average outweigh the costs incurred on particular occasions by not performing the action that maximizes utility at that instant (Frank 1988; 2003). The latter seems to me highly plausible, but neither view needs exploring here. There is, however, a third sort of rationalization to which I want to draw attention; one of my great regrets is that nobody (myself included) has ever managed to provide a philosophically convincing elaboration of the thought that animates much of the work of Erving Goffman. That is the thought that we spend a great deal of our time literally enacting ourselves; this is a process of a slightly peculiar kind inasmuch as we devote our efforts both to reassuring other people that we can cooperate and do exactly as we ought, but also to showing them how much more of us there is than that part of ourselves that we devote to social cooperation (Goffman 1963; 1971). Acting out a drama in which we indicate our reliability and knowability, while at the same time suggesting that we may be quite other than we appear is a very distinctive but wholly commonplace human skill.

The surgeon who asks his nurse for a "niblick" when she and he know full well that he needs a favourite scalpel is showing that he is a wholly competent surgeon, reassuring his surgical team that he is fully in command in a tricky situation, and showing them that he has plenty of sangfroid – if that is the right word in the context – in reserve. The surgeon shows without insisting that he is wholly in command; put otherwise, he affirms his identity as a wholly competent surgeon. Goffman was very good at the topic of preserving our "real" identity – an issue that might arise when a father plays with his child too unself-consciously, and casts doubt upon his commitment to grown-up existence. He invented the concept of "role-distance" to pick out this phenomenon. Goffman saw that fathers taking their children on fairground rides and in similar situations acted as though they knew that they might be watched by grown-up observers and adjusted their conduct in ways that indicated "yes, I am playing here, but I am also the grown-up whom you see in the office." Voting behavior may be illuminated by the same observation. A person who votes, whether in the direction of self-interest, or as some norm dictates, may also seek to *enact* his commitment. Modern voting is not the most obvious arena for this, since the fashion for the secret ballot means that people don't have a public place in which to act out their allegiances. Nonetheless, one can see how a person might think that voting was a form of *avowing* commitments. I could not vote for a Conservative, no matter how excellent, not because Conservatives are wicked, nor because a Conservative government would be less in my interest than a Labour government, but because it would declare to an invisible audience that I was the person I grew up determined not to be – whether the red-faced hearty in the local golf club or the weasel-faced racist who runs the dodgy car dealership round the corner.

Expressive rationality is genuinely explanatory, but parasitic on other explanatory considerations. That is, we have to express something, and the something that we express must in logic be prior to the expression of it. I give my love a bunch of red roses to express my affection, but the affection gets the whole thing going. My desire not to be a gin-swiller or a dealer in dodgy automobiles must antedate my sense that voting for someone other than the Tory candidate will express that desire. An operatic aria allows us to express rage and jealousy with more verve than everyday life permits, but the rage and jealousy have to be there to be expressed, even if in due course there will be feedback such that the *way* we feel rage and jealousy will reflect the available modes of expression of the emotions. Once there is opera, I can behave operatically when I flounce out of department meetings. It is at least possible that there is some constraint on our feelings imposed by the need that they be *expressible* in ways we have some natural aptitude for or inclination towards.

Interestingly, we get three distinct sorts of mistake or misfires in action corresponding to the three sorts of rationalization I have described. The simplest mistake is that made by someone aiming to maximize his pay-off in the standard rational choice situation; he will fail to do what actually maximizes his payoff and will do what does not achieve it. If *A* aims at *Y* by doing *X* and *X* won't bring about *Y*, *A* has made a mistake. Either his factual beliefs are wrong or he has made some error of calculation. The normative case is more complicated, inasmuch as there are several different slants on it. One is simply that we think the agent's normative attachments are wrong; *he* behaves as a conscientious head-hunter should in slaughtering the family of his victim, and *we* think he is wicked. His mistake is to be wicked. But, less starkly, he might have made a mistake about the implications of attachments with which we would not quarrel, or, as with the ordinary rational choice instance he might have made a mistake about the likelihood of the action bringing about the normatively required situation. (His friends may in fact regard voting as a matter of individual conscience and not think his solidarity an appropriate virtue.) Lastly, the gesture may be theatrically wrong in all sorts of ways: it might be inappropriate, over-blown, or self-indulgent. The interesting thing about the last case, however, is that it shares with the second but not the first a mode of mistake that is interesting in its own right, that of inaptness. It is an important fact, for what seems the enactment of sincerity to one audience can readily seem the demonstration of blind stupidity to another – whence my view that we must attend *both* to the role of the expressive and to the question of what it expresses. Of course, there are depths here worth exploring in their own right; Sen's famous essay on "Rational Fools" is but one example of the insights to be had by considering what happens if we cannot change our ends as well as the means to previously adopted ends (Sen 1979; 2002).

> 2. *That explanation is only* prima facie *rational actor explanation needs elucidation; I want to insist that we can always, but not always usefully or interestingly, ask why one sort of reason rather than another motivated an agent in a given context; sometimes this will take the form of asking why she or he believed or valued what she or he did, but sometimes it can take the form of asking why the agent thought in self-interested terms rather than normative ones or vice versa, or why they chose the particular mode of dramatization that they did. On other occasions, the question will undercut their avowed reasons in search of their real reasons – readily in the case of simple lying, but more complicatedly in the case of self-deception.*

We should begin by observing that the priority of rational explanation so construed gives no priority to rational choice theory. Rational choice theory is committed to the view that explanations should be couched in terms of the interactions of rational utility maximizers. Here we duck the familiar findings of Daniel Kahneman and his collaborators (2000) that show how impossible it is for us to rank alternatives consistently or in such a way as to provide the completeness that theorists assume for model-building purposes. By the same token "framing" problems will be tackled only in passing and casually when we discuss the ways in which stimuli other than those internal to rational choice affect the way we rank alternatives and assess how much we prefer one to another. The first issue to examine is the rationality of maximization. In the ordinary sense of the word, we rarely try to maximize our acquisition of what we are seeking. We do not eat the largest possible meal, or drink the largest possible drink. It might be retorted that the largest possible meal is not the same thing as the best meal under the circumstances. That only gets us so far. We do not seek out the very best car, or the very best television set, or even the very best meal under the circumstances. What we mostly do is "satisfice." But this is an area where ordinary habits of speech are at odds with the technicalities of the economist; the plain man's understanding of maximization is not that of the economist. What we do not do in the economist's sense is "optimize"; we do not usually seek, let alone usually find, the absolutely best alternative. The economist insists that we must nonetheless maximize, which is to say that we must choose what is – given what we want to achieve – an alternative that is no worse than any of the others available. On this view, Buridan's Ass, who starved to death between two equidistant bundles of hay, suffered from an inability to distinguish between optimizing, as a search for the absolutely best – which did not exist – and maximizing, as a search for an alternative as good as the others in the neighborhood such as starving to death (Sen 2002: 42–47).

The greater difficulty lurks in the word "utility," which has always covered a multitude of sins. The common complaint is that it means only "whatever it is that a rational actor is maximizing," and thus renders rational choice theory vacuous. The better thought is that we must be careful to employ rational choice theory in only one of the two distinct ways in which it can be employed. That is, if we set it as an *axiom* that some person or group of persons are "maximizing something," *and* it is an axiom that they are fully informed, suffer no weakness of will or defects in processing information and the like, then we can analyze their behavior to discover *what* it is they are maximizing. In the alternative, if we know in fact what they are trying to maxmize, and they are fully informed, suffer no weakness of will, and so on and so forth, we can predict what they will do. In general, we can, and do, hold some parts of the framework constant to illuminate the state of the other parts – such as defects in information, weakness of will, etc. The gratuitious (and false) assumption is that what they are trying to maximize can only be their own, narrowly selfish interests. If rationality requires them to maxmize in the economist's sense, it does not require them to adopt as their ends only their own well-being.

This is sometimes thought to be a sentimental or moralizing view; it is not. "What's in it for me?" is a question we can always ask, but frequently do not. In some contexts even asking the question would display a low character. The converse is more interesting. In many contexts we behave in a rationally self-interested fashion not because this consorts with human nature, but because we are morally obliged to or legally required to. A trustee of a charity, for instance, is legally required to maximize the charity's resources. The notion that the self-interest oriented explanation is the most basic explanation is not plausible. If the interest of anything is basic, it is the survival interest of our genes, and they have neither selves, nor the capacity to be moved by self-interest, *pace* my colleague Richard Dawkins. It is not easy to provide a general account of the contexts in which rational choice theory complete with the priority of selfish motives is most illuminating, but it is tempting to say that it is those contexts where most things that human beings enjoy have been excluded. Holding a stock for moral reasons is likely to be expensive, and I'd be well advised to exclude such thoughts from my motivation. Again, treating my role as CEO as the field for operatic displays of courage without thinking quite hard about the underlying profitability of the company may work for a little but probably not for long. (Operatic displays intended to reinforce the point that I really am the boss, that I really take hard decisions, and that I really have a good eye for the main chance may all be quite important, as suggested above.)

The Weberian view that rational choice theory becomes useful at the point where we create institutions that provide many payoffs for so organizing our conduct and very little emotional, aesthetic, or normative satisfaction is right. There are constraints on the creation of such institutions, too, such as the obvious constraint that the institutions that buy and sell in order to make a profit – and not to glorify God, serve the poor, or enhance the glory of the local prince – must not be doing something obviously immoral. The prostitute who said that she went on the streets rather than sit on her assets was making an improper joke, but she neatly called attention to the contestability of the line between those areas in which we may decently consult the promptings of cost-benefit analysis and those where we may not. The fact that there are not merely large numbers of jokes along the same general lines, but a whole film starring

Julia Roberts, suggests that this is well understood by the majority of the human race.

Equally importantly, there will be many occasions when we need to distinguish between a person's "reasons" and their "real reasons" for action. The latter are those that are causally efficacious in the context in question, whereas the former merely "rationalize" the action without being causally efficacious; they are reasons which would be good enough to explain the action, but in fact do not, because they were not the reasons that got the agent to act. Absent this distinction, both self-deception and ideological blandishment are unintelligible. We constantly employ the distinction when we think that someone is lying about what they are up to; if we think that nobody could believe what, say, Paul Wolfowitz said that he believed about the dangers to the United States posed by Iraq, we look for something that he might believe that could provide reasons – reasons that are logically and causally efficacious whatever their moral qualities – for invading Iraq. But we need not think that he was lying. The interesting cases are those where people believe the account they give but are not properly informed about their own motives or beliefs or both. We might think that a politician who always adopted the more aggressive or risky of the policies available, when the more and less risky were equally likely to succeed, was trying to show his father that he wasn't a wimp. Or we might think that such a person had swallowed the ethos of the social class that had hitherto run the country, and was demonstrating that he shared their outlook – he might be a *parvenu* impressing a military aristocracy, for instance. The sociologically more complex distinction between reasons and real reasons would include some familiar Marxist themes, such as the suggestion that the working-class voter who votes Tory is mystified into thinking that by engaging in middle-class activity, he demonstrates that he is a middle-class person.

Although in everyday life we operate perfectly satisfactorily with the distinction between avowed reasons and real reasons without begging the question whether real reasons must be deeply suspect, social scientists tend to get into a tangle by assuming that real reasons can only be of one sort and must be unavowed because they are discreditable. But there is no reason to believe either of these things. When Hobbes gave some money to a beggar, a friend asked him whether he would have done it if Our Lord had not commanded us to be charitable. "Certainly," said Hobbes, "that man's distress distressed me and in easing him I eased myself." Hobbes was right to insist that unselfish behavior – that is, devoting himself to someone else's welfare – could have a self-regarding payoff – a kindly person like Hobbes is upset by the misfortune of others and helps himself by helping them. Self-regarding is importantly not the same thing as selfish; that is, there *must* on Hobbes's account be a motivating payoff to the agent, but the route to that motivating payoff can be something quite other than the agent's own welfare. Hobbes' account is a precursor of Robert Frank's. What Hobbes gets from the encounter is the reduction of his distress at someone else's distress, and this is not in the abusive sense a "purely instrumental" act. People frequently give a less self-praising account of their behavior than their friends would come up with, and explain themselves in more self-deprecating fashion than Hobbes. What goes for motives also goes for compulsion; people behave well as often as badly in a quite compulsive fashion – helping others when they are determined not to or contributing to charity when they had decided not to. Succumbing to temptation can readily mean succumbing to the temptation to behave well.

It is unclear how much clarification of the different sorts of reason for action is possible. Weber's distinction of *Zweck-* and *Wertrational* action is famous: the distinction between means-end rationality and "value-implementing" rationality. The second embraces both the sort of normative rationalization that I mentioned above and the expressive dimension I added in homage to Goffman. I incline to the view that dimensions may be added *ad libitum* for analytical purposes. One thing to emphasize is that the issues that come up when we deepen or undermine proferred rationalizations have nothing to do with this form of explanation, but with any kind of explanation. The temptation to think otherwise stems from the fact that to call an action "rational" is often to commend it; but "rational" for our purposes only means "rationalizable." Consider an example that Martin Hollis made famous: a man looking for a large piece of buttered toast to sit on doesn't reduce our bewilderment when he explains that he is a poached egg. Hollis wanted to preserve the thought that to call an action rational is to commend it and so to say that even though a piece of buttered toast is the right place for a poached egg to be, the action couldn't be baptized as "rational" in the commendatory sense. The deeper explanation is that the absurdity of the belief is so salient that it overwhelms one's grip on the explanation. We want to know why on earth the man thinks he is a poached egg, and what on earth it might be like to think anything of the sort.

So, although rational action is basic in social science explanation, we in practice spend very little time elaborating rational actor explanations, unless we are obsessive practitioners of rational choice theory. More often, we want to know the things that allow us to fill in – should we wish to cross our Ts and dot our Is – the details of their decisionmaking. So, we spend our time wondering *why* people have the weird beliefs they do – that the UN is sending black helicopters to spy on them, for instance – and how they come by the strange values they seem to hold; and we shall spend almost no time thinking about what follows from those beliefs and values. What we are puzzled by and what we look for to provide an explanation depends on what we antecedently think it plausible that people *might* believe and value; we don't wonder why people believe that a lack of health care imperils their health, but we do wonder why they think that rubbing a weak magnet over a painful part of their body will do them good.

> *3. Finally, the approach sketched here may seem to slight – in no particular order – holistic, historical, and functional explanation. That is, it might seem that an insistence on the role of rationalization explanations, even with no priority given to rational choice explanations as conventionally understood may seem to be too individualistic, too given to syllogistic explanation and not enough to narrative, and always seeking Humean causes to the exclusion of an understanding of social function. There is a response to the first two complaints, and none is necessary to the third.*

"Holistic" explanations in the Marxian or Durkheimian mode rely in the last resort on the "logic of the situation" mode of analysis offered above. This is not to say that such explanations *look* like situational logic explanations on the page, only to say that when the explanations are properly unpicked, their dependence on situational logic becomes evident. The point is this: suppose we think of Marx's claim that the fate of capitalism is dictated with an iron necessity. Part of this claim rests on a philosophical conviction about the rationality of history that is closer to a religious conviction than anything one might prove on the basis of empirical evidence. But the basic machinery of a Marxian explanation does not rely on a belief in the overall rationality of the process. It relies on a familiar pattern: the situation at time T_1 gives people overwhelming reason to behave in ways *a,b,c . . . x,y,z*, and the consequences of their behaving in those ways will be to create at T_2 a new situation that gives them overwhelming reason to behave in ways a_2 to z_2, and the consequences will be . . . The way this is set out has to include various items that will have the effect of preventing people radically revising their values, changing their beliefs, or devising new strategies, but in well-organized old-fashioned Marxism this was taken care of. At the end of the day, T_n, what resulted from the consequences of all the prior situations and people's reactions to those situations was Marx's inevitable catastrophe. Marx was, of course, wrong. But he was wrong about the facts, not about the logic of explanation; what holists believe is that the intervening processes of perception, evaluation, and decision can be so taken for granted that a "situation-situation" explanation will suffice. This claim is false in fact, but not offensive in logic.

There is, for different reasons, no conflict between situational logic and explanatory narrative and therefore no tension between historical and rational explanation. Historical explanations depend on "logic of the situation" analysis, and a good narrative is one where the causal chain from beginning to end is well put together in a series of rationalizing links together with an account of how actions created new situations to which appropriate responses took place, and so on down the chain. There is no conflict between the "generalizing" mode that underlies a lot of rationalizing explanation and historical explanation. Indeed, it is truer to say that all explanation is historical and that theories provide idealized histories than that situational logic explanation is unhistorical. On this view, economics is a box from which we can extract idealized histories of the behavior of interacting agents. In practice, of course, we do not bother to do it – we are more interested in the box of tricks from which we can extract the histories or in the mathematical formulation of those histories. Still, as a matter of logic, that is what we do, so there is no general tension between rationalizing and historical explanations.

Finally, and dogmatically, because I have argued the point at length elsewhere, "functional" explanations turn out to be one of three things: not explanations at all, explanations in terms of purpose, and explanations in terms of situational logic. They are not explanations at all when all they do is point to the agreeable, useful, benign or otherwise helpful effects of whatever it is we are functionally explaining. Robert Merton's famous and wonderful essay on "Manifest and Latent Functions" (Merton 1957) is a very good example of all of these possibilities. Manifest functions are what institutions and practices do because people set them up for that purpose. The purposes of the agents who set them up provide the crucial causative factors. They are good causal explanations in terms of purpose – the purposes of human agents, and by extension the purpose of those agents' creations. In the case of the latent functions of the boss system – a wonderful piece of observation – where it is observed that the boss system substituted for the missing welfare state, it is perfectly proper to observe that the boss system provided benefits to newly arrived immigrants that were useful to the immigrants and whose usefulness explains the immigrants' readiness to support the bosses against middle-class reformers. That does not explain why there was a boss system; effects cannot explain the events or institutions whose effects they are. What explains the existence of the boss system is the entrepreneurial opportunity available to quick and unscrupulous folk, given a broad electorate, feeble policing of corruption, and a large, underresourced immigrant population whose votes could be used to get control of the political system and therewith of its revenues.

The good effects of this corrupt system do not explain its genesis, but they are the beginnings of an explanation of something else, namely the difficulty of stamping out the system. For, given that the bosses provided benefits to their clients – at a very high cost to the wider community and indeed at a higher cost to their clients than would have been charged by a cleaner system – their clients had no reason to support middle-class reformers who would leave them without any benefits at all. Since, for all the reasons explored by Mancur Olson many years ago, the bosses were likely to be better organized and to have greater stamina than the reformers, their clients were likely to remain loyal to them. The demise of the boss system was signaled by the arrival of the welfare state on the one hand and more effective employment arrangements on the other. Once the city job was unattractive compared to what one might earn elsewhere, one of the major benefits that the boss could offer was of much less value; and once the welfare state could do what sporadic deliveries of coal and food in return for one's vote had done before, the old bargain lost its value. In short, when there is a real explanation of something identifiable as the *explanandum*, a situational logic explanation is what we want. In this case, moreover, something very like an explanation in terms of rational self-interest does as well as can be imagined. Still, it is not impossible that something else would have been the right explanation – for instance, that as the immigrant poor became more assimilated they changed their moral perspective and simply sided with respectable America and against both the bosses and their earlier selves. Happily, American politics has never been a healthy environment for respectability, and that explanation, though impeccable in logic, has no credibility in fact.

REFERENCES

Davidson, Donald. 1980. *Essays on Actions and Events*. Oxford: The Clarendon Press.

De Luca, Tom. 1996. *The Two Faces of Apathy*. Philadelphia: Temple University Press.

Elster, Jon. 1989a. *Nuts and Bolts for the Social Sciences*. Cambridge: Cambridge University Press.

1989b. *The Cement of Society*. Cambridge: Cambridge University Press.

Frank, Robert H. 1988. *Passions within Reason*. New York: W.W. Norton.

2003. *Microeconomics and Behavior*, 4th edn. Boston: McGraw-Hill.

Goffman, Erving. 1963. *Stigma: Notes on the Management of Spoiled Identity* Englewood Cliffs, NJ: Prentice-Hall.

1971. *Relations in Public*. London: Allen Lane.

Hollis, Martin. 1994. *The Philosophy of Social Science: An Introduction*. Cambridge University Press.
Kahneman, Daniel and Amos Tversky. 2000. *Choices, Values and Frames*. Cambridge: Cambridge University Press.
Lazarsfeld, Paul, Bernard Berelson, and Hazel Gaudet. 1948. *The People's Choice*. New York: Columbia University Press.
Lesnoff, Michael. 1974. *The Structure of Social Science*. London: Allen and Unwin.
Merton, Robert K. 1957. *Social Theory and Social Structure*. Glencoe, IL: The Free Press.
Sen, Amartya. 1979. "Rational Fools," in Frank Hahn and Martin Hollis (eds.). *Philosophy and Economic Theory*. Oxford: Clarendon Press.
2002. *Rationality and Freedom*. Cambridge, MA : Harvard University Press.

10 An analytic narrative approach to puzzles and problems

Margaret Levi

The false dichotomy between problem-oriented and method-driven work is belied by most of the best extant social science research. Occasionally, there may be research motivated only by the desire to improve a technique, but virtually all work of significant interest is trying to untangle a theoretical or empirical puzzle. Not all are puzzles of general public or contemporary interest, however, and this may be what generates much of the current concern about the "relevance" of contemporary social science. But this, too, is often an issue that is more apparent than real.

Many of the major topical questions, especially the Big Problems, are not easily susceptible to social science analysis. We can speculate and postulate and write op eds about how to end racism, poverty, or dictatorship and how to improve the well-being and voice of the neglected. We can seldom offer definitive solutions. Sometimes this is because the causes are distant and the outcomes cumulative and slow moving (Pierson 2003). Often it is because the problems themselves are inadequately specified and the information social scientists need for analysis unavailable or even unobtainable.

Despite these and other limitations, social scientists can inform public debate and policy. There are excellent examples of scholars who have, among them James Scott (1985, 1998) on the green revolution and on state actions, Frances Fox Piven and Richard Cloward (1977, 1993 [1971]) on welfare systems and other policies meant to aid the poor, Doug McAdam, Sidney Tarrow, and Charles Tilly, separately and together (2001) on contentious politics, Fritz Scharpf (1991, 1999) on governance and democracy in the new Europe, and David Laitin

This chapter draws on earlier work (Bates *et al.* 1998, 2000a, b) co-authored with my *Analytic Narrative* collaborators: Robert Bates, Avner Greif, Jean-Laurent Rosenthal, and Barry Weingast and on an earlier published paper (Levi 2002). For their tough, smart, and helpful critiques I thank Turan Kayaoglu, Kevin Quinn, Renate Mayntz, Sidney Tarrow, Charles Tilly, and the two commentators at the Yale Conference, Arjun Appadurai and James Scott.

(1992, 1998) on language choice. What distinguishes their distinguished work is that they frame their research in terms of enduring theoretical issues and use rigorous analytic techniques – although of quite different kinds – to make their case. They frame the puzzle in terms that make it possible, using their methods of choice, to actually provide compelling answers to the questions they raise. The variety of approaches they use suggests that "doubly engaged" (Skocpol 2003) social scientists emerge from all the competing methodological persuasions.

Their methods divide them, but they share some commonalities. All ground their arguments in a model of the interaction of actors given structural constraints. This model need be neither explicit nor formalized, and the key actors may be collectivities rather than individuals. They also identify causal mechanisms (even if only some of these authors label them as such) that are generalized from or generalizable to other cases. For example, in *Weapons of the Weak* (1985), Scott emphasizes how outrage at the injustice of the transformed social contract motivates behavior. In both *Regulating the Poor* (1993 [1971]) and *Poor People's Movements* (1977), mass mobilizations generate electoral tensions that lead to policy concessions to the poor to achieve peace. For Scharpf, and for Laitin, there are a variety of mechanisms, depending on the specific puzzle, actors, and structural constraints. McAdam, Tarrow, and Tilly (2001) are by far the most explicit and most developed in their use of a mechanisms-based approach.

All of these works focus on processes and develop qualitative case studies to explore them. In that sense all use narrative techniques, at least as understood by Tim Büthe (2002: 481–2 and n. 2), and all are analytic in their efforts to provide a model of action, to theorize about the relationships among the actors, institutions, and structures that are the constituent parts of a complex phenomenon. Only Scharpf and Laitin, however, rely heavily on rational choice and game theory, which makes their work closest to that of rational choice analytic narrative. But this is only one of the flavors of "analytic narratives." Ira Katznelson (1997) frequently uses the label to describe the method for taking up large-scale macro-historical questions with a configurative approach derived from Barrington Moore (1966). Theda Skocpol and Margaret Somers (1980) label their preferred comparative method as "macro-analytic comparative history" and certainly rely both on narrative and on iteration in their research (also see Skocpol 2000). Others use the term "causal narrative" to describe a process by which the sequences and variables in an historical narrative are disaggregated in a way that allows cross-case comparisons (Mahoney 2003: 365–7; Sewell 1996). Most recently, Sidney Tarrow (2003) describes the *Dynamics of Contention* project as "'analytic narratives' more deeply rooted in context . . . than the parallel effort" of the approach presented here.

There are various routes to a social science ultimately capable of addressing the major problems of the day. However, this chapter eschews the huge comparisons of big structures and large processes taken on by the earlier Charles Tilly (1984) in favor of single cases or small-N comparisons that permit the elaboration of precise models, which generate testable implications and whose domain of operation can be clearly delimited. Scholars who share this taste can still ask big questions but only certain kinds of big questions, ones that can be framed in ways that lend themselves to one form or other of methodological rigor (Geddes 2003). By analyzing language choice (Laitin 1998), the role of emotional mechanisms (Petersen 2002), electoral strategies (Chandra 2004), or social boundary mechanisms (Tilly forthcoming), it is possible to address important facets of ethnic conflict. By considering the relationship between rural and urban populations and their relative bargaining power with the state (Bates 1981, 1983; 1991), the constraints bureaucrats impose on political reform (Geddes 1994), or the deals between long dead rulers and legislators (e.g., Hoffman and Norberg 1994; Hoffman and Rosenthal 1997; Kiser and Barzel 1991; Levi 1988; North and Weingast 1989; Root 1994), it is possible to speak to central issues of political and economic development.

Ultimately, social science should assist in the resolution of pressing problems. This is not to suggest that all social scientists must concern themselves with relevance. There is a division of labor in social science as in all else. Rather, the aim is to develop a social science built on solid enough theoretical and empirical foundations to ensure that applied research is in fact grounded in what we authoritatively know about the world. Unfortunately, we still know relatively little. As many have long claimed, we are far from achieving even the level of science attained in physics or biology, which are themselves far from immune from contests and challenge. For some this signals the need to admit the impossibility of achieving anything more than sensible, reasonable judgments grounded in a sophisticated understanding of the world and the ways of power (Flyvbjerg 2001). For others it means that we should develop clear standards for good social science research and thus remain committed to scientific progress (Laitin 2003). Given the level of maturity of social science *qua* science, the steps may have to be small and incremental. The puzzles need not be tiny and esoteric, but they do need to be susceptible to solution with the finest analytic and methodological tools available. Moreover, in the best of all possible worlds, their resolutions should serve as building blocks for further and possibly more elaborate puzzle solving.

Another feature of pressing contemporary problems is that each tends to be unique. There may be and usually are comparable issues in the past or in other parts of the world, but they take place at specific moments and in particular places that can make them distinct in quite important ways. A recent example is the terrorist attack on the United States on September 11, 2001. Comparisons with Pearl Harbor have a certain resonance but ultimately fall flat. The conflict between security and civil liberties recalls the Cold War, but we live under different threats with modified institutions and significant differences in political organizations and mobilization.

The analytic narratives project (Bates *et al.* 1998, 2000a, b) – possibly better labeled the rational choice analytic narrative project – represents one attempt to improve explanations of unique events and outcomes, unravel particular puzzles, and at the same time construct the basis for a social science capable of addressing significant questions of the past and present. The approach is "problem driven, not theory driven" or method driven (Bates *et al.* 1998: 11): the motivation for each of the studies is substantive interest by the author in what took place and why. However, our version of analytic narrative is clearly informed by theory, specifically rational choice theory, and by the conjoined methodologies of historical analysis and in-depth case studies.

The analytic narrative approach

The analytic narratives project attempts to make explicit the approach adopted by numerous scholars trying to combine historical and comparative research with rational choice models in order to understand institutional formation and change. It is important to be clear. These essays do not represent a methodological breakthrough. There are many comparativists engaged in similar enterprises. What distinguishes the essays in *Analytic Narratives* is not the approach *per se* but rather the presentation of the material so as to emphasize the steps involved in assembling an analytic narrative. In the original book and in responses to critics, the authors have begun the process of outlining the key elements of a widely shared approach; there is systematization still to be done.

The book, *Analytic Narratives*, brings together five scholars who share a commitment to understanding institutional change and variation. Influenced by the work of Douglass C. North (1981, 1990, 1996 [1993]), to whom the book is dedicated, the aim is to use the tools of the new economic institutionalism to investigate such enduring questions of political economy as political order, governance of the economy and polity, and interstate relations. The pieces in the book represent an extension of

North's conception of institutions as rules, both formal and informal, that influence behavior by means of constraints and incentives.

The authors emphasize a view of institutions derived from the theory of games, especially extensive form games and the associated equilibrium concept of subgame perfection. Choices are coordinated, regularized, stable, and patterned – institutionalized, if you will – because they are made in equilibrium. From this perspective, the key questions become how that particular equilibrium emerged in a world of multiple equilibria and why it changes. Relative price changes and variations in distributions of bargaining power and resources among the participating actors (Knight 1992; Levi 1990) suggest the outlines of answers that only a detailed investigation of the case can illuminate.

Thinking about institutional and other forms of social change as problems of multiple equilibria imposes certain boundaries on the scholarly endeavor. Whereas North's programmatic books may consider the whole history of the Western world, his research tends to take on narrower questions, such as the origin of constitutional constraints on the English monarch following the Glorious Revolution (North and Weingast 1989). The essays in *Analytic Narratives* are similarly focused on specific institutions in particular times and places.

The project also reflects the recognition of the extent to which the authors combine rational choice with inductive practices. Even the formulation of the model is an inductive process, requiring some initial knowledge of the case; the refinement of the model depends on iteration with the narrative.

There are potential gains and losses from the combination of rational choice and game theory with historical analysis. The potential gains come from the capacity to model the sequencing and dynamics of the narrative, to have a technique for specifying the temporal ordering.[1] In addition, the reliance on game theory offers a means to model "constellation effects" (Mayntz forthcoming), that is the relationships and interactions among corporate actors whose individual choices have consequences at the macro level. The application to history and narrative provides a rich empirical account that allows for process tracing (George and Bennett forthcoming; George and McKeown 1985) and evaluation of the model.

The potential losses are of two sorts. The first is the danger (which the *Analytic Narrative* authors like to believe they have avoided) of failing to meet criteria of good research design. David Laitin (2002, 2003) advocates the "tripartite method of comparative research," which integrates

[1] Büthe makes this point while also criticizing game theory for being "based on a truncated conception of temporality" (2002: 485).

statistical analysis, formal theory, and narrative. However, for studies undertaken as analytic narratives, the point is to establish the basic causal mechanisms at work; only once this goal is accomplished can an author even begin to consider statistical analysis.[2] Furthermore, for some problems statistics are either unavailable or inadequate for getting at the key questions and the number of observations too few for anything approaching what most social scientists consider a test. While paying as much attention as possible to statistical rules of causal inference (King, Keohane, and Verba 1994), analytic narrativists are searching for a means to understand important instances of institutional change not always susceptible to many standard techniques. Generalization is more difficult, but satisfying causal explanation should be more possible.

The second limitation comes from the reliance on rational choice and game theory. Formal models may encourage a search for immediate causes that link the strategic behavior of the key individuals and groups, thus losing sight of the causes situated in the more distant past (Pierson 2003). Firm grounding in economic history and its sensitivity to the *longue durée* tend to obviate this problem, albeit sometimes making the modeling more difficult and complex. Increasingly, as behavioral economists and others (Jones 2001) have made us aware, assumptions of full rationality are problematic, but this need not mean, as Jon Elster (2000) claims, that there is little or no value in using rational choice models in studying complex processes. Such models still offer powerful accounts of certain kinds of phenomena – even if these models may one day, even soon, be superseded by better ones. Nor does it mean that uncertainty, contingency, unintended consequences and the like have no role to play in rational choice analytic narratives. This issue will be more fully addressed below.

There is a tendency for rational choice scholars to focus almost exclusively on cognitive mechanisms at the individual level. This can (but need not) lead to insufficient attention to emotional (Elster 2000: 692, 694) or to relational and environmental mechanisms (McAdam, Tarrow, and Tilly 2001; 2001: 22–3). To some extent the emphasis on institutions and explicit elaboration of structural constraints permits some movement in the direction of achieving an explanation concerned with macro- as well as micro-processes. Nonetheless, the analytic narrative project, at least in its current state of development, is less attentive to the whole range of mechanisms and processes than are others (see, esp., Mayntz forthcoming; Tarrow 2003; Tilly forthcoming) who also focus on developing an analytic but historically grounded social science.

[2] This is a major distinction from the prescriptions of King, Keohane, and Verba (1994) only because they assume the causal mechanism is already understood.

What rational choice analytic narratives promise are explanatory accounts of how structured choices arise in complex historical situations, how and why certain choices are made, and their consequences for actors and institutional arrangements. It is to the efforts to achieve these objectives that we now turn.

The essays

The substantive chapters in *Analytic Narratives* explore institutional change in a wide range of places and times. All focus on a specific historic puzzle, sometimes taking place only in one country. The primary aim is to understand a particular set of institutions, but the combination of approach and findings do have implications for a wider set of issues. Because of the desire to systematize the approach of analytic narrative, the essays are written in a way that attempts to reveal the skeleton of the reasoning and decisions that the authors make in building their models, selecting what is essential from the larger history, and devising their explanations. This makes some of the writing more pedagogical than literary.[3] After all, it is the quality of the theory and its confirmation in the essays on which the project rests.

Avner Greif (1998) accounts for the origins of the twelfth-century Genovese *podestà*, a ruler with no military power. In explicating an exotic institution in an interesting moment of history, Greif constructs an argument with significant implications for theorizing the relationship between factional conflict and political order. Jean-Laurent Rosenthal (1998) compares rulers' capacities to raise revenue and wage wars in France and England in the seventeenth and eighteenth centuries. He argues that the distribution of fiscal authority is a major explanatory variable. Rosenthal speaks to the sources of regime variation and change as well as the relationship between domestic political structures and war making. Margaret Levi (1998) investigates the institutional bases for variation in government policies and citizen responses to conscription in France, the United States, and Prussia. Levi's finding that revised norms of fairness, resulting from democratization, influence the timing and content of institutional change suggests the importance of normative considerations and the institutional bases of legitimacy in accounting for citizen compliance with government and regulatory agencies more generally. Barry Weingast's (1998) focus is the balance rule, the compromise over the

[3] This produces the criticism (Carpenter 2000: 656–7) that the analytic narrative project authors tell a story in the way that fails to fit the aesthetic criteria claimed by some historical institutionalists.

admission of slave states, how it promoted ante-bellum American political stability, and how its breakdown was a critical factor in precipitating the Civil War in the United States. Weingast advances the program of understanding the institutional foundations and effects of federalism. Robert Bates (1998) also addresses one specific institution, the International Coffee Agreement. He explains why it rose and fell and why the United States, a principal coffee consumer, cooperated with the cartel to stabilize prices during World War II and the Cold War. Bates offers a significant contribution to understanding the circumstances under which a political basis for organization will trump economic competition in an international market.

Of analytics and narratives

Analytic narratives involve choosing a problem or puzzle, then building a model to explicate the logic of the explanation and to elucidate the key decision points and possibilities, and finally evaluating[4] the model through comparative statics and the testable implications the model generates. As an approach, analytic narrative is most attractive to scholars who seek to evaluate the strength of parsimonious causal mechanisms. The requirement of explicit formal theorizing (or at least theory that could be formalized even if it is not) compels scholars to make causal statements and to identify a small number of variables.

Analytics here refers to the use of models derived from rational choice, particularly the theory of extensive form games. The narrative refers to the detailed and textured account of context and process, with concern for both sequence and temporality. It is not used in the postmodern sense of a master- or meta-narrative. Rather, it refers to research grounded in traditional historical methods.

The approach requires, first, extracting from the narratives the key actors, their goals, and their preferences and the effective rules that influence actors' behaviors. Second, it means elaborating the strategic interactions that produce an equilibrium that constrains some actions and facilitates others. The emphasis is on identifying the reasons for the shift from an institutional equilibrium at one point in time to a different institutional equilibrium at a different point in time. By making the assumptions and reasoning clear and explicit, it is then possible to pose a challenge

[4] In *Analytic Narratives*, the authors use the language of theory testing. However, as Kevin Quinn has pointed out, it is not really clear that testing is at issue in most analytic narratives, where the number of observations is so few. This is an issue that deserves considerable further exploration. But see Büthe (2002) for a different perspective.

that might produce new insights and competitive interpretations of the data.

The model must also entail comparative static results. The comparative statics are crucial for comparative research since they are the basis for hypotheses of what could have taken place under different conditions. Comparative statics clarify the effects of the key exogenous variables on the endogenous variables and offer yet another source of hypothesis building. The consideration of off the equilibrium path behavior can reveal the reasoning behind why actors took one path and not another. Indeed, what actors believe will happen should they make a different choice might determine what choices they do make.

The narrative of analytic narratives establishes the actual and principal players, their goals, and their preferences while also illuminating the effective rules of the game, constraints, and incentives. Narrative is "a useful tool for assessing causality in situations where temporal sequencing, particular events, and path dependence must be taken into account" (Mahoney 1999: 1164): The narrative provides the necessary information for causal assessments.

First, it offers a means to arbitrate among possible explanations for instances of observational equivalence, when either of two distinct processes could be producing the outcome under investigation. For example, in the illustrative game in the Appendix of *Analytic Narratives* (Bates *et al.* 1998), in equilibrium the opposition does not attack a country with a large army. Is having a large army when there is no attack the very reason for peace or is it a waste of resources? Different people have different beliefs that can only be understood contextually:

> the observationally equivalent interpretations rest on markedly different theories of behavior. To settle upon an explanation, we must move outside the game and investigate empirical materials. We must determine how the opponent's beliefs shape their behavior. This blend of strategic reasoning and empirical investigation helps to define the method of analytic narratives . . . (241)

In Levi's chapter (1998), for example, democratization changes both norms of fairness and institutions. The first alters the ranking of preferences by key actors and the second their beliefs about the effects of their actions.

Sometimes the narrative is insufficient to arbitrate between two alternative explanations, and the theory specifies the conditions that must obtain to ascertain which is correct. This was the case in Greif's analysis of political order in pre-1164 Genoa (1998: 35–6). The narrative then offers the key for causal assessment.

Case selection

The analytic narrative authors combine a commitment to rational choice, a deep interest in a particular case, and a desire to elaborate more general conditions for institutional change. The standard approaches to case selection emphasize the bases for choice among a sample of cases which are informative about the causal chain of interest, either because of the absence, presence, or extreme values of key variables (Van Evera 1997: 49–88). While these may be critical criteria for selecting cases to test or generate a general theory, the criteria used by the analytic narrativists are sometimes closer to that of the historians than of generalizing social scientists. There is attention to the properties of causal inference embodied in statistical analysis (King, Keohane, and Verba 1994), but the choice of cases is not drawn from the conceivable sample but because of actual puzzles that have attracted the authors.

Sometimes these puzzles are generated by the availability of an exciting new data set. This was part of what started Greif in his explorations of the Magrhibi traders; his interest in the Genovese derived from that initial enterprise, which generated additional questions about the changing bases of international trade and commerce, cultural beliefs, and institutional arrangements for coordinating social order (Greif 1989, 1993, 1994a, b, 1995a, b). Sometimes the motivation comes from long-term commitment to understanding a particular part of the world in a particular era. Bates' interest in the coffee cartel arose from his investigations of postcolonial African markets (Bates 1981, 1983, 1997). Rosenthal's focus on French and English absolutism builds on his years of scholarship on the development of markets in France, prior to and after its Revolution (Hoffman, Postel-Vinay, and Rosenthal 2000; Rosenthal 1992). Sometimes the case is inherently interesting and significant; the limited success of the huge literature on the timing of the Civil War inspired Weingast to reconsider the evidence from a new perspective. Sometimes the puzzle is generated by a larger theoretical project, in which a particular case provides a particularly nice window. Levi was not attracted to questions of conscription because of their inherent interest to her or a long-term investment in their investigation. She chose that subject because it allows her to investigate the conditions under which citizens give or withdraw consent from government's extractive demands. To some extent such a motivation existed for all the authors, for all have a long-term theoretical project.

Reliance on analytic narratives can occur because there is no way to accumulate the kind of data required for good statistical analysis; either the problem itself or the period under study is not amenable to quantitative research. Analytic narratives may also complement statistical evidence; the construction of analytic narratives is not an alternative to Laitin's tripartite method. However, all analytic narrative cases must include features that make them amenable to modeling, which not all puzzles or problems are. In addition, they must provide an opportunity to get at an important process or mechanism not easily accessible through other means. Finally, the causal mechanisms and the structures or relationships must be generalizable to other cases, once they are modified for the specifics of the new case.

Strategic choice situations

Essential to the model-building is the choice of cases in which there are strategic interactions among the key actors. That is, the choice of one depends on the choice of the other. By considering situations that can be modeled as extensive form games in which there is a subgame-perfect equilibrium, self-enforcement of the institution becomes a matter of credible commitments. In the equilibrium of a game, it is in the interest of the players to fulfill their threats or promises against those who leave the equilibrium path.

This does not mean that the game – and its incentives – cannot change. The comparative static results suggest where it might. Indeed, understanding sources of change is critical to understanding institutional transformation. However, it is also important to comprehend the reasons for institutional stability. Thus, Weingast's analysis of the balance rule accounts for the mechanics of how actors, with conflicting interests in the perpetuation of slavery, benefited from a compromise on slavery, why they stopped benefiting, and therefore why the institution changed and with what consequences.

Contingency and uncertainty

The existence of multiple equilibria makes both contingency and uncertainty crucial features of analytic narratives. If the starting point is fortuitous, then the resulting equilibrium is effectively determined by contingency. Whatever the starting point, there is uncertainty about the outcome. It is indeterminate *ex ante* since it depends on the strategic choices of actors, who vary in the kind of information they possess, their discount rates, and other constraints acting upon them. The narrative is key to unraveling the probable choices of the actors, but the narrative also reveals the range of characterizations of the situation and conceptualizations of their positions that key individuals possess.

Levi's research on labor union leadership (in process) illuminates the contingency of the starting point. She finds that union members have a dominant preference for competent representation, leaders who can solve the initial strategic dilemmas of organizing a union, winning strikes, attaining the right to bargain collectively, and winning decent contracts. They tend to give their support to whoever effectively solves those problems, whoever happens to be in the right place at the place time. The Teamsters of the 1930s offer an example. In Minneapolis, members coordinated around Farrell Dobbs, a Trotskyite. In Detroit, they turned to James S. Hoffa, who was relative conservative politically. Both Hoffa and Dobbs, who mentored Hoffa, were effective in gaining employer compliance to their demands for union recognition and improved pay and hours, but Hoffa also relied on strong-arm tactics and third-party enforcers, such as the *Cosa Nostra*. The choice of the leader had consequences for the governance arrangements of the union. If members knew and understood all the implications of their original choice, they might make a different one – if they could.

The cases must include problems of randomness or contingency but not if they are too extreme. Again, the example of the unions makes the point. Members solved their leadership problem in the face of uncertainty about the occurrence of strikes and only partial information about the reaction of employers to their demands. Because the interactions between unions and managers are unpredictable and leaders cannot always deliver what they promise, members might have responded by regularly replacing leaders. Where that is true, union leadership is not as good a subject for an analytic narrative. It might make more sense to treat each election independently and thus use quantitative methods to develop and test models of leadership selection and retention.

Even when uncertainty about outcomes appears to be minimal, as in instances where there are clear focal points and strategies, factors in the situation can change unexpectedly. Some contextual changes may have clear and significant consequences, others have butterfly effects, and others little or no effect. The narrative is crucial here for sorting out what matters for what. In Rosenthal's chapter, the potential birth of a Catholic heir to James II affects the calculations of both monarch and elites, but its importance lies in how it changes the strategies of the elites even unto the point of revolution (92). Why they had to resort to revolution rather than peaceful institutional change becomes apparent through the narrative.

Building models

The analytic narratives privilege parsimonious models, ones where the number of exogenous factors are sufficiently few that it is possible to know how changes in their value can affect the institutional equilibrium. This affects the narrative by reducing the importance of other variables for the story. For instance, for Bates one sort of uncertainty was critical: movements in the price of coffee. Variations in US economic activity – however important they might be to the world economy – were much less salient. All narratives have to have an anchor (or set of anchors). Analytic narratives make the theoretical anchor more explicit (and thereby easier to criticize) than in more configurative accounts. Analytic narratives require a thicker account than a formal model alone, but the use of game theory ensures that the account will be as thin as is essential to produce an explanatory story.

In building a model, it is advisable to avoid using off-the-shelf models unless they demonstrably enhance the explanatory project. Empirical social scientists often appeal to one of a small number of models (prisoner's dilemma, battle of the sexes, principal agent with moral hazard, principal agent with adverse selection), even when these models do no more than redescribe the situation in slightly different terms or illuminate only a small part of what is under investigation. Their analysis depends on the context; each of these models can either lead to an efficient or inefficient solution, to a problem solved, or to a problem not solved. Which it is depends on detailed knowledge of the context. Once the context is sufficiently understood, the researcher can build a model that fits the particular case better and that captures actual institutional constraints. The institutional constraints illuminate the set of possible outcomes that are possible, and they suggest how the particular problem faced by a society can be solved.

Deduction, induction, and iteration

In analytic narratives the narrative and the analytics are very intertwined. For analytic narratives the models used to elucidate the causal connections among variables are iterative and inductive although the initial intuitions may have been deduced. This is a different strategy than the establishment of a general model from which are derived testable hypotheses, explored with appropriate cases.[5] The assumptions of rational choice and the logic of game theory generate hypotheses, but the models are refined in interplay with the detailed elements of the narrative. While the claim to generalizability of findings is clearer when hypotheses are deduced from general theory, the explanations of specific instances may be less compelling and realistic. This has long been a critique of the rational choice

[5] This is a common practice among rational choice scholars and one that has led to some very important insights and findings (Golden 1997; Kiser 1994; Levi 1988).

program in comparative and historical politics and one the analytic narrative project attempts to address.

All of the *Analytic Narrative* authors rely on rational choice to derive hypotheses and provide theoretical leverage, and each has training and experience in historical research, fieldwork, or both. The research began with some basic information and some theoretical priors, but the next step is to accumulate new information and formulate new models. Bates exemplifies this process. In moving from a model of oligopoly to a model of political economy, he started from one clearly articulated vantage point, confronted it with the evidence, and then selected a new one.

Iteration between theory and data also has implications for the conduct of research. Each new model adopted should be consistent with the facts on hand and reveal what new data still needs to be acquired. When constructing the theory, social scientists often already know a lot about the data/problem/case that they study. While this is true generally in social science, it is even more obvious in an approach like analytic narrative where the theory must be imbedded in the narrative.

Path dependence

Attention to the narrative ensures that, to the extent possible, the authors can reconstruct the points in the strategic interactions when contingency and uncertainty have an impact on the outcome. The model suggests and the narrative explicates why certain paths were chosen, others purposely forgone, and some not considered at all. The analytic narrativist generally selects cases where there is path dependence (David 1985, 1994) but not necessarily complete certainty about the outcome. There is nothing about the approach that limits its use to cases of certainty or modest uncertainty, despite Elster's claim otherwise (2000: 693). Extensive form games have long been useful in studying settings of high uncertainty and contingency. Path dependence requires more than identifying the constraints that derive from past actions or the incentives that are built into new institutions. The sequence in which events occur is causally important; events in the distant past can initiate particular chains of causation that have effects in the present. Game theory is particularly well suited to modeling path dependence since so many of its outcomes depend on the sequence of moves.

Path dependence, as understood through the prism of the paths not taken, means more than "history matters." This is trivially true. The starting point of the game affects and often determines the end point but only once the proper payoffs are incorporated. Certain institutions in certain contexts become self-enforcing in the sense that the alternatives continue to appear unattractive. Beliefs by the players then matter as much as history. While beliefs are certainly affected by historical experience, they also are affected by what actors know of the other players within the current context.

Path dependence in analytic narratives also implies that once certain institutional arrangements are in place – and with them certain distributions of power and authority – it becomes more difficult to reverse or change course. Rosenthal makes this very clear in accounting for the divergence between French and English political institutions in the seventeenth century. Are these the feedback effects that Paul Pierson (2000) emphasizes in his influential article (also, see Thelen 1999: 392–6)? This is not so clear. The analytic narratives approach shares features with this formulation, but the extent of difference and similarity remains to be fully explored.

Greif's (1996) research on the Commercial Revolution is illustrative. Greif identifies various forms of expectations that coordinate action and, in some instances, give rise to organizations that then influence future economic development. The expectations arise out of the complex of economic, social, political, and cultural – as well as technological – features of a society. The existence of a coordination point in itself makes change difficult since it requires considerable effort to locate and then move enough others to a different coordination point (Hardin 1999). When organization develops, the path is even more firmly established, for organization tends to bring with it vested interests who will choose to maintain a path even when it is not or is no longer optimal.

Comparative statics generate hypotheses about the relationship between exogenous changes and endogenous variables. They provide a way to understand why institutions shift and change. At a minimum comparative statics help us understand the choice of path within a given context by spelling out how a shock or other external variable will alter the choices of the actors. The Weingast and Levi essays in *Analytic Narratives* use comparative statics to help account for the switch from one equilibrium to another within a given society, creating a new starting point for institutional evolution. There remains a question about the extent to which comparative statics can capture the big foundational moments,[6] which the historical institutionalist literature on critical junctures[7] attempts to explain. However, comparative statics may also illuminate even those big one-off events labeled as critical junctures (see, e.g.,

[6] This is a somewhat contested point among the five authors of *Analytic Narratives*.
[7] For an excellent review of the literature on critical junctures, see Thelen (1999: 388–92).

Bates, de Figueiredo, and Weingast 1998; de Figueiredo and Weingast 1999; Weingast in-process).[8]

Evaluation

There are a variety of criteria for evaluating implementations of the analytic narrative approach. Several of these are logical, but others involve issues of confirmation and generalizability. Most are in the standard toolbox of the well-trained social scientist and are used in evaluation of virtually all research.

Jon Elster (2000) claims that rational choice analytic narratives are only "just so stories." He echoes the more generic complaint lodged against rational choice by Donald Green and Ian Shapiro (1994; also, see Shapiro 2004). The analytic narrative project members strongly reject this accusation. A major point of the enterprise is to self-consciously adhere to standards of narrative construction, problem formation, and assessment that prevent fashioning "just so stories." Nonetheless, it is fair to claim, as James Mahoney (2000: 86) does, that "advocates of the new rational choice theory have still not said enough about their methods of hypothesis testing." What follows below is the humblest of beginnings of a corrective.

Logic

One of the advantages of analytic narratives is the possibility of assessing the argument according to rigorous and, often, formal logic. Conclusions must follow from the premises. If the reasoning is wrong or even insufficiently precise, then the account lacks credibility. Logical consistency disciplines both the causal chain and the narrative. This is especially true where there is explicit formal and mathematical reasoning employed. The math is either right or wrong.

However, correct math is hardly the only criterion of assessment. A given narrative suggests a model which, when explicated, should have implications for choices, behaviors, and strategic interactions among the players. Those implications force the scholar to reconsider the narrative and then to reevaluate the extent to which key elements of the

[8] Barry Weingast (in process) points out that while scholars from multiple perspectives have identified mechanisms that generate path dependence, the same cannot be said for the vaguer term, critical junctures. Both of his claims found considerable confirmation in the second meeting of the Russell Sage Foundation workshop on "Preferences in Time," an interdisciplinary group representing a variety of approaches. It was organized by Weingast with Ira Katznelson.

narrative lie outside of the proposed theory. If one must appeal too often to forces outside the model, then the theory must be rejected. At the same time, the model clarifies those issues that cannot be resolved logically and can only be resolved through narrative materials.

Confirmation

There are various means to assess the extent to which the authors have offered an account confirmed by the data and superior to the alternatives. Not only must the assumptions and causal mechanisms fit the facts, but the model must also have generated testable implications that no alternative would.

Explicit assumptions assist appraisal. However, explicitness puts a burden on the analyst to demonstrate that the assumptions are reasonable. Are the actors identified as the key players in fact the key players? Have the authors provided a plausible and defensible definition of the content and order of preferences as well as of the nature of the beliefs and goals of the actors? The reliance on simplifying assumptions tends to prompt challenges from those who are steeped in the history and the context of the place and period, but it also provokes challenges from those who do not find the assumptions logical or consistent with the material presented in the narrative. The logic embedded in analytic narratives and the empirical material used to create, on the one hand, and evaluate, on the other, soon become enmeshed. Both must meet high standards.

Producing testable implications from the model and then subjecting them to disconfirmation is as critical to the evaluation of the analytic narrative as it is for other means of generating hypotheses and implications. However, the iterative feature of analytic narratives adds another dimension. As the authors of *Analytic Narratives* note,

> We stop iterating when we have run out of testable implications. An implication of our method is that, in the last iteration, we are left uncertain; ironically perhaps, we are more certain when theories fail than when they fit the case materials. (17)

Analytic narratives share with process tracing (George and Bennett forthcoming) the commitment to modification of the models and explanations as the data reveal new possibilities. There is a difference, however. By explicitly using rational choice and especially when relying on extensive form games, there is an even clearer and more rigorous delineation of the process. Both process tracing and analytic narratives easily degenerate into curve-fitting exercises if improperly done. One of the analytic narratives' greatest strengths may be that the combination of game theory and iteration compels the researcher to search for novel facts that the old model neither recognized nor captured. This then makes the refinement of the model part of a progressive research program in the Lakatosian sense.[9]

The final criterion of adequate confirmation is the extent to which the theory offers a more powerful account than other plausible explanations. Sometimes, there are alternatives against which an author is contesting or subsuming. This was true for the issues Weingast and Levi study. Sometimes, there are obvious alternatives that derive from a general framework. For example, Greif demonstrates that his political economic theory is superior to a cultural theory (1998; also see 1994a, 1995b). To some extent, the comparative static results offer some competitive explanations. However, often, the generation of an alternative account must come from someone on the other side of the scholarly debate and not from the author of the original analytic narrative. Authors usually become committed to their own interpretations and their own ways to discipline the narrative even when they engage in very strenuous efforts to remain objective. Thus, real social scientific progress comes when different scholars with different perspectives attempt to offer more powerful explanations of the same phenomenon (Büthe 2002: 489; Fiorina 1995).

Generalizability

The Achilles' heel of analytic narratives, indeed of most historically specific accounts and small-N case study research (Rueschemeyer 2003), is in the capacity to generalize. These are, after all, efforts to account for a particular puzzle in a particular place and time with a model and theory tailored to that situation. Even so, it is possible to use the cases to make some more general points.

The reliance on rational choice, which is after all a general theory of how structures shape individual choices and consequently collective outcomes,[10] assists in the portability of findings. The additional reliance on game theory highlights certain properties of the structure and strategic choices that then arise: "whereas the specific game may not be portable ... they may yield explanations that can be tested in different settings" (Bates *et al.* 1998: 234). By identifying the specific form of collective action problems, principal–agent issues, credible commitments, veto points, and the like, analytic narratives provide a way to suggest the characteristics of

9 Turan Kayaoglu suggested this point to me.

10 This is not quite the same claim as that made by Kiser and Hechter (1991, 1998), who are rigorously deductive. The difference lies not only in the emphasis on the relative roles of deduction and induction but equally on the extent of portability of the findings. Nor do they find game theory useful in their practice or their theory (Hechter 1990, 1992).

situations to which these apply and in what ways. For example, the models of federalism, as initially delineated by William Riker (1964) and developed by Barry Weingast and his collaborators (Weingast 1995; Weingast, Montinolo, and Qian 1995), are useful in explicating a large number of problems in a wide range of countries, including the case Weingast addresses in his *Analytic Narrative* chapter.

What makes possible the application of the logic of one setting to another is the existence of identifiable causal mechanisms, one of the strengths of rational choice. Roger Petersen (1999: 66) offers a primer for linking mechanisms and structures by means of game theoretic analyses:

> When the structures can be identified a priori, that is independently from outcomes, prediction becomes possible. Secondly, the decision structures may connect individual actions to aggregate level phenomena. Through its specification of causal linkages across levels of analysis, game theory can provide individual level prediction from existing aggregate level theory.

Mechanisms then do the work Stinchcombe (1991) wants of them; they "generate new predications at the aggregate or structural level."

Conceptions of path dependence, an attribute of nearly all institutional approaches, also rely on causal mechanisms to account for the adaptations and reproduction of institutions. There is a tension here, however, and it takes a similar form for both historical comparative and rational choice institutionalists. If the presumption is that different societies encounter similar problems, it is possible to posit a limited set of mechanisms and solutions. Context influences the outcome but is not totally determinative, and some level of generalization is possible. However, if the presumption is that in some cases the micro details of the situation matter a lot for the form of institutional evolution, and in other cases not at all, then generalization is far less viable.[11]

The literature on mechanisms is currently in a state of development; there is little consensus on the definition of mechanisms, let alone how to best use them in explanatory accounts (see, e.g. Hedstrom and Swedberg 1998; Johnson 2002; Mahoney 2001; Mayntz forthcoming; McAdam, Tarrow, and Tilly 2001; Tarrow 2003; Tilly 2001, forthcoming). However, if they are to be useful as a tool of causal assessment, they cannot be simply what Elster suggests (1998; 1999): a list of mechanisms, such as emotions, resentment, and the like, that offer a fine-grained explanation of the link between actions and alternatives.

[11] Jean-Laurent Rosenthal made this point to me. He used North (1981) as exemplar of a conception of path dependence that permits generalization and North (1990) as exemplar of a conception that does not. This distinction is, however, widespread among comparative and historical scholars.

Elster himself admits his enterprise borders on "explanatory nihilism" (Elster 1999: 2). A more useful view is that mechanisms are processes that causally connect initial conditions and the outcomes of interest and which may occur at the macro and at the relational as well as the cognitive or emotional levels (Mayntz forthcoming; McAdam, Tarrow, and Tilly 2001).

To further establish the generalizability of the theory out-of-sample tests are necessary. The presumption today in social science research is that the authors will provide those tests themselves, and many attempt to, including Levi and Rosenthal in their chapters in the book. However, seldom does the level of knowledge for the out-of-sample case rival the detailed understanding of the original case that puzzled the author. While it can be argued that it is incumbent upon an author to have equal authority over a wide range of cases, this is seldom realistic for area specialists, historians, and others who must conquer languages, archives, and other sources to acquire the in-depth command of the subject matter and the narrative detail essential for an analytic narrative. The demonstration of generalizability may rest ultimately on a larger community of scholars who take the findings applicable to one place and time to illuminate a very different place and time.

Conclusion

The rational choice analytic narrative approach stands in sharp opposition to views of history that would make the outcomes of events totally systematic or unsystematic in the extreme. It is the claim of the analytic narrativists that understanding the institutional context within which events occur can explain outcomes in certain (but not all) important historical situations: where structured choices have significant consequences.[12] This is an ideological position rather than a methodological position because there is nothing *per se* in game theory that rules out complete uncertainty. However, it is not a claim that the approach reveals all the effects of that structured choice; unintended consequences and unforeseen contingencies are beyond the scope of the enterprise.

Use of the rational choice analytic narrative approach has implications for developing a social science capable not only of explaining the past but also of the far more difficult task of advising about the present. By

[12] I thank Chuck Tilly for urging me to be clearer about the ontological claims of the approach. Further elaboration of the status of this claim and specification of the conditions under which it can operate will have to await a future paper, however.

providing a way to model and explicate a complex but relatively unique event, the combination of rational choice and narrative offers a plausibility test for some kinds of proposed policies. By focusing on the institutional details that affect strategic interactions, choices, and outcomes, it can be useful in suggesting likely outcomes under given initial conditions. In using comparative statics, it can help predict the likely consequences of certain kinds of changes and shocks. The analytic narrative approach is applicable only for a relatively narrow range of problems, perhaps, but it offers some promise of revealing important relations and probable outcomes for those kinds of problems for which it is suited. It may provide insight into a piece of a larger puzzle.

It is only one of many tools, however, available to social scientists and not always the most appropriate one. As we progress toward a social science better capable of diagnostics and prediction, we may find better and more powerful tools. In the meantime, we must use what we have to do what we can.

REFERENCES

Bates, Robert H. 1981. *Markets and States in Tropical Africa*. Berkeley: University of California Press.

1983. *Essays on the Political Economy of Rural Africa*. Berkeley: University of California Press.

1991. *Beyond the Miracle of the Market*. New York: Cambridge University Press.

1997. *Open Economy Politics: The Political Economy of the World Coffee Trade*. Princeton: Princeton University Press.

1998. "The International Coffee Organization: An International Institution," in *Analytic Narratives*, R. H. Bates, A. Greif, M. Levi, J.-L. Rosenthal, and B. Weingast (eds.). Princeton: Princeton University Press.

Bates, Robert H., Jr., Rui J. P. de Figueiredo, and Barry R. Weingast. 1998. "The Politics of Interpretation: Rationality, Culture and Tradition." *Politics & Society* 26(4): 603–42.

Bates, Robert H., Avner Greif, Margaret Levi, Jean-Laurent Rosenthal, and Barry R. Weingast. 1998. *Analytic Narratives*. Princeton: Princeton University Press.

2000a. "The Analytic Narrative Project." *American Political Science Review* 94(3): 696–702.

2000b. "Analytic Narratives Revisited." *Social Science History* 24(4): 685–96.

Büthe, Tim. 2002. "Taking Temporality Seriously: Modeling History and Use of Narratives as Evidence." *American Political Science Review* 96(3): 481–93.

Carpenter, Daniel P. 2000. "What is the Marginal Value of Analytic Narratives?" *Social Science History* 24(4): 653–68.

Chandra, Kanchan. 2004. *Why Ethnic Parties Succeed: A Comparative Study*. New York: Cambridge University Press.

David, Paul. 1985. "Clio and the Economics of QWERTY." *American Economic Review* 75(2): 332–37.

1994. "Why are Institutions the 'Carriers of History'?: Path Dependence and the evolution of Conventions, Organizations, and Institutions." *Structural Change and Dynamics* 5: 205–20.

de Figueiredo, Jr., Rui J. P., and Barry R. Weingast. 1999. "The Rationality of Fear: Political Opportunism and Ethnic Conflict," in *Civil Wars, Insecurity and Intervention*, B. Walter and J. Snyder (eds.). New York: Columbia University Press.

Elster, Jon. 1998. "A Plea for Mechanisms," in *Social Mechanisms*, P. Hedstrom and R. Swedberg (eds.). New York: Cambridge University Press.

1999. *Alchemies of the Mind: Rationality and the Emotions*. New York: Cambridge University Press.

2000. "Rational Choice History: A Case of Excessive Ambition." *American Political Science Review* 94(3): 685–95.

Fiorina, Morris P. 1995. "Rational Choice, Empirical Contributions, and the Scientific Enterprise." *Critical Review* 9(1–2): 85–94.

Flyvbjerg, Bent. 2001. *Making Social Science Matter: Why Social Science Inquiry Fails and How it Can Succeed Again*. Cambridge: Cambridge University Press.

Geddes, Barbara. 1994. *The Politicians' Dilemma*. Berkeley: University of California Press.

2003. *Paradigms and Sand Castles in Comparative Politics of Developing Areas*. Ann Arbor: University of Michigan.

George, Alexander and Andrew Bennett. forthcoming. *Case Study and Theory Development*. Cambridge, MA: MIT Press.

George, Alexander, and Timothy McKeown. 1985. "Case Studies and Theories of Organizational DecisionMaking." *Advances in Information Processing in Organizations* 2: 21–58.

Golden, Miriam. 1997. *Heroic Defeats: The Politics of Job Loss*. New York: Cambridge University Press.

Green, Donald, and Ian Shapiro. 1994. *The Pathologies of Rational Choice*. New Haven: Yale University Press.

Greif, Avner. 1989. "Reputation and Coalitions in Medieval Trade: Evidence on the Maghribi Traders." *Journal of Economic History* 49(4): 857–82.

1993. "Contract Enforceability and Economic Institutions in Early Trade: Evidence on the Maghribi Traders' Coalition." *American Economic Review* 83(3): 525–548.

1994a. "Cultural Beliefs and the Organization of Society: A Historical and Theoretical Reflection on Collectivist and Individualist Societies." *Journal of Political Economy* 102(5): 912–50.

1994b. "On the Political Foundations of the Late Medieval Commercial Revolution: Genoa During the Twelfth and Thirteenth Centuries." *Journal of Economic History* 54(4): 271–87.

1995a. "The Institutional Foundations of Genoa's Economic Growth: Self-Enforcing Political Relations, Organizational Innovations, and Economic Growth During the Commerical Revolution."

1995b. "Micro Theory and Recent Developments in the Study of Economic Institutions Through Economic History," in *Advances in Economics and Econometrics: Theory and Applications*, D. M. Krep (eds.). New York: Cambridge University Press.

1996. "On the Study of Organizations and Evolving Organizational Forms Through History: Reflections from the Late Medieval Firm." *Industrial and Corporate Change* 5(2): 473–502.

1998. "Self-Enforcing Political Systems and Economic Growth: Late Medieval Genoa," in *Analytic Narratives*, R. Bates, A. Greif, M. Levi, J.-L. Rosenthal, and B. Weingast (eds.). Princeton: Princeton University Press.

Hardin, Russell. 1999. *Liberalism, Constitutionalism, and Democracy*. Oxford: Oxford University Press.

Hechter, Michael. 1990. "On the Inadequacy of Game Theory for the Solution of Real-World Collective Action Problems," in *The Limits of Rationality*, K. S. Cook and M. Levi (eds.). Chicago: University of Chicago Press.

1992. "The Insufficiency of Game Theory for the Resolution of Real-World Collective Action Problems." *Rationality & Society* 4: 33–40.

Hedstrom, Peter and Richard Swedberg (eds.). 1998. *Social Mechanisms: An Analytical Approach to Social Theory*. Cambridge: Cambridge University Press.

Hoffman, Philip T. and Kathryn Norberg. 1994. *Fiscal Crises and the Growth of Representative Institutions*. Stanford: Stanford University Press.

Hoffman, Philip T., and Jean-Laurent Rosenthal. 1997. "The Political Economy of Warfare and Taxation in Early Modern Europe: Historical Lessons for Economic Development," in *The Frontiers of the New Institutional Economics*, J. N. Drobak and J. V. Nye (eds.). San Diego: Academic Press.

Hoffman, Philip T., Gilles Postel-Vinay, and Jean-Laurent Rosenthal. 2000. *Priceless Markets: The Political Economy of Credit in Paris, 1660–1870*. Chicago: University of Chicago Press.

Johnson, James. 2002. "How Conceptual Problems Migrate: Rational Choice, Interpretation, and the Hazards of Pluralism." *Annual Review of Political Science* 5: 223–48.

Jones, Bryan D. 2001. *Politics and the Architecture of Choice*. Chicago: University of Chicago Press.

Katznelson, Ira. 1997. "Structure and Configuration in Comparative Politics," in *Comparative Politics: Rationality, Culture and Structure*, M. Lichbach and A. Zuckerman (eds.). New York: Cambridge University Press.

King, Gary, Robert Keohane, and Sidney Verba. 1994. *Scientific Inference in Qualitative Research*. Princeton: Princeton University Press.

Kiser, Edgar. 1994. "Markets and Hierarchies in Early Modern Tax Systems: A Principal–Agent Analysis." *Politics & Society* 22(3): 285–316.

Kiser, Edgar and Yoram Barzel. 1991. "The Origins of Democracy in England." *Rationality & Society* 3: 396–422.

Kiser, Edgar, and Michael Hechter. 1991. "The Role of General Theory in Comparative-Historical Sociology." *American Journal of Sociology* 97: 1–30.

1998. "The Debate on Historical Sociology." *American Journal of Sociology* 104(3): 785–816.

Knight, Jack. 1992. *Institutions and Social Conflict*. New York: Cambridge University Press.
Laitin, David D. 1992. *Language Repertoires and State Construction in Africa*. New York: Cambridge University Press.
1998. *Identity in Formation: The Russian-speaking Populations in the Near Abroad*. Ithaca, NY: Cornell University Press.
2002. "Comparative Politics: The State of the Discipline." In *The State of the Discipline: Power, Choice, and the State*, I. Katznelson and H. Milner (eds). New York: W. W. Norton.
2003. "The Perestroikan Challenge to Social Science." *Politics & Society* 31(1): 163–84.
Levi, Margaret. 1988. *Of Rule and Revenue*. Berkeley: University of California Press.
1990. "A Logic of Institutional Change," in *The Limits of Rationality*, K. S. Cook and M. Levi (eds.). Chicago: University of Chicago Press.
1998. "The Price of Citizenship: Conscription in France, the United States, and Prussia in the Nineteenth Century," in *Analytic Narratives*, R. H. Bates, A. Greif, M. Levi, J.-L. Rosenthal, and B. Weingast (eds.). Princeton: Princeton University Press.
2002. "Modeling Complex Historical Processes with Analytic Narratives," in *Akteure, Mechanismen, Modelle: Zur Theoriefahigkeit makrosozialer Analyse*, R. Mayntz (eds.). Frankfurt/Main: Campus Verlag.
in process. "Inducing Preferences within Organizations: The Case of Unions." In *Preferences in Time*, I. Katznelson and B. Weingast (eds.). New York: Russell Sage Foundation.
Mahoney, James. 1999. "Nominal, Ordinal, and Narrative Appraisal in Macro-causal Analysis." *American Journal of Sociology* 104(4): 1154–96.
2000. "Rational Choice Theory and the Comparative Method: An Emerging Synthesis?" *Studies in Comparative International Development* 35(2): 83–94.
2001. "Beyond Correlational Analysis: Recent Innovations in Theory and Method." *Sociological Review* 16(3): 575–93.
2003. "Strategies of Causal Assessment in Comparative Historical Analysis," in *Comparative Historical Analysis in the Social Sciences*. New York: Cambridge University Press.
Mayntz, Renate. forthcoming. "Mechanisms in the Analysis of Macro-Social Phenomena.
McAdam, Doug, Sidney Tarrow, and Charles Tilly. 2001. *Dynamics of Contention*. New York: Cambridge University Press.
Moore, Barrington. 1966. *Social Origins of Dictatorship and Democracy*. Boston: Beacon Press.
North, Douglass C. 1981. *Structure and Change in Economic History*. New York: Norton.
1990. *Institutions, Institutional Change, and Economic Performance*. New York: Cambridge University Press.
1996 [1993]. "Economic Performance Through Time," in *Empirical Studies in Institutional Change*, L. J. Alston, T. Eggertsson, and D. C. North (eds.). New York: Cambridge University Press.

North, Douglass C. and Barry R. Weingast. 1989. "Constitutions and Commitment: The Evolution of Institutions Governing Public Choice in Seventeenth Century England." *Journal of Economic History* 49(4): 803–32.

Petersen, Roger. 1999. "Mechanisms and Structures in Comparisons," in *Critical Comparisons in Politics and Culture*, J. Bowen and R. Petersen (eds.). New York: Cambridge University Press.

2002. *Understanding Ethic Violence*. New York: Cambridge University Press.

Pierson, Paul. 2000. "Path Dependency, Increasing Returns, and the Study of Politics." *American Political Science Review* 94(2): 251–67.

2003. "Big, Slow-Moving, and . . . Invisible," in *Comparative Historical Analysis in the Social Sciences*, J. Mahoney and D. Rueschemeyer (eds.). New York: Cambridge University Press.

Piven, Frances Fox, and Richard Cloward. 1977. *Poor People's Movements*. New York: Vintage Books.

1993 [1971]. *Regulating the Poor*. New York: Pantheon.

Riker, William. 1964. *Federalism*. Boston: Little, Brown.

Root, Hilton L. 1994. *The Fountain of Privilege: Political Foundations of Markets in Old Regime France and England*. Berkeley: University of California Press.

Rosenthal, Jean-Laurent. 1992. *The Fruits of Revolution*. New York: Cambridge University Press.

1998. "The Political Economy of Absolutism Reconsidered," in *Analytic Narratives*, R. H. Bates, A. Greif, M. Levi, J.-L. Rosenthal, and B. Weingast (eds.). Princeton: Princeton University Press.

Rueschemeyer, Dietrich. 2003. "Can One or a Few Cases Yield Theoretical Gains?" In *Comparative Historical Analysis in the Social Sciences*, J. Mahoney and D. Rueschemeyer (eds.). New York: Cambridge University Press.

Scharpf, Fritz W. 1991. *Crisis and Choice in European Social Democracy*. Ithaca, NY: Cornell University Press.

1999. *Governing in Europe: Effective and Democratic?* New York: Oxford University Press.

Scott, James C. 1985. *Weapons of the Weak*. New Haven: Yale University Press.

1998. *Seeing Like a State*. New Haven: Yale University Press.

Sewell, William H., Jr. 1996. "Three Temporalities: Toward an Eventful Sociology," in *The Historic Turn in the Human Sciences*, T. J. McDonald (eds.). Ann Arbor: University of Michigan Press.

Shapiro, Ian. 2004. "Problems, Methods, and Theories in the Study of Politics, or: What's Wrong with Political Science and What to Do About It?" In *Problems and Methods in the Study of Politics*, I. Shapiro, R. Smith, and T. Masoud (eds.). New York: Cambridge University Press.

Skocpol, Theda. 2000. "Commentary: Theory Tackles History." *Social Science History* 24(4): 669–76.

2003. "Doubly Engaged Social Science: The Promise of Comparative Historical Analysis," in *Comparative Historical Analysis in the Social Sciences*, J. Mahoney and D. Rueschemeyer (eds.). New York: Cambridge University Press.

Skocpol, Theda and Margaret R. Somers. 1980. "The Uses of Comparative History in Macrosociological Inquiry." *Comparative Studies in Society and History* 22(2): 174–97.

Stinchcombe, Arthur L. 1991. "On the Conditions of Fruitfulness of Theorizing about Mechanisms in the Social Science." *Philosophy of the Social Sciences* 21: 367–88.

Tarrow, Sidney. 2003. "Confessions of a Recovering Structuralist." *Mobilization* 8(1): 134–146.

Thelen, Kathleen. 1999. "Historical Institutionalism in Comparative Politics." *Annual Review of Political Science* 2: 369–404.

Tilly, Charles. 1984. *Big Structures, Large Processes, and Huge Comparisons*. New York: Russell Sage Foundation.

2001. "Mechanisms in Political Processes." *Annual Review of Political Science* 4: 21–41.

forthcoming. "Social Boundary Mechanisms." *Philosophy of the Social Sciences*.

Van Evera, Stephen. 1997. *Guide to Methods for Students of Political Science*. Ithaca, NY: Cornell University Press.

Weingast, Barry R. 1995. "The Economic Role of Political Institutions: Market-Preserving Federalism and Economic Development." *International Political Science Review* 11 (Spring): 1–31.

1998. "Political Stability and Civil War: Institutions, Commitment, and American Democracy," in *Analytic Narratives*, R. H. Bates, A. Greif, M. Levi, J.-L. Rosenthal, and B. R. Weingast (eds.). Princeton: Princeton University Press.

in process. *Institutions and Political Commitment: A New Political Economy of the American Civil War Era*.

Weingast, Barry R., Gabriella Montinolo, and Yingyi Qian. 1995. "Federalism, Chinese Style: The Political Basis for Economic Success in China." *World Politics* 48: 50–81.

11 The methodical study of politics

Bruce Bueno de Mesquita

Ecclesiastes teaches that, "To everything there is a season and a time for every purpose under the heaven (Book of Ecclesiastes 3:1)." Although methodological seasons come and go, I believe most students of politics are united in their purpose. We want to understand how the world of politics works. Some may be motivated in this purpose by a desire to improve, or at least influence, the world, others to satisfy intellectual curiosity, and still others by mixes of these and other considerations. Within the generally agreed purpose for studying politics, however, there are important differences in emphasis that lead to variations in methodological choices. For instance, those concerned to establish the verisimilitude and precision of specific conjectures about particular political events are likely to choose the case study method and archival analysis as the best means to evaluate the link between specific events and particular explanations. Those concerned to establish the general applicability or predictive potential of specific conjectures are likely to choose experimental designs or large-N, statistical analysis as the best means to evaluate the link between independent and dependent variables. Those concerned to probe normative arguments or to engage in social commentary will find other methods, such as poststructuralism and some forms of constructivism, more fruitful.

Whatever the explanatory goal and whatever the substantive concern, research benefits from efforts to establish the logical foundation of propositions as it is exceedingly difficult to interpret and build upon commentary or analysis that is internally inconsistent. The concern for logical rigor to accompany empirical rigor encourages some to apply mathematics to the exploration of politics. Mathematical models – whether they are structural, rational choice, cognitive, or take other forms – provide an especially useful set of tools for ensuring logical consistency or for uncovering inconsistencies in complex arguments. As most of human interaction is complex – and certainly political choices involve complex circumstances – mathematical models provide a rigorous foundation for making clear precisely what is being claimed and for discovering unanticipated contingent predictions or relations among variables. By linking careful analysis or social commentary to logically grounded empirical claims we improve the prospects of understanding and influencing future politics.

The mathematical development of arguments is neither necessary nor sufficient for getting politics right, but it does greatly improve the prospects of uncovering flaws in reasoning. As such, mathematics does not, of course, provide a theory of politics any more than natural language discourse is itself a theory of politics. Rather, each – mathematics and natural language – is simply a tool for clarifying the complex interactions among variables that make up most prospective explanations of politics. The advantage of mathematics is that it is less prone to inconsistent usage of terms than is ordinary language. Therefore, it reduces the risk that an argument mistakenly is judged as right when it turns out to rely on changes in the meaning of terms.

Some draw a distinction between those who choose problems based on methodological tastes and those who choose based on substantive concerns. This seems to me to be a false distinction. Rather, it seems likely to me that substantive interests propel researchers to rely more on some methods than on others. Of course, all researchers are limited in their skills in using some tools of research as compared to others. This too surely plays a role in the selection of methods that are applied to substantive problems. It is, for instance, particularly difficult for those lacking the requisite technical training in a method, whether it be mathematical modeling, archival research, experimental design, or statistical analysis, to adapt to the use of unfamiliar, technically demanding tools. We need not concern ourselves here with this limitation since its remedies are straightforward. Researchers can acquire the methodological skills needed for a research problem or they can select collaborators with the appropriate technical know-how. Naturally, the costlier retooling is in time and effort, the less likely it is that a researcher will pay the price and the more likely it is that there will be a division of labor among specialized co-authors. Co-authorship is an increasingly common practice in political science presumably for this reason. With these few caveats in mind, I put the distinction between methodological tastes and substantive interests aside and focus instead on what I believe are the linkages between substantive issues and methodological choices.

When researchers fuse evidence and logic, they improve our ability to judge whether their claims are plausible on logical grounds and are refuted or supported empirically. This combination of logic and evidence has been the cornerstone of studies of politics over more than two millennia. Such an approach typifies the writings of Aristotle, Thucydides, Kautilya, Sun Tzu, Aquinas, Machiavelli, Hobbes, Hume, Locke,

Madison, Montesquieu, Marx, Weber, Dahl, Riker, Arrow, Black and so on. Reliance on the tools of rhetoric or on advocates' briefs without logic and balanced evidence probably makes evaluating research more difficult and probably subject to higher interpersonal variation than when relying on logic and evidence. With this in mind I begin by considering what forms of argument and evidence are most likely to advance knowledge and persuade disinterested observers of the merits or limitations of alternative contentions about how politics works. As I proceed, I will occasionally offer illustrative examples. When I offer such examples they will be drawn primarily from research on international relations since this is the aspect of political research best known to me.

The case study method of analysis

One path to insight into politics is through the detailed analysis of individual events such as characterizes many applications of the case study method. This technique, often relying on archival research, proves to be a fertile foundation from which new and interesting ideas germinate, ideas that suggest hypotheses about regularities in the world that are worthy of being probed through close analysis of individual events and through careful repetition across many events. *The Limits of Coercive Diplomacy* (George, Hall, and Simons 1971), for instance, uses five case studies to infer an interesting hypothesis that ties conflict outcomes to asymmetries in motivation across national decisionmakers. Subsequent research has replicated the basic insights from that work in applications to other case histories of conflict, though I am not aware of applications yet to decisions to avoid conflict based on an anticipated motivational disadvantage.

Case studies also are used to evaluate the empirical plausibility of specific theoretical claims derived from sources other than the case histories themselves. Kiron Skinner, Serhiy Kudelia, Condoleezza Rice, and I (2004), for instance, investigate William Riker's (1996) theory of rhetoric and heresthetic political maneuvering by using archival records to probe decisionmaking strategies during campaigns led by Boris Yeltsin and Ronald Reagan. The case studies in this instance are examined to see whether details of predictions about "off the equilibrium path" expectations influenced Reagan's and Yeltsin's strategies in the manner anticipated by Riker's theory.

The nuances of case histories uncover both the strengths and weaknesses of this method. Case studies promote close probing of the details behind an event. This probing enhances the prospects of achieving internal validity. By bringing theory as close as possible to the many small events and variations in circumstance that characterize real-world politics, the researcher enhances the prospects of providing a convincing account of the event or events in question. However, to the extent the researcher's primary interest is in explaining and predicting other events in the class to which the case history belongs, a focus on the unique nuances of the individual event is an impediment to reliable generalization. The unique features of the case frequently are permitted to become a part of the explanation rather than elements in control variables that do not make up part of a general hypothesis applicable to other events in the "class."

Case studies can provide existence proofs for some theoretical claims. For instance, a case history and archival research on the Seven Weeks War provide an existence proof that a seemingly small war can lead to fundamental changes in the organization of international politics, contrary to claims in theories of hegemonic war or of power transition wars (Bueno de Mesquita 1990a). Assuming no measurement error for argument's sake, a case study can also falsify a strong theoretical claim – rare in political science – that some factor or set of factors is necessary, sufficient, or necessary and sufficient for a particular outcome to occur. A single case study, however, cannot provide evidence that the specific details are germane to other, similar occurrences. For those who envision progress through the accumulation of knowledge in the manner suggested by theorists like Laudan (1977) or Lakatos (1976, 1978), a single case study by itself cannot help choose among competing theoretical claims. In these approaches a theory's superiority is judged by its ability to remove more anomalies or explain more facts than a rival theory. That is, these approaches require a theory to account for the findings regarding previously known cases, plus patterns that emerge from other cases, thereby requiring more than a single case or even very few cases to judge that one explanation is superior to another. Nor can individual case studies evaluate the merits of arguments that involve a probability distribution over outcomes. Even when arguments are deterministic, replication of case study quasi-experiments is important for building confidence in a putative result. Such replication naturally leads to a second method of investigation.

Large-N, usually statistical analysis

Replication of quasi-experiments to test the accuracy of a general claim helps us discern whether specific hypothesized general patterns hold up within cases that fall within a class of situations. Alexander L. George, David K. Hall, and William E. Simons (1971, 1994), for instance, examine variations in motivations across disputants in case after case. Each time they are probing the same hypothesis: a motivational advantage can help overcome a power disadvantage in disputes. The second edition takes the theory of the first edition and adds new case histories to include foreign policy developments through the 1980s, thereby testing and retesting their main hypothesis.

As the number of cases used for testing a theoretical claim increases, the general patterns – rather than particular details – are efficiently discerned through statistical analysis. This is especially true when the predicted pattern involves a probability distribution across possible outcomes. Few expectations in political science, at least at the current stage of theory development, are deterministic. Statistical studies uncover ideas about the general orderliness of the world of politics through the assessment of regular patterns tying dependent variables to independent variables. While the case study approach provides confidence in the internal workings of specific events, statistical analysis probes the generality or external validity of the hypotheses under investigation. Statistical patterns offer evidence about how variables relate to one another across a class of circumstances, providing clues about general causal mechanisms, but generally fail to illuminate a specific and convincing explanation of any single case. Of course, statistical analysis is not intended as a method for elaborating on details within a case just as an individual case study is not designed to expose general patterns of action.

Statistical analysis in political science has opened the doors to successful electoral predictions, as well as to the development of insights into virtually every aspect of politics. Within the field of international relations, statistical methods have been used to investigate such important questions as the causes and consequences of war, changes in trade policy, alterations in currency stability, alliance reliability, and so forth. The contributions of statistical studies in American politics and comparative politics are probably even greater.

Consider, for instance, the fruitful statistical research into the links between regime type and war proneness. Research about whether there is a democratic peace and, if so, why it persists (Russett 1993; Maoz and Abdolali 1989; Bremer 1993) illustrates an instance in which a seemingly compelling pattern of variable interaction has led to what may be a law of nature, a law that could not be discerned by observing only a few individual examples. The exact statement of that law remains open to debate. Is the apparent relative peace among democracies a consequence of shared values and norms, as some argue, or the product of specific constraints on action that make it particularly difficult for leaders who depend on broad support to wage war, as I and others have maintained, or is there some other, as yet uncovered explanation that best accounts for the numerous regularities that collectively are drawn together under the label, the democratic peace? So strong are the statistical patterns that policymakers as well as scholars have embraced the idea that a democratic peace exists. Indeed, Bill Clinton in 1994 specifically cited his belief in the democratic peace findings as a basis for American action in restoring Bertrand Aristide to power in Haiti.

Policymakers appear to have such a strong belief in the democratic peace that it is sometimes argued that an all-democratic world, if it is ever attained, will be a peaceful world. These sanguine beliefs require a note of caution. Structural models and accompanying statistical tests also suggest that the impact on peace of spreading democracy is nonlinear (Crescenzi, Kadera, and Shannon 2003) while another statistically observed and logically deduced pattern shows that democracies are especially prone to overthrow the regimes of defeated rivals through war and, in doing so, are prone to construct new autocracies (Bueno de Mesquita, Smith, Siverson, and Morrow 2003). Leaders pursue war for different reasons, but one reason apparently is the domestic inducement to change a foe's policies by altering its regime. Such a motivation is common among leaders in democracies. This is readily shown statistically and is illustrated by regime changes imposed by the democratic victors in World War II. A glance at the Thirty Years War or postcolonial African history makes apparent that such a motivation was remote in a world led by monarchs or by petty dictators. The statistical evidence for the generalization that democracies fight more often than autocracies to overthrow regimes and impose new leaders is strong. Autocrats fight for spoils, to gain territory and riches, and only rarely to overthrow their foreign foes.

Experimental design

A necessary condition for an explanation to be true is that it passes appropriate empirical tests of both its internal and external validity. Experiments provide the best controlled setting in which to conduct tests of political theory, but in practice laboratory experiments with human subjects throw up impediments not usually confronted by physical scientists. Randomization in case selection is one element in designing reasonable quasi-experiments based on data from the past. Another approach, about which I will have more to say later, is to do predictive studies about events whose outcome is not yet known. These studies are natural "out of sample" experiments. Nature randomizes the value of variables not explicitly within the theoretical framework while the researcher chooses cases without bias with respect to whether they turn out to fit or refute the hypotheses being tested. The absence of such a bias is assured by the fact that the outcomes of the cases are unknown at the time the cases enter the experimental frame. Such predictive case studies are becoming more common among international relations researchers (a small sampling includes Bueno de Mesquita 1984, 2002; Bueno de Mesquita and Stokman 1994; James 1998; James and Lusztig 1996, 1997a, 1997b, 2000, 2003; Williams and Webber 1998; Friedman 1997; Kugler and Feng 1997; and Achen forthcoming). These studies are instances of repeated prospective natural experiments in international relations in which the explanatory variables and their interrelationships are not altered from test case to test case. Of course, in American politics such natural experiments are undertaken all of the time in the context of elections. True laboratory experiments are also growing in prominence in political science. Within the international relations context, the best-known studies relate to experiments about repeated prisoner's dilemma situations (e.g., Rapoport and Chammah 1965; Rapoport, Guyer, and Gordon 1976; Axelrod 1984), though there is also a literature on weapons proliferation and other conflict-related experiments (Brody 1963). And, of course, the Defense Department routinely runs war game experiments to help train officers for combat. These experiments, however, rarely are designed to test competing hypotheses about the politics of war.

Mathematic modeling

Whatever the source of data for testing hypotheses, empirical methods alone are insufficient to establish that a given conjecture, hypothesis, hunch, or observation provides a reliable explanation. After all, correlation does not prove causation, though a strong correlation certainly encourages a quest for causality and the absence of correlation often provides evidence against causality. In both instances, in the pursuit of internal and external validity, we grow confident in our observations and the prospective causal ties between independent and dependent variables when our tests explore the deductively derived *logic* of action. This non-empirical methodological approach – the logic of action – furnishes a proposed explanation for the regularities we discern and for previously unobserved regularities implied by the logic of action. The logic of action establishes the internal consistency of the claims we make or the observations we report. The test of logical consistency establishes a coherent link between those observations and evidence that *describes* the world and the theory purported to explain why our observations are expected to look as they do.

Consider some of the empirical regularities proposed through formal logical analysis, some of which have been tested and others of which remain for future statistical and case study evaluation. Repeated game theory offers fundamental insight into the conditions under which to expect – or not expect – cooperation. The prisoner's dilemma is a model situation in which the absence of trust or credible commitment can prevent cooperation. Through repeated play, however, we know that cooperation can be achieved between patient decisionmakers (Taylor 1976; Axelrod 1984). This insight is important in uncovering prospective solutions to enduring conflicts and it is supported in laboratory experiments, case histories, and statistical evaluations.

Equally significant, however, we also know through the logic of certain repeated games that patience does not always encourage cooperation. Indeed, the chance for cooperation can be diminished by patience; that is, through a long shadow of the future. When costs endured now in preparation for war provide improved prospects for gains later through military conquest, then patient decisionmakers more keenly appreciate the future benefits of war than do their impatient counterparts. The result is that it is more difficult to deter patient leaders when current costs presage greater subsequent gains. Aggression rather than cooperation is the prospective consequence, as has been indicated in studies of spending on guns or butter (Powell 1999). So, when costs precede benefits, patience is not a virtue that stimulates cooperation; it can be a liability that promotes conflict.

The sequence in which costs and benefits are realized apparently alters the prospects of peace. This is an important conceptual insight for those who shape agendas in international negotiations, an insight that was not revealed either through case analysis or statistical assessment. It was uncovered through the rigorous, formal logic of game theory. This insight may help to provide a key to better addressing the fractious issues that threaten relations between India and Pakistan, North and South Korea, Israel and the Palestinians, China and Taiwan, Northern Irish Catholics and Protestants and so on. Finally, I should note that rational actor models have not only influenced academic understanding, but also practical policymaking. America's nuclear policy through much of the Cold War traces its roots to the strategic reasoning of Thomas Schelling (1960). More recently, a formal game theoretic model of bargaining has been described by the United States government as producing: "forecasts . . . given to the President, Congress, and the US government." These forecasts are described as "a substantial factor influencing the elaboration of the country's foreign policy course" (Izvestiya 1995). Contrary to what we sometimes hear, policymakers in the United States and elsewhere rely in part on inferences drawn from mathematical models of politics to choose their course of action. I could go on with innumerable examples of rational choice studies that yield counterintuitive, but empirically valid, deductions, such as the benefits for foreign policy of political partisanship (Schultz 2001) and "yes men" (Calvert 1985) or explanations of alliance behavior (Smith 1995), but I believe the point is clear. We have learned much about politics through mathematical reasoning, particularly in a rational choice setting, as well as from statistical assessments, and that knowledge spans a wide array of substantively important questions.

Research through multiple methods

Convincing, cumulative progress in knowledge is bolstered by and may in fact require the application of *ex post* case histories, *ex post* statistical studies, and *ex ante* natural and laboratory experiments, though the order in which such research is done may be irrelevant to success in uncovering covering laws of politics if they exist. Let me illustrate the benefits of combining methods with a brief foray into a crucial political debate of the early fourteenth-century, one that remains pertinent today to those in political science who dismiss this or that method out of hand or out of personal predilections rather than from solid theoretical or empirical foundations.

The use of logic, case evidence, and repeated observations within a class of cases to explain behavior and to influence policy was at the heart of the fourteenth-century essay *De Potestate Regia et Papali*, that is, *On Royal and Papal Power*, written by Jean Quidort (known today as John of Paris) around 1302 (1971). John challenged Pope Boniface VIII on behalf of Philip the Fair, King of France, in what was the main international conflict of the day, a conflict that resulted in war, in the seizure, imprisonment, and death of the pope, and in the gradual decline of the Catholic Church as the hegemonic power in Europe. In defending the claims of French sovereignty, John of Paris resorted to three methods of analysis: logical exegesis, detailed reference to evidence in history and scriptures, and an appeal to the general pattern of relations between the Church and monarchs over many centuries in his effort to show that the pope had no authority to depose kings, contrary to Boniface VIII's claims. Indeed, John, foreshadowing modern debate over methods of analysis, warned of the dangers of generalizing *solely* from a single case. The pope had defended his right to overthrow kings by pointing to the deposition of Childeric III by Pope Zachary centuries earlier. Jean Quidort contested the claim of precedent, noting that, "It is inappropriate to draw conclusions from such events which . . . are of their nature unique, and not indicative of any principle of law" (167). He went on to state principles of law and to document them empirically. Furthermore, he went on to argue about when evidence is to be taken as unbiased and when it is biased, foreshadowing contemporary debate about both selection bias and selection effects. He noted:

> When it is the power of the pope in temporal affairs that is at issue, evidence for the pope from an imperial source carries weight, but evidence for the pope from a papal source does not carry much weight, unless what the pope says is supported by the authority of Scripture. What he says on his own behalf should not be accepted especially when, as in the present case, the emperor manifestly says the contrary and so too do other popes. (John of Paris: 170)

John clearly understood that a case history or example by itself is insufficient for stating general principles and that such evidence, selected to make the case, is likely to be biased by design. In fact, the case of Childeric and his actual dethroning by Pipin III for domestic reasons and not by Pope Zachary, illustrates the dangers of too readily generalizing from a singular event.

John of Paris relied exclusively on ordinary language to make his case for the illogic of Boniface's claims and he did so rather effectively. Still, he faced the problem that often haunts ordinary language arguments today. His arguments against the pope were interpreted in different (generally self-serving) ways by different listeners or readers. Indeed, phenomenology, postmodernism, discourse analysis, and a variety of other contemporary approaches to research are grounded in this very characteristic of ordinary language. With ordinary language, there is a great risk that arguments that *seem* to make sense, when subjected to the closer scrutiny of formal, mathematical logic, fail the test of reason. Yet, today, there are still many who denounce the application of mathematical reasoning and quantitative assessments as conceits that cannot illuminate understanding. Often they argue that the interesting problems of politics are too complex to be reduced to a few equations even though it is *exactly* when dealing with complex problems that mathematics becomes an attractive substitute for ordinary language as it is in *complex problems* that errors in informal logic are most easily made and hardest to discover. It is in just such circumstances that we too often utter phrases such as "It stands to reason" or "It is a fact that . . ." without pausing to demonstrate that these assertions are true.

The study of politics affords a particularly fertile ground for the application of multiple methods and complementary theoretical approaches. For instance, constructivism and game theory are generally cast as competing approaches to politics and yet there is a substantial, but largely overlooked, complementarity between them. Consider two central questions about politics: (1) what do people want and (2) what do they do to fulfill their wants? Game theory provides no means to answer the first question except in special circumstances. Constructivism generally does not provide an answer to the second question.

Game theory offers a theory of action. The primary solution concept used in contemporary game theory is the notion of Nash equilibrium or any of a host of refinements of that idea (e.g., subgame perfection, Markov perfect equilibrium, Bayesian perfect equilibrium). The theory of strategic interaction posited by game theory seems especially pertinent to politics. It stipulates that people do what they believe is in their best interest given the constraints they face. Strategies (i.e., complete plans of action for every contingency that could arise in the situation) constitute an equilibrium when no one in the strategic situation has a unilateral incentive to switch to a different course of action. As such, game theory emphasizes counterfactual reasoning; that is, decisionmakers choose their actions with an eye to what they expect will happen to them if they choose some alternative course of action. They rarely get to do what they most would like to do, but they always choose to do what they think will give them the best return under the constrained circumstances in which their actions take place. Thus, game theory relies on two critical considerations: beliefs and desires. Beliefs and desires (or at least some types of desires) can change because of endogenous pressures within the circumstances of the game, but initial beliefs and preferences over outcomes are taken as given or determined non-strategically.

Constructivism, research on socialization, and many features of psychological research address the very matters that game theory takes as given by assumption. Constructivism, in particular, is concerned with the origins of beliefs and desires. From a constructivist perspective, beliefs and desires are socially constituted. Authority, reflection, established norms, culture, and other social elements give shape to people's self-identity, including their preferences (desires) and beliefs about what they or others can or cannot do. As such, constructivism provides a plausible theoretical framework for empirical research into the endogenous generation of preference profiles and structures of beliefs. Yet game theorists have yet to capitalize on constructivism's theory of identity formation just as constructivists have not yet tested their conjectures against historical or experimental evidence regarding differences in choices of action across alternative constructed realities.

Allow me to suggest an experiment that would explore the complementarities between game theory and constructivism. One well-known empirical challenge for game theorists is to predict decisions in games with multiple equilibria. Even many simple games like chicken or battle of the sexes produce three equilibria (two in pure strategies and one mixed strategy). In infinitely repeated games almost any (but not all) combination of actions can be supported as a Nash equilibrium. At present, the best that can be done empirically in such circumstances is to examine the distribution of observed strategies (or at least actions) given different constraints to see if the distribution matches expectations, or one can do tests to see whether non-equilibrium strategies occur. If constructivist claims are correct and if game theory's theory of action is correct, then it should be possible to do much better in identifying expected equilibrium strategies in different circumstances. For example, if one culture or other factor determining *ex ante* preferences and beliefs favors a given equilibrium strategy (or the outcome it leads to) while another culture prefers a different equilibrium approach in a game with multiple equilibria, then identity formation provides the means to refine the predicted course of action in different settings. By combining a theory of constituted identities with game theory, then, it may become possible to improve the answers to the two questions posed earlier: (1) what do people want and (2) what do they do to fulfill their wants?

With consideration of multiple methods in mind (and mindful of John of Paris's warnings) I focus the remainder of my discussion on the application of a small set of mathematical models to numerous predictive case studies and to statistical assessments. The applied models discussed here are intended to illuminate a body of research that explicitly combines mathematical reasoning, case analysis, and statistical assessment and that focuses both on processes and outcomes in politics, doing so in a real-time, predictive setting.

The case for more predictive testing of theories of politics

In recent years many in political science have questioned the value of rational choice models (Cohn 1999; Lowi 1992; Green and Shapiro 1994). Elsewhere I have expressed my views on what constitutes some of the important limitations of rational choice approaches (Bueno de Mesquita 2003). Here I focus on some practical contributions of two sets of applied rational choice models.

Research on decisionmaking has progressed over the past few decades from individual case analyses that lack the guidance of an explicit theoretical perspective to theoretically informed, genuinely predictive assessments of decisions and the process leading to them in complex settings. Much of this work draws upon elements of social choice theory, game theory, or combinations of the two. Some of this research represents efforts in pure formal theorizing, while others attempt to examine the retrospective, statistical interplay between formal models of decisionmaking and the empirical environment. Still others apply insights from formal models in developing practical tools for use in predicting the outcomes and bargaining dynamics of future decisions. These latter efforts provide evidence regarding the prospects of such models as tools for explaining and predicting political behavior.

In this section I discuss what I term applied formal models. That is, I examine the empirical application of rational actor models as opposed to reviewing the more formal literature that is concerned with deriving theorems that elucidate analytic solutions to abstract political questions. By applied models I mean rational choice models designed to provide computational solutions to complex, multiactor real-world problems. These models are embedded in, build on, and sometimes contribute to the development of broad empirical generalizations, but their primary purpose is to bridge the gulf between basic theorizing, logically sound explanation, and social engineering.

Although there is a growing body of applied models, especially in economics, I draw attention here to models developed and applied by Frans Stokman and his colleagues – loosely referred to as the Groningen Group – and a model I developed, as well as its applications by myself and others. I do so because these two groups have been sufficiently active in their effort to convert theoretical insights from rational choice models into tools for real-world experiments and problem solving that they have extensive, often externally audited, track records from which we can assess the potential benefits of such models as tools that help explain political choices, both in terms of process and outcomes, and as a help to guide decisionmaking. Consequently, I ignore here other valuable developments related to prediction, including the use of neural network models by Beck, King, and Zeng (2000) or the innovative research on fair division by Steven Brams and Alan Taylor (1996). These promising areas of inquiry are just beginning to build track records that demonstrate their value as tools for practical problem solving about politics.

Ultimately, the best way we have to evaluate the explanatory power of a model is to assess how well its detailed analysis fits with reality. This is true whether the reality is about a repeated event, like the price of a commodity, or about important, one-time political decisions. This can be done with retrospective data or prospective data. In the latter case, when actual outcomes are unknown at the time of investigation, there is a pure opportunity to evaluate the model, independent of any possibility that the data have been made to fit observed outcomes. Indeed, perhaps the most difficult test for any theory is to apply it to circumstances in which the outcome events have not yet occurred and where they are not obvious.

The prediction of uncertain events and the processes leading to them is demanding precisely because the researcher cannot fit the argument to the known results. This is a fundamental difference between real-time prediction and post-diction.

The models discussed here have been tested thousands of times against problems with unknown and uncertain outcomes. Frans Stokman, his colleagues, and his students have done extensive testing of the reliability of their log-rolling models. A special issue of *Rationality and Society* (2003) provides an excellent overview of their modeling and of their track record. Their log-rolling models – called the exchange model and the compromise model – have undergone extensive testing in the context of decisionmaking in the European Union. They generally achieve accuracy levels on their predictions that range between about 80 to 90 percent. What is more, a Dutch company, Decide BV, has used these models in corporate and government projects in Europe for more than a decade. They are credited with playing a significant part in several significant labor-management negotiations in the Netherlands. In one particular instance, Stokman and others associated with Decide bv provided Dutch journalists with advance notice of predicted outcomes of a major Dutch railway negotiation. Their predictions were substantially different from the common journalistic view at the time and from the Dutch government's expectations and yet the log-rolling, rational choice models proved correct in virtually all of the details of the resolution of the negotiations. The details of this and other genuinely predictive experiences using the Groningen Group's models can be found at http://www.decide.nl/.

Many of the Groningen Group's applications have been done in the context of the consulting company they established in cooperation with the University of Groningen. They have, in addition, published all of their theoretical research and many applications in the academic literature. Likewise, many of the applications of the forecasting and bargaining model I developed have been conducted in a consulting setting. Few applications from the company I helped establish have been widely available for public scrutiny. As with studies by the Groningen Group, others and I have published the theoretical research and many applications in academic journals and books (see Bueno de Mesquita 2002, 2003b for a list of relevant publications). Even some of the commercial applications are publicly available and so can be used to evaluate the accuracy of these models and their helpfulness in shaping policy decisions.

The Central Intelligence Agency (CIA) declassified its own evaluation of the accuracy of the forecasting model I developed, a model it refers to as Policon and which it has used in various forms since the early 1980s. According to Stanley Feder (1995), then a national intelligence officer at the CIA and now a principal in another company (Policy Futures) that uses a version of one of the models I developed:

> Since October 1982 teams of analysts and methodologists in the Intelligence Directorate and the National Intelligence Council's Analytic Group have used Policon to analyze scores of policy and instability issues in over 30 countries. This testing program has shown that the use of Policon helped avoid analytic traps and improved the quality of analyses by making it possible to forecast specific policy outcomes and the political dynamics leading to them. . . . Political dynamics were included in less than 35 percent of the traditional finished intelligence pieces. Interestingly, forecasts done with traditional methods and with Policon were found to be accurate about 90 percent of the time. Thus, while traditional and Policon-based analyses both scored well in terms of forecast accuracy, Policon offered greater detail and less vagueness. Both traditional approaches and Policon often hit the target, but Policon analyses got the bull's eye twice as often. (Feder 1995: 292)

Feder goes on to report that, "Forecasts and analyses using Policon have proved to be significantly more precise and detailed than traditional analyses. Additionally, a number of predictions based on Policon have contradicted those made by the intelligence community, nearly always represented by the analysts who provided the input data. In every case, the Policon forecasts proved to be correct" (Feder 1995).

The modeling method can address diverse questions. Analysts have examined economic, social, and political issues. They have dealt with routine policy decisions and with questions threatening the survival of particular regimes. Issues have spanned a variety of cultural settings, economic systems, and political systems. Feder cites numerous examples of government applications in which the model and the experts who provided the data disagreed. His discussion makes clear that the model is not a Delphi technique – that is, one in which experts are asked what they believe will happen and then these beliefs are reported back in the form of predictions. Rather, this model and the Groningen models provide added insights and real value above and beyond the knowledge of the experts who provide the data.

Government assessments are not the only basis on which to evaluate the predictions made by these models. Christopher Achen and others continue testing begun by Stokman and me (1994) in the context of the European Union, a repeated "game" environment better suited to Stokman's and Achen's models than to mine. They find strong predictive accuracy from these rational actor models (Torenvlied 1996; Van der Knoop, and Stokman 1998; Stokman, van Assen, van der Knoop, and van Oosten 2000; Achterkamp 1999, 2000; Achen n.d). These empirical results encourage growing confidence that the study of international negotiations has reached a stage where several practical and reliable rational actor tools are becoming available to help in real-world policy formation.

Additional evidence for the reliability of the so-called Policon model can be found in the published articles that contain predictions based on it. James Ray and Bruce Russett (1996) have evaluated many of these publications to ascertain their accuracy. They maintain that:

> The origins of the political forecasting model based on rational choice theory can be traced to *The War Trap* by Bruce Bueno de Mesquita. The theory introduced there was refined in 1985, and served in turn as the basis for a model designed to produce forecasts of policy decisions and political outcomes in a wide variety of political settings . . . This "expected utility" forecasting model has now been tried and tested extensively. [T]he amount of publicly available information and evidence regarding this model and the accuracy of its forecasts is sufficiently substantial, it seems to us, to make it deserving of serious consideration as a "scientific" enterprise . . . [W]e would argue in a Lakatosian fashion that in terms of the range of issues and political settings to which it has been applied, and the body of available evidence regarding its utility and validity, it may be superior to any alternative approaches designed to offer specific predictions and projections regarding political events. (1996: 1569)

They go on to provide numerous examples of predictions made by that model, especially predictions that were contrary to conventional expectations at the time of their publication. Other predictions over the years can be found in articles dealing with a peace agreement in the Middle East, political instability in Italy over the budget deficit, the dispute over the Spratly Islands, Taiwanese nuclear weapons capabilities, bank debt management in several Latin American countries, the content of NAFTA, the Maastricht referendum in Europe, admission of the two Koreas to the United Nations, policy changes in Hong Kong leading up to and following the return to China in 1997, the leadership transition in China, resolution of the Kosovo War, implementation of the Good Friday Agreement in Northern Ireland, among many other topics.[1]

Each of these instances includes accurate predictions that were surprising to many at the time. For example, in 1990 the model predicted that Yasser Arafat would shift his stance to reach an agreement with the next Labour government in Israel by accepting very modest territorial concessions that would not include autonomy (Bueno de Mesquita 1990b). The prediction was made and published well before the Gulf War and yet foreshadowed the steps Arafat took. Francine Friedman has

[1] The bibliography of such studies is too copious to list everything here. I would be happy to provide citations to all or any such studies on request.

demonstrated that the model indicated in December 1991 that the Croat–Serbian conflict "would in the end settle for Croatian independence with some Croation (territorial) concessions to Serbia but also that the Bosnian Muslims were the next likely target for extremist Serbian forces" (1997: 55). At the time, this seemed an unlikely event.

The forecasting model has also been pitted against prospect theory in the context of predictions regarding the implementation of the Good Friday Agreement in Northern Ireland (Bueno de Mesquita, McDermott, and Cope 2001). The predictions were made well in advance and were designed to predict and explain the detailed political dynamics through 2001. The prospect theory predictions, undertaken by a former student of Amos Tversky, achieved about 50–60 percent accuracy while the forecasting model and its bargaining component proved accurate in this instance in 100 percent of the cases (eleven distinct issues), including that the resumption of violence would not lead to a breakdown in further negotiations toward implementation of the agreement (while the prospect theory prediction indicated renewed violence would terminate further efforts to implement the Good Friday Agreement), and that the IRA would do only token decommissioning of weapons and the Protestant paramilitaries would do none.

To be sure, some predictions have been wrong or inadequate. The forecasting and bargaining model successfully predicted the break of several East European states from the Soviet Union, but failed to anticipate the fall of the Berlin Wall. The model predicted that the August 1991 Soviet coup would fail quickly and that the Soviet Union would unravel during the coming year, but it did not predict the earlier, dramatic policy shifts introduced by Mikhail Gorbachev. To be fair, the expected utility model was not applied to that situation so that such predictions could not have been made. This, of course, is an important difference between prediction and prophecy. The first step to a correct – or incorrect – prediction is to ask for a prediction about the relevant issue.

Both my and the Groningen Group's predictive models and bargaining components have important limitations, of course. They are inappropriate for predicting market-driven events not governed by political considerations. They can be imprecise with respect to the exact timing behind decisions and outcomes. The models can have timing elements incorporated into them by including timing factors as contingencies in the way issues are constructed, but the models themselves, though dynamic, are imprecise with regard to time. The dynamics of the models indicate whether a decision is likely to be reached after very little give and take or after protracted negotiations. Most importantly, the models by themselves are of limited value without the inputs from area or issue experts. While the views of experts are certainly valuable without these models, in combination the two are substantially more reliable than are the experts alone (Feder 1995). The limitations remind us that scientific and predictive approaches to politics are in their infancy and underscore the benefits of collaboration among researchers with different areas of specialization. Some encouragement can be taken from the fact that in many domains it has already proved possible to make detailed, accurate predictions accompanied by detailed explanations.

Conclusion

Existing applied models seem to produce accurate and reliable predictions (Feder 1995; Ray and Russett 1996; Thomson 2000). The predictions include detailed explanations of why choices are likely to turn out in particular ways and how each individual decisionmaker involved in a decision is likely to behave. What is more, some of these models have proven themselves not only in the social science laboratory, but also in the perhaps more demanding domain of the market place where survival depends on producing information valued by clients. The track record amassed by these models provides encouragement for the belief that even after relaxing assumptions to turn concepts into practical tools, rational choice models of decisionmaking are effective in predicting and helping to steer complex negotiations. Of course, confidence in such tools must also be tempered by the need for still more extensive and thorough tests of reliability. These and other models point to the benefits that may be developed by linking mathematical logic, game theoretic reasoning, area expertise, case studies, and statistical analysis in the pursuit of internally and externally valid accounts of important political phenomena.

REFERENCES

Achen, Christopher. H. Forthcoming. "Forecasting European Union Decisionmaking," in Frans Stokman and Robert Thomson (eds.).

Achterkamp, Marjolein. 1999. "Influence Strategies in Collective Decision Making: A Comparison of Two Models," Ph.D. Dissertation, University of Groningen.

Akkerman, Agnes. 2000. "Trade Union Competition and Strikes," Ph.D. Dissertation, University of Groningen.

Axelrod, Robert. 1984. *The Evolution of Cooperation*. New York: Basic Books.

Beck, Nathaniel, Gary King, and Langche Zeng. 2000. "Improving Quantitative Studies of International Conflict: A Conjecture." *American Political Science Review* 94: 21–35.

Brams, Steven J. and Alan Taylor. 1996. *Fair Division: From Cake-cutting to Dispute Resolution*. New York and Cambridge: Cambridge University Press.

Bremer, Stuart. 1993. "Democracy and Militarized Interstate Conflict, 1816–1965." *International Interactions* 18: 231–49.
Brody, Richard A. 1963. "Some Systemic Effects of the Spread of Nuclear Weapons Technology," *Journal of Conflict Resolution* 7: 663–753.
Bueno de Mesquita, Bruce. 1990a. "Pride of Place: The Origins of German Hegemony." *World Politics* (October): 28–52.
1990b. "Multilateral Negotiations: A Spatial Analysis of the Arab–Israeli Dispute." *International Organization* 44: 317–40.
2002. *Predicting Politics*. Columbus: Ohio State University Press.
2003a. "Ruminations on Challenges to Prediction with Rational Choice Models." *Rationality and Society* 15, 1: 136–47.
2003b. *Principles of International Politics*, 2nd ed. Washington, DC: CQ Press.
Bueno de Mesquita, Bruce and Frans Stokman. 1994. *European Community Decision Making*. New Haven: Yale University Press.
Bueno de Mesquita, Bruce, Rose McDermott, and Emily Cope. 2001. "The Expected Prospects for Peace in Northern Ireland." *International Interactions* 27, 2: 129–67.
Bueno de Mesquita, Bruce, Alastair Smith, Randolph M. Siverson and James D. Morrow. 2003. *The Logic of Political Survival*. Cambridge, MA: MIT Press.
Calvert, Randall L. 1985. "The Value of Biased Information: A Rational Choice Model of Political Advice." *Journal of Politics* 47: 530–55.
Cohn, Jonathan. 1999. "Revenge of the Nerds: When Did Political Science Forget About Politics?" *The New Republic* (October 25): 25–32.
Crescenzi, Mark, Kelly Kadera, and Megan Shannon. 2003. "Democratic Survival, Peace and War in the International System." *American Journal of Political Science*, forthcoming.
Feder, Stanley. 1995. "Factions and Policon: New Ways to Analyze Politics," in *Inside CIA's Private World: Declassified Articles from the Agency's Internal Journal, 1955–1992*. H. Bradford Westerfield (ed.). New Haven: Yale University Press.
Friedman, Francine. 1997. "To Fight or Not to Fight: The Decision to Settle the Croat–Serb Conflict." *International Interactions* 23: 55–78.
George, Alexander L., David K. Hall, and William E. Simons. 1971. *The Limits of Coercive Diplomacy*. Boston: Little, Brown.
1994. *The Limits of Coercive Diplomacy*, 2nd edn. Boulder: Westview Press.
Green, Donald and Ian Shapiro. 1994. *Pathologies of Rational Choice*. New Haven: Yale University Press.
James, Patrick. 1998. "Rational Choice?: Crisis Bargaining over the Meech Lake Accord." *Conflict Management and Peace Science* 16: 51–86.
James, Patrick and Michael Lusztig. 1996. "Beyond the Crystal Ball: Modeling Predictions about Quebec and Canada." *American Review of Canadian Studies* 26: 559–75.
1997a. "Assessing the Reliability of Prediction on the Future of Quebec." *Quebec Studies* 24: 197–210.
1997b. "Quebec's Economic and Political Future with North America." *International Interactions* 23: 283–98.

2000. "Predicting the Future of the FTAA." *NAFTA: Law and Business Review of the Americas* 6: 405–20.

2003. "Power Cycles, Expected Utility and Decision Making by the United States: The Case of the Free Trade Agreement of the Americas." *International Political Science Review* 24: 83–96.

Kugler, Jacek and Yi Feng (eds.). 1997. *International Interactions*.

Lakatos, Imre. 1976. *Proofs and Refutations: The Logic of Mathematical Discovery*. Cambridge: Cambridge University Press.

1978. *The Methodology of Scientific Research Programmes*. Vol. I. Cambridge: Cambridge University Press.

Laudan, Lawrence. 1977. *Progress and its Problems*. Berkeley: University of California Press.

Lowi, Theodore J. 1992. "The State in Political Science: How We Become What We Study." *American Political Science Review* 86: 1–7.

Maoz, Zeev, and Maoz and Nazrin Abdolali. 1989. "Regime Type and International Conflict, 1816–1976." *Journal of Conflict Resolution* 33: 3–36.

Powell, Robert. 1999. *In the Shadow of Power: States and Strategy in International Politics*. Princeton: Princeton University Press.

Quidort, Jean (John of Paris). 1302 [1971]. *On Royal and Papal Power*. J. A. Watt (trans.). Toronto: The Pontifical Institute of Mediaeval Studies.

Rapoport, Anatol and A. M. Chammah. 1965. *The Prisoners' Dilemma*. Ann Arbor: University of Michigan Press.

Rapoport, Anatol, Melvin Guyer, and David Gordon. 1976. *The 2X2 Game*. Ann Arbor: University of Michigan Press.

Ray James L. and Bruce M. Russett. 1996. "The Future as Arbiter of Theoretical Controversies: Predictions, Explanations and the End of the Cold War." *British Journal of Political Science* 25: 1578.

Riker, William H. 1996. *The Strategy of Rhetoric*. New Haven: Yale University Press.

Russett, Bruce M. 1993. *Grasping the Democratic Peace*. Princeton: Princeton University Press.

Schelling, Thomas. 1960. *Strategy of Conflict*. Cambridge, MA: Harvard University Press.

Schultz, Kenneth A. 2001. *Democracy and Coercive Diplomacy*. New York: Cambridge University Press.

Skinner, Kiron, S. Kudelia, B. Bueno de Mesquita, and C. Rice. 2004. *The Strategy of Campaigning: Ronald Reagan and Boris Yeltsin*. Ann Arbor: University of Michigan Press.

Smith, Alastair. 1995. "Alliance Formation and War." *International Studies Quarterly* 39: 405–25.

Stokman, Frans, Marcel van Assen, J. van der Knoop, and R. C. H. van Oosten. 2000. "Strategic Decision Making." *Advances in Group Processes* 17: 131–53.

Taylor, Michael. 1976. *Anarchy and Cooperation*. New York: John Wiley.

Thomson, Robert. 2000. "A Reassessment of the Final Analyses in European Community Decision Making." Unpublished ms, ICS, University of Groningen.

Torenvlied, Rene. 1996. "Decisions in Implementation: A Model-Guided Test of Implementation Theories Applied to Social Renewal Policy in Three Dutch Municipalities." Ph.D. Dissertation, University of Groningen.

Van der Knoop, J. and F. Stokman. 1998. "Anticiperen op Basis van Scenarios." *Gids voor Personeelsmanagement* 77: 12–15.

Williams, John H. P., and Mark J. Webber. 1998. "Evolving Russian Civil Military Relations: A Rational Actor Analysis." *International Interactions* 24: 115–50.

Part III

Possibilities for pluralism and convergence

12 The illusion of learning from observational research

Alan S. Gerber, Donald P. Green, and Edward H. Kaplan

Introduction

Empirical studies of cause and effect in social science may be divided into two broad categories, experimental and observational. In the former, individuals or groups are randomly assigned to treatment and control conditions. Most experimental research takes place in a laboratory environment and involves student participants, but several noteworthy studies have been conducted in real-world settings, such as schools (Howell and Peterson 2002), police precincts (Sherman and Rogan 1995), public housing projects (Katz, Kling, and Liebman 2001), and voting wards (Gerber and Green 2000). The experimental category also encompasses research that examines the consequences of randomization performed by administrative agencies, such as the military draft (Angrist 1990), gambling lotteries (Imbens, Rubin, and Sacerdote 2001), random assignment of judges to cases (Berube 2002), and random audits of tax returns (Slemrod, Blumenthal, and Christian 2001). The aim of experimental research is to examine the effects of random variation in one or more independent variables.

Observational research, too, examines the effects of variation in a set of independent variables, but this variation is not generated through randomization procedures. In observational studies, the data generation process by which the independent variables arise is unknown to the researcher. To estimate the parameters that govern cause and effect, the analyst of observational data must make several strong assumptions about the statistical relationship between observed and unobserved causes of the dependent variable (Achen 1986; King, Keohane, and Verba 1994). To the extent that these assumptions are unwarranted, parameter estimates will be biased. Thus, observational research involves two types of

The authors are grateful to the Institution for Social and Policy Studies at Yale University for research support. Comments and questions may be directed to the authors at alan.gerber@yale.edu, donald.green@yale.edu, or edward.kaplan@yale.edu.

uncertainty, the statistical uncertainty given a particular set of modeling assumptions and the theoretical uncertainty about which modeling assumptions are correct.

The principal difference between experimental and observational research is the use of randomization procedures. Obviously, random assignment alone does not guarantee that an experiment will produce unbiased estimates of causal parameters (cf. Cook and Campbell 1979, ch. 2, on threats to internal validity). Nor does observational analysis preclude unbiased causal inference. The point is that the risk of bias is typically much greater for observational research. This chapter characterizes experiments as unbiased and observational studies potentially biased, but the analytic results we derive generalize readily to situations in which both are potentially biased.

The vigorous debate between proponents of observational and experimental analysis (Cook and Payne 2002; Heckman and Smith 1995; Green and Gerber 2003; Weiss 2002) raises two meta-analytic questions. First, under what conditions and to what extent should we update our prior beliefs based on experimental and observational findings? Second, looking to the future, how should researchers working within a given substantive area allocate resources to each type of research, given the costs of each type of data collection?

Although these questions have been the subject of extensive discussion, they have not been addressed within a rigorous analytic framework. As a result, many core issues remain unresolved. For example, is the choice between experimental and observational research fundamentally static, or does the relative attractiveness of experimentation change depending on the amount of observational research that has accumulated up to that point in time? To what extent and in what ways is the tradeoff between experimental and observational research affected by developments in "theory" and in "methodology"?

The analytic results presented in this chapter reveal that the choice between experimental and observational research is fundamentally dynamic. The weight accorded to new evidence depends upon what methodological inquiry reveals about the biases associated with an estimation procedure as well as what theory tells us about the biases associated with our extrapolations from the particularities of any given study. We show that the more one knows *ex ante* about the biases of a given research approach, the more weight one accords the results that emerge from it. Indeed, the analytics presented below may be read as an attempt to characterize the role of theory and methodology within an observational empirical research program. When researchers lack prior information about the biases associated with observational research, they will assign observational findings zero weight and will never allocate future resources to it. In this situation, learning is possible only through unbiased empirical methods, methodological investigation, or theoretical insight. These analytic results thus invite social scientists to launch a new line of empirical inquiry designed to assess the direction and magnitude of research biases that arise in statistical inference and extrapolation to other settings.

Assumptions and notation

Suppose you seek to estimate the causal parameter M. To do so, you launch two empirical studies, one experimental and the other observational. In advance of gathering the data, you hold prior beliefs about the possible values of M. Specifically, your prior beliefs about M are distributed normally with mean μ and variance $\sigma^2{}_M$. The dispersion of your prior beliefs ($\sigma^2{}_M$) is of special interest. The smaller $\sigma^2{}_M$, the more certain you are about the true parameter M in advance of seeing the data. An infinite $\sigma^2{}_M$ implies that you approach the research with no sense whatsoever of where the truth lies.

You now embark upon an experimental study. Before you examine the data, the central limit theorem leads you to believe that your estimator, X_e, will be normally distributed. Given that $M = m$ (the true effect turns out to equal m) and that random assignment of observations to treatment and control conditions renders your experiment unbiased, X_e is normal with mean m and variance $\sigma^2{}_{Xe}$. As a result of the study, you will observe a draw from the distribution of X_e, the actual experimental value x_e.

In addition to conducting an experiment, you also gather observational data. Unlike randomized experimentation, observational research does not involve a procedure that ensures unbiased causal inference. Thus, before examining your observational results, you harbor prior beliefs about the bias associated with your observational analysis. Let B be the random variable that denotes this bias. Suppose that your prior beliefs about B are distributed normally with mean β and variance $\sigma^2{}_B$. Again, smaller values of $\sigma^2{}_B$ indicate more precise prior knowledge about the nature of the observational study's bias. Infinite variance implies complete uncertainty.

Further, we assume that priors about M and B are independent. This assumption makes intuitive sense: there is usually no reason to suppose *ex ante* that one can predict the observational study's bias by knowing whether a causal parameter is large or small. It should be stressed, however, that independence will give way to a negative correlation once the experimental and observational results become known.[1] The analytic results we present here are meant to describe what happens as one moves from prior beliefs to posterior views based on new information. The results can also be used to describe what happens after one examines an entire literature of experimental and observational studies. The precise sequence in which one examines the evidence does not affect our conclusions, but tracing this sequence does make the analytics more complicated. For purposes of exposition, therefore, we concentrate our attention on what happens as one moves from priors developed in advance of seeing the results to posterior views informed by all the evidence that one observes subsequently.

The observational study generates a statistical result, which we denote X_o (*o* for observational). Given that $M = m$ (the true effect equals m) and $B = b$ (the true bias equals b), we assume that the sampling distribution of X_o is normal with mean $m + b$ and variance $\sigma^2{}_{Xo}$. In other words, the observational study produces an estimate (x_o) that may be biased in the event that b is not equal to *0*. Bias may arise from any number of sources, such as unobserved heterogeneity, errors in variables, and other well-known problems. The variance of the observational study ($\sigma^2{}_{Xo}$) is a function of sample size, the predictive accuracy of the model, and other features of the statistical analysis used to generate the estimates.

Finally, we assume that given $M = m$ and $B = b$, the random variables X_e and X_o are independent. This assumption follows from the fact that the experimental and observational results do not influence each other in any way. In sum, our model of the research process assumes (1) normal and independently distributed priors about the true effect and the bias of observational research and (2) normal and independently distributed sampling distributions for the estimates generated by the experimental and observational studies. We now examine the implications of this analytic framework.

The joint posterior distribution of *M* and *B*

The first issue to be addressed is how our beliefs about the causal parameter M will change once we see the results of the experimental and observational studies. The more fruitful the research program, the more

[1] This negative correlation results from the fact that the experiment provides an unbiased estimator of M, whereas the observational study provides an unbiased estimator of $M + B$. As we note below, once these findings become known, higher estimates of M from the experimental study imply lower values of B when the experimental result is subtracted from the observational result.

our posterior beliefs will differ from our prior beliefs. New data might give us a different posterior belief about the location of M, or it might confirm our prior belief and reduce the variance (uncertainty) of these beliefs.

Let $f_X(x)$ represent the normal probability density for random variable X evaluated at the point x, and let $f_{X|A}(x)$ be the conditional density for X evaluated at the point x given that the event A occurred. Given the assumptions above, the joint density associated with the compound event $M = m$, $X_e = x_e$, $B = b$, and $X_o = x_o$ is given by

$$f_M(m) \times f_{X_e|M=m}(x_e) \times f_B(b) \times f_{X_o|M=m,B=b}(x_o). \tag{1}$$

What we want is the joint posterior distribution of M, the true effect, and B, the bias associated with the observational study, given the experimental and observational data. Applying Bayes' rule we obtain:

$$f_{M,B|X_e=x_e,X_o=x_o}(m,b)$$
$$= \frac{f_M(m) \times f_{X_e|M=m}(x_e) \times f_B(b) \times f_{X_o|M=m,B=b}(x_o)}{\int_{m=-\infty}^{\infty} \int_{b=-\infty}^{\infty} f_M(m) \times f_{X_e|M=m}(x_e) \times f_B(b) \times f_{X_o|M=m,B=b}(x_o)\,db\,dm}. \tag{2}$$

Integrating over the normal probability distributions (cf. Box and Tiao 1973) produces the following result.

Theorem 1: The joint posterior distribution of M and B is bivariate normal with the following means, variances, and correlation.

The posterior distribution of M is normally distributed with mean given by

$$E(M \mid X_e = x_e, X_o = x_o) = p_1\mu + p_2 x_e + p_3(x_o - \beta)$$

and variance

$$\sigma^2_{M|x_e,x_o} = \frac{1}{\dfrac{1}{\sigma^2_M} + \dfrac{1}{\sigma^2_{X_e}} + \dfrac{1}{\sigma^2_B + \sigma^2_{X_o}}},$$

where

$$p_1 = \frac{\sigma^2_{M|x_e,x_o}}{\sigma^2_M}, \quad p_2 = \frac{\sigma^2_{M|x_e,x_o}}{\sigma^2_{X_e}}, \quad \text{and} \quad p_3 = \frac{\sigma^2_{M|x_e,x_o}}{\sigma^2_B + \sigma^2_{X_o}}.$$

The posterior distribution of B is normally distributed with mean

$$E(B \mid X_e = x_e, X_o = x_o) = q_1\beta + q_2(x_o - \mu) + q_3(x_o - x_e)$$

and variance

$$\sigma^2_{B|x_e,x_o} = \frac{1}{\dfrac{1}{\sigma^2_B} + \dfrac{\dfrac{1}{\sigma^2_{X_o}}\left(\dfrac{1}{\sigma^2_M} + \dfrac{1}{\sigma^2_{X_e}}\right)}{\dfrac{1}{\sigma^2_M} + \dfrac{1}{\sigma^2_{X_e}} + \dfrac{1}{\sigma^2_{X_o}}}},$$

where

$$q_1 = \frac{\sigma^2_{B|x_e,x_o}}{\sigma^2_B},$$

$$q_2 = \frac{\sigma^2_{B|x_e,x_o}\left(\dfrac{1}{\sigma^2_M \sigma^2_{X_o}}\right)}{\dfrac{1}{\sigma^2_M} + \dfrac{1}{\sigma^2_{X_e}} + \dfrac{1}{\sigma^2_{X_o}}},$$

$$q_3 = \frac{\sigma^2_{B|x_e,x_o}\left(\dfrac{1}{\sigma^2_{X_e} \sigma^2_{X_o}}\right)}{\dfrac{1}{\sigma^2_M} + \dfrac{1}{\sigma^2_{X_e}} + \dfrac{1}{\sigma^2_{X_o}}}.$$

The correlation between M and B after observing the experimental and observational findings is given by $\rho \leq 0$, such that

$$1 - \rho^2 = \frac{\dfrac{1}{\sigma^2_M} + \dfrac{1}{\sigma^2_{X_e}} + \dfrac{1}{\sigma^2_B + \sigma^2_{X_o}}}{\dfrac{1}{\sigma^2_M} + \dfrac{1}{\sigma^2_{X_e}} + \dfrac{1}{\sigma^2_{X_o}}}.$$

This theorem reveals that the posterior mean is an average (since $p_1 + p_2 + p_3 = 1$) of three terms: the prior expectation of the true mean effect (μ), the observed experimental value (x_e), and the observational value corrected by the prior expectation of the bias ($x_o - \beta$). This analytic result parallels the standard case in which normal priors are confronted with normally distributed evidence (Box and Tiao 1973). In this instance, the biased observational estimate is recentered to an unbiased estimate by subtracting off the prior expectation of the bias. It should be noted that such recentering is rarely, if ever, done in practice. Those who report observational results seldom disclose their priors about the bias term, let alone correct for it. In effect, researchers working with observational data routinely, if implicitly, assume that the bias equals zero and that the uncertainty associated with this bias is also zero.

To get a feel for what the posterior distribution implies substantively, it is useful to consider several limiting cases. If prior to examining the data one were certain that the true effect were μ, then $\sigma^2{}_M = 0$, $p_1 = 1$, and $p_2 = p_3 = 0$. In this case, one would ignore the data from both studies and set $E(M \mid X_e = x_e, X_o = x_o) = \mu$. Conversely, if one had no prior sense of M or B before seeing the data, then $\sigma^2{}_M = \sigma^2{}_B = \infty$, $p_1 = p_3 = 0$, and $p_2 = 1$, in which case the posterior expectation of M would be identical to the experimental result x_e. In the less extreme case one has some prior information about M such that $\sigma^2{}_M < \infty$, p_3 remains zero so long as one remains completely uninformed about the biases of the observational research. In other words, in the absence of prior knowledge about the bias of observational research, one accords it zero weight. Note that this result holds even when the sample size of the observational study is so large that $\sigma^2{}_{Xo}$ is reduced to zero.

For this reason, we refer to this result as the Illusion of Observational Learning Theorem. If one is entirely uncertain about the biases of observational research, the accumulation of observational findings sheds no light on the causal parameter of interest. Moreover, for a given finite value of $\sigma^2{}_B$ there comes a point at which observational data cease to be informative and where further advances to knowledge can come only from experimental findings. The illusion of observational learning is typically obscured by the way in which researchers conventionally report their nonexperimental statistical results. The standard errors associated with regression estimates, for example, are calculated based on the unstated but often implausible assumption that the bias associated with a given estimator is known with perfect certainty before the estimates are generated. These standard errors would grow much larger were they to take into account the value of $\sigma^2{}_B$.

The only way to extract additional information from observational research is to obtain extrinsic information about its bias. By extrinsic information, we mean information derived from inspection of the observational procedures, such as the measurement techniques, statistical methodology, and the like. *Extrinsic information does not include the results of the observational studies and comparisons to experimental results.* If all one knows about the bias is that experimental studies produced an estimate of 10 while observational studies produced an estimate of 5, one's posterior estimate of the mean will not be influenced at all by the observational results.

To visualize the irrelevance of observational data with unknown biases, consider a hypothetical regression model of the form

$$Y = a + bX + U,$$

where Y is the observed treatment effect across a range of studies, X is a dummy variable scored 0 if the study is experimental and 1 if it is observational, and U is an unobserved disturbance term. Suppose that we have non-informative priors about a and b. The regression estimate of a provides an unbiased estimate of the true treatment effect.[2] Similarly, the regression estimate of b provides an unbiased estimate of the observational bias. Regression of course generates the same estimates of a and b regardless of the order in which we observe the data points. Moreover, the estimate of a is unaffected by the presence of observational studies in our dataset. This regression model produces the same estimate of a as a model that discards the observational studies and simply estimates

$$Y = a + U.$$

This point warrants special emphasis, since it might appear that one could augment the value of observational research by running an observational pilot study, assessing its biases by comparison to an experimental pilot study, and then using the new, more precise posterior of $\sigma^2{}_B$ as a prior for purposes of subsequent empirical inquiry. The flaw in this sequential approach is that conditional on seeing the initial round of experimental and observational results, the distributions of M and B become negatively correlated. To update one's priors recursively requires a different set of formulas from the ones presented above. After all is said and done, however, a recursive approach will lead to exactly the same set of posteriors. As demonstrated in the Appendix, the formulas above describe how priors over M and B change in light of *all* of the evidence that subsequently emerges, regardless of the sequence in which these studies become known to us.

Although there are important limits to what observational findings can tell us about the causal parameter M, the empirical results may greatly influence posterior uncertainty about bias, $\sigma^2{}_B$. This posterior distribution is a weighted sum of three quantities: the prior belief about the location of β, the difference between the observational finding and the expected true effect, and the observed gap between the experimental and observational results. If the researcher enters the research process with diffuse priors about M and B such that $\sigma^2{}_M = \sigma^2{}_B = \infty$, the weights q_1

[2] We are grateful to Doug Rivers, who suggested this analogy.

and q_2 will be zero, and the posterior will reflect only the observed discrepancy between experimental and observational results. In this instance, the posterior variance reduces to the simple quantity

$$\sigma^2_{B|x_e,x_o;\sigma^2_M=\sigma^2_B=\infty} = \sigma^2_{X_e} + \sigma^2_{X_o},$$

which is the familiar classical result concerning the variance of the difference between two independent estimates. Note that this result qualifies our earlier conclusion concerning the futility of gathering observational data when one's priors are uninformative. When analyzed in conjunction with experimental results, observational findings can help shed light on the biases associated with observational research. Of course, this comes as small consolation to those whose primary aim is to learn about the causal parameter M.

A numerical example

A simple numerical example may help fix ideas. Consider two studies of the effects of face-to-face canvassing on voter turnout in a particular election in a particular city. The observational study surveys citizens to assess whether they were contacted at home by political canvassers and uses regression analysis to examine whether reported contact predicts voting behavior, controlling for covariates such as political attitudes and demographic characteristics (Kramer 1970; Rosenstone and Hansen 1993). The key assumption of this study is that reported contact is statistically unrelated to unobserved causes of voting. This assumption would be violated if reported contact were an imprecise measure of actual contact or if political campaigns make a concerted effort to contact citizens with unusually high propensities to vote. The majority of published studies on the effects of voter mobilization use some version of this approach.

The experimental study of face-to-face canvassing randomly assigns citizens to treatment and control groups. Canvassers contact citizens in the treatment group; members of the control group are not contacted.[3] This type of fully randomized experiment dates back to Eldersveld (1956)

[3] As Gerber and Green (2000) explain, complications arise in this type of experiment when canvassing campaigns fail to make contact with those subjects assigned to the treatment group. This problem may be addressed statistically using instrumental variables regression, although as Heckman and Smith (1995) note, the external validity of this correction requires the assumption that the canvassing has the same effect on those who could be reached as it would have had among those the canvassing campaign failed to contact. This assumption may be tested by randomly varying the intensity of the canvassing effort. Below, we take up the question of how our conclusions change as we take into account the potential for bias in experimental research.

and currently enjoys something of a revival in political science (Gerber and Green 2000).

Suppose for the sake of illustration that your prior beliefs about M, the effect of canvassing on voter turnout, were centered at 10 with a variance of 25 (or, equivalently, a standard deviation of 5). These priors imply that you assign a probability of about 0.95 to the conjecture that M lies between 0 and 20. You also hold priors about the observational bias. You suspect that contact with canvassers is measured unreliably, which could produce an underestimate of M, but also that contact with canvassers is correlated with unmeasured causes of voting, which may produce an overestimate of M. Thus, your priors about the direction of bias are somewhat diffuse. Let us suppose that B is centered at 2 with a variance of 36 (standard deviation of 6). Finally, you confront the empirical results. The experimental study, based on approximately 1,200 subjects divided between treatment and control groups, produces an estimate of 12 with a standard deviation of 3. The observational study, based on a sample of 10,000 observations, produces an estimate of 16 with a standard deviation of 1.

With this information, we form the posterior mean and variance for M:

$$E(M \mid X_e = 12, X_o = 16) = p_1(10) + p_2(12) + p_3(16 - 2) = 11.85$$

$$\sigma^2_{M|x_e,x_o} = \frac{1}{\frac{1}{25} + \frac{1}{9} + \frac{1}{1+36}} = 5.61$$

$$p_1 = \frac{5.61}{25} = .23, \quad p_2 = \frac{5.61}{9} = .62, \quad p_3 = \frac{5.61}{36+1} = .15.$$

Notice that although the observational study has much less sampling variability than the experimental study, it is accorded much less weight. Indeed, the experimental study has four times as much influence on the posterior mean as the observational study. The observational study, corrected for bias, raises the posterior estimate of M, while the prior lowers it, resulting in a posterior estimate of 11.9, which is very close to the experimental result. The prior variance of 25 has become a posterior variance of 5.6, a reduction that is attributable primarily to the experimental evidence. Had the observational study contained 1,000,000 observations instead of 10,000, thereby decreasing its standard error from 1 to 0.1, the posterior variance would have dropped imperceptibly from 5.61 to 5.59, and the posterior mean would have changed only from 11.85 to

11.86. A massive investment in additional observational data produces negligible returns.

One interesting feature of this example is that the experimental and observational results are substantively rather similar; both suggest that canvassing "works." Sometimes when experimental results happen to coincide with observational findings, experimentation is chided for merely telling us what we already know (Morton 2002: 15), but this attitude stems from the illusion described above. The standard error associated with the N = 10,000 observational study would conventionally be reported as 1, when in fact its root mean squared error is 6.1. In this situation, what we "already know" from observational research is scarcely more than conjecture until confirmed by experimentation. The posterior variance is four times smaller than the prior variance primarily because the experimental results are so informative.

The empirical results also furnish information about the bias associated with the observational data. The posterior estimate of B is 4.1, as opposed to a prior of 2. Because the experimental results tell us a great deal about the biases of the observational study, the posterior variance of B is 6.3, a marked decline from the prior value of 36. In advance of seeing the data, our priors over M and B were uncorrelated; afterwards, the posterior correlation between B and M becomes -0.92. This posterior correlation is important to bear in mind in the event that subsequent evidence becomes available. The estimating equations presented above describe how uncorrelated priors over M and B change in light of all of the evidence that emerges subsequently, regardless of the order in which it emerges. Thus, if a second observational study of 10,000 observations were to appear, we would recalculate the results in this section on the basis of the cumulative N = 20,000 observational dataset.

Allocating resources to minimize the posterior variance of *M*

Above we considered the case in which a researcher revises prior beliefs after encountering findings from two literatures, one experimental and the other observational. Now we consider a somewhat different issue: how should this researcher allocate scarce resources between experimental and observational investigation?

Suppose the research budget is R. This budget is allocated to experimental and observational studies. The marginal price of each experimental observation is denoted π_e; the price of a non-experimental observation is π_o. Let n_e be the size of the experimental study; n_o the size of the observational study. The budget is allocated to both types of research subject to the constraint $\pi_e n_e + \pi_o n_o = R$.

Let the variance[4] of the experimental study equal $\sigma^2{}_{Xe} = \sigma^2{}_e/n_e$, and the variance of the observational study equal $\sigma^2{}_{Xo} = \sigma^2{}_o/n_o$. Using the results in Theorem 1, the aim is to allocate resources so as to minimize the posterior variance of M, subject to the budget constraint R.

Theorem 2. The optimal allocation of a budget R, given prices π_e and π_o, disturbance variances $\sigma^2{}_e$ and $\sigma^2{}_o$, and variance of priors about observational bias $\sigma^2{}_B$, takes one of three forms depending on the values of the parameters:

Case 1. For $\sigma_o^2(\pi_o/\pi_e) \geq \sigma_e^2$, allocate $n_e = R/\pi_e$ and $n_o = 0$.

Case 2. For $\sigma_o^2(\pi_o/\pi_e)\left(1 + \dfrac{R\sigma_B^2}{\pi_o\sigma_o^2}\right)^2 \geq \sigma_e^2 \geq \sigma_o^2(\pi_o/\pi_e)$,

allocate $n_O^* = \left[\left(\dfrac{\pi_e\sigma_e^2}{\pi_o\sigma_o^2}\right)^{1/2} - 1\right]\left(\dfrac{\sigma_o^2}{\sigma_B^2}\right)$ and $n_e = \dfrac{R - \pi_o n_o^*}{\pi_e}$

Case 3. For $\sigma_e^2 \geq \sigma_o^2(\pi_o/\pi_e)\left(1 + \dfrac{R\sigma_B^2}{\pi_o\sigma_o^2}\right)^2$, allocate $n_o = R/\pi_o$ and $n_e = 0$.

The implications of the Research Allocation Theorem in many ways parallel our earlier results. When allocation decisions are made based on uninformative priors about the bias ($\sigma^2{}_B = \infty$), no resources are ever allocated to observational research. As budgets approach infinity, the fraction of resources allocated to experimental research approaches 1. When $\sigma^2{}_B$ is zero, resources will be allocated entirely to either experiments or observational studies, depending on relative prices and disturbance variances.

The most interesting case is the intermediate one, where finite budgets and moderate values of $\sigma^2{}_B$ dictate an apportioning of resources between the two types of studies. Here, possible price advantages of observational research are balanced against the risk of bias. Particularly attractive, therefore, are observational studies that are least susceptible to bias, such as those based on naturally occurring randomization of

[4] The notation used in this section has been selected for ease of exposition. When the population variance for the subjects in an experimental study equals v, and n experimental subjects are divided equally into a treatment and control group, the variance for the experiment is $4v/n$. To convert from these units to the notation used here, define π_e as the cost of adding four subjects.

the independent variable (Imbens, Rubin, and Sacerdote 2001), near-random assignment (McConahay, 1982), or assignment that supports a regression-discontinuity analysis (Cook and Campbell 1979: ch. 3).

Notice that the allocation decision does not depend on σ^2_M, that is, prior uncertainty about the true causal parameter. How much one knows about the research problem before gathering data is irrelevant to the question of how to allocate resources going forward. Experimentation need not be restricted, for example, to well-developed research programs.

One further implication deserves mention. When the price-adjusted disturbance variance in observational research is greater than the disturbance variance in experimental research (see Case 1), all of the resources are allocated to experimental research. Reduction in disturbance variance is sometimes achieved in highly controlled laboratory settings or through careful matching of observations prior to random assignment. Holding prices constant, the more complex the observational environment, the more attractive experimentation becomes.

The Research Allocation Theorem provides a coherent framework for understanding why researchers might wish to conduct experiments. Among the leading justifications for experimentation are (1) uncertainty about the biases associated with observational studies; (2) ample resources; (3) inexpensive access to experimental subjects; and (4) features of experimental design that limit disturbance variability. Conversely, budget constraints, the relative costliness of experimental research, and the relative precision of observational models constitute leading arguments in favor of observational research. It should be emphasized, however, that the case for observational research hinges on prior information about its biases.

Discussion

In this concluding section, we consider the implications of these two theorems for research practice. We consider in particular (i) the possibility of bias in experiments; (ii) the value of methodological inquiry; (iii) the conditions under which observational data support unbiased causal inference, and (iv) the value of theory.

What about bias in experiments? In the preceding analysis, we have characterized experiments as unbiased and observational studies as potentially biased. It is easy to conceive of situations where experimental results are potentially biased as well. The external validity of an experiment hinges on three factors: whether the subjects in the study are as strongly influenced by the treatment as the population to which a generalization is made, whether the treatment in the experiment corresponds to the treatment in the population of interest, and whether the response measure used in the experiment corresponds to the variable of interest in the population. In the example mentioned above, a canvassing experiment was conducted in a given city at a given point in time. Door-to-door canvassing was conducted by certain precinct workers using a certain type of get-out-the-vote appeal. Voter turnout rates were calculated based on a particular source of information. Extrapolating to other times, places, and modes of canvassing introduces the possibility of bias.

A straightforward extension of the present analytic framework could be made to cases in which experiments are potentially biased. Delete the expressions related to the unbiased experiment, and replace them with a potentially biased empirical result akin to the observational study. The lesson to be drawn from this type of analysis parallels what we have presented here. Researchers should be partial to studies, such as field experiments, which raise the fewest concerns about bias. The smaller the inferential leap, the better.

This point has special importance for the distinction between laboratory experiments and field experiments. Although lab experiments are often less costly than field experiments, the inferential leap from the laboratory to the outside world increases the risk of bias. Experiments often involve convenience samples, contrived interventions, and response measures that do not directly correspond to the dependent variables of interest outside the lab. These are more serious drawbacks than the aforementioned problems with field experiments. Field experiments may be replicated in an effort to sample different types of interventions and the political contexts in which they occur, thereby reducing the uncertainty associated with generalizations beyond the data. Replication lends credibility to laboratory experiments as well, but so long as the outcome variable observed in the lab (e.g., stated vote intention) differs from the variable of interest (actual voter turnout rates), and so long as there is reason to suspect that laboratory results reflect the idiosyncrasies of an artificial environment, the possibility of bias remains acute.

Disciplines such as medicine, of course, make extensive and productive use of laboratory experimentation. In basic research, animals such as rats and monkeys are used as proxies for human beings. One might suspect that the idiosyncratic biology and social environments of human beings would render this type of animal research uninformative, but in fact the correspondence between results obtained based on animal models and those based on human beings turns out to be substantial. Even stronger is the empirical correspondence between laboratory results involving non-random samples of human subjects and outcomes that occur when medical treatments are deployed in the outside world. As Achen points out, when experience shows that results may be generalized readily from a laboratory setting, "even a tiny randomized experiment may be better than a large uncontrolled experiment" (1986: 7). Our point is not that laboratory experiments are inherently flawed; rather, the external validity of laboratory studies is an empirical question, one that has been assessed extensively in medicine and scarcely at all in political science.

All things being equal, the external validity of field experimentation exceeds that of laboratory experimentation. However, the advantages of field experimentation in terms of external validity may be offset by threats to internal validity that arise when randomized interventions are carried out in naturalistic settings. Heckman and Smith (1995) note that the integrity of random assignment is sometimes compromised by those charged with administering treatments, who deliberately or unwittingly divert a treatment to those who seem most deserving. They note also the complications that arise when people assigned to the control group take it upon themselves to obtain the treatment from sources outside the experiment. To this list may be added other sources of bias, such as problems of spillover that occur when a treatment directed to a treatment group affects a nearby control group or the experimenter's failure to measure variations in the treatment that is actually administered.

Whether a given field experiment confronts these difficulties, of course, depends on the nature of the intervention and the circumstances in which the experiment is conducted. When these threats to unbiasedness do present themselves, valid inference may be rescued by means of statistical correctives that permit consistent parameter estimation. When treatments are shunted to those who were assigned to the control group, for example, the original random assignment provides a valid instrumental variable predicting which individuals in the treatment and control groups actually received the treatment. Despite the fact that some members of the control group were treated inadvertently, the causal parameters of interest may be obtained using instrumental variables regression so long as the assigned treatment group was more likely to receive the treatment than the assigned control group (Angrist, Imbens, and Rubin 1996).[5]

Although statistical correctives are often sufficient to mend the problems that afflict field experiments, certain potential biases can only be

[5] The interpretation of these estimates is straightforward if one assumes (as researchers working with observational data often do) that the treatment's effects are the same for all members of the population. Strictly speaking, the instrumental variables regression provides an estimate of the effect of the treatment on those in the treatment group who were treated (or, in the case of contamination of the control group, those treated at different rates in the treatment and control groups).

addressed by adjusting the experimental design. For example, Howell and Peterson's (2002) experiments gauge the effects of private school vouchers on student performance by assigning vouchers to a random subset of families that apply for them. This design arguably renders a conservative estimate of the effects of vouchers, inasmuch as the competitive threat of a private voucher program gives public schools in the area an incentive to work harder, thereby narrowing the performance gap between the treatment group that attends private schools and the control group that remains in public school.[6] In order to evaluate the systemic effects of vouchers, it would be useful to perform random assignment at the level of the school district rather than at the level of the individual. In this way, the analyst could ascertain whether the availability of vouchers improves academic performance in public schools. This result would provide extrinsic evidence about the bias associated with randomization at the individual level.

Field experiments of this sort are of course expensive. As we point out in our Research Allocation Theorem, however, the uncertainties associated with observational investigation may impel researchers to allocate resources to field experimentation in spite of these costs. Observational inquiry may involve representative samples of the population to which the causal generalization will be applied and a range of real-world interventions, but the knowledge it produces about causality is more tenuous. The abundance of observational research and relative paucity of field experimentation that one currently finds in social science – even in domains where field experimentation is feasible and ethically unencumbered – may reflect excessive optimism about what is known about the biases of observational research.

The value of methodological inquiry. Our analytic results underscore not only the importance of unbiased experimental research but also the value of basic methodological inquiry. To the extent that the biases of observational research can be calibrated through independent inquiry, the information content of observational research rises. For example, if through inspection of its sampling, measurement, or statistical procedures the bias associated with a given observational study could be identified more precisely (lowering $\sigma^2{}_B$), the weight assigned to those observational findings would go up. The social sciences have accumulated enormous amounts of data, but like ancient texts composed in some inscrutable language, these data await the discovery of insights that will enable them to become informative.

6 The magnitude of this bias is likely to be small in the case of the Howell and Peterson interventions, which affected only a small percentage of the total student-age population.

Unfortunately, the *ex ante* prediction of bias in the estimation of treatment effects has yet to emerge as an empirical research program in the social sciences. To be sure, methodological inquiry into the properties of statistical estimators abounds in the social sciences. Yet, we know of no study that assesses the degree to which methodological experts can anticipate the direction and magnitude of biases simply by inspecting the design and execution of observational social science research.[7] The reason that this literature must be empirically grounded is that observational data analysis often involves a variety of model specifications; the statistical attributes of the final estimator presented to the reader may be very different from the characteristics of the circuitous estimation procedure that culminated in a final set of estimates. One potential advantage of experimentation is that it imposes discipline on data analysis because the statistical model is implied by the experimental design. Since data-mining potentially occurs in experimental analysis as well; the relative merits of experimentation remain an open research question. This type of basic empirical investigation would tell us whether, as a practical matter, the uncertainty associated with observational bias is so great that causal inference must of necessity rely on experimental evidence.

A model for this type of literature may be found in survey analysis. The estimation of population parameters by means of random sampling is analogous to the estimation of treatment effects by means of randomized experimentation. Convenience and other non-random samples are analogous to observational research. In fields where random sampling methods are difficult to implement (e.g., studies of illicit drug use), researchers make analytic assessments of the biases associated with various non-random sampling approaches and test these assessments by making empirical comparisons between alternative sampling methodologies. The upshot of this literature is that biased samples can be useful so long as one possesses a strong analytic understanding of how they are likely to be biased.

When is observational learning not illusory? Sometimes researchers can approximate this kind of methodological understanding by seizing upon propitious observational research opportunities. For example, students of public opinion have for decades charted movements in presidential popularity. As a result, they have a clear sense of how much presidential popularity would be expected to change over the course of a few

[7] In the field of medicine, where plentiful experimental and observational research makes cross-method comparisons possible, scholars currently have a limited ability to predict *ex post* the direction and magnitude of observational bias (cf. Heinsman and Shadish 1996; Shadish and Ragsdale 1996). We are aware of no study showing that medical researchers can predict the direction of biases *ex ante*.

days. Although the terrorist attacks of September 11, 2001 were not the product of random assignment, we may confidently infer that they produced a dramatic surge in presidential popularity from the fact that no other factors can plausibly account for a sudden change of this magnitude. Expressed in terms of the notation presented above, the size of the observational estimate (x_o) dwarfs the bias term (B) and the uncertainty associated with it ($\sigma^2{}_B$), leaving little doubt that the terrorist attacks set in motion a train of events that increased the president's popularity.

Notice that the foregoing example makes use of substantive assumptions in order to bolster the credibility of a causal inference. In this case, we stipulate the absence of other plausible explanations for the observed increase in popularity ratings. Were we able to do this routinely in the analysis of observational data, the problems of inferences would not be so daunting. Lest one wonder why humans were able to make so many useful discoveries prior to the advent of randomized experimentation, it should be noted that physical experimentation requires no explicit control group when the range of alternative explanations is so small. Those who strike flint and steel together to make fire may reasonably reject the null hypothesis of spontaneous combustion. When estimating the effects of flint and steel from a sequence of events culminating in fire, $\sigma^2{}_B$ is fairly small; but in those instances where the causal inference problem is more uncertain because the range of competing explanations is larger, this observational approach breaks down. Pre-experimental scientists had enormous difficulty gauging the efficacy of medical interventions because the risk of biased inference is so much greater, given the many factors that plausibly affect health outcomes.

This analytic framework leaves us with an important research principle: In terms of mean-squared error, it may be better to study fewer observations, if those observations are chosen in ways that minimize bias. To see this principle at work, look back at our numerical example and imagine that our complete data set consisted of 11,200 observations, 1,200 of which were free from bias. If we harbor uninformative priors about the bias in the observational component of the data set, the optimal weighting of these observations involves placing zero weight on the 10,000 bias-prone observations and focusing exclusively on the 1,200 experimental observations. The implication for large-N studies of international relations or comparative politics is that it may be better to focus on a narrow but uncontaminated portion of a larger data set.

How can one identify the unbiased component of a larger data set? One way is to generate the data through unbiased procedures, such as random assignment. Another is to look for naturally occurring instances where similar cases are confronted with different stimuli, as when defendants are assigned by lot to judges with varying levels of punitiveness. More tenuous but still defensible may be instances where the processes by which the independent variables are generated have no plausible link to unobserved factors that affect the dependent variable. For example, if imminent municipal elections cause local officials to put more police on the streets, and if the timing of municipal elections across a variety of jurisdictions has no plausible relationship to trends in crime rates, the municipal election cycle can be used as an instrumental variable by which to estimate the effects of policing on crime rates (Levitt 1997). More tenuous still, but still arguably more defensible than an indiscriminant canvass of all available cases, is an attempt to match observations on as many criteria as possible prior to the occurrence of an intervention. For certain applications, this is best done by means of a panel study in which observations are tracked before and after they each encounter an intervention, preferably an intervention that occurs at different points in time. These methods are not free from bias, but the care with which the comparisons are crafted reduces some of the uncertainty about bias, which makes the results more persuasive.

The value of theory. The more theoretical knowledge the researcher brings to bear when developing criteria for comparability, the smaller the $\sigma^2{}_B$ and the more secure the causal inference. Thus, in addition to charting a new research program for methodologists, our analysis calls attention to an underappreciated role of theory in empirical research. The point extends beyond the problem of case selection and internal validity. Any causal generalization, even one based on a randomized study that takes place in a representative sample of field sites, relies to some degree on extrapolation. Every experiment has its own set of idiosyncrasies, and we impose theoretical assumptions when we infer, for example, that the results from an experiment conducted on a Wednesday generalize readily to Thursdays. Just as methodological insight clarifies the nature of observational bias, theoretical insight reduces the uncertainty associated with extrapolation.

But where does theoretical insight come from? To the extent that what we are calling theories are testable empirical propositions, the answer is some combination of priors, observational data, and experimental findings. Theory development, then, depends critically on the stock of basic, carefully assessed empirical claims. Social scientists tend to look down on this type of science as narrow and uninspiring, grasping instead for weak tests of expansive and arresting propositions. Unlike natural scientists, therefore, social scientists have accumulated relatively few secure empirical premises from which to extrapolate. This deficiency is unfortunate, because developing secure empirical premises speeds the rate at which learning occurs as new experimental and observational results become available.

The framework laid out here cannot adjudicate the issue of how a discipline should allocate its resources to research questions of varying substantive merit, but it does clarify the conditions under which a given research program is likely to make progress. Our ability to draw causal inferences from data depends on our prior knowledge about the biases of our procedures.

Appendix: Proof of the proposition that the sequence of the empirical results does not affect the analytic results presented above

We seek to prove the following: if instead of observing the single experimental value x_e and the single observational value x_o we instead observe the series of experimental and observational sample mean values $x_e^{(1)}, x_e^{(2)}, \ldots x_e^{(k_e)}$ and $x_o^{(1)}, x_o^{(2)}, \ldots x_o^{(k_o)}$ such that

$$\frac{\sum_{j=1}^{k_e} n_e^{(j)} x_e^{(j)}}{\sum_{j=1}^{k_e} n_e^{(j)}} = x_e \quad \text{and} \quad \frac{\sum_{j=1}^{k_o} n_o^{(j)} x_o^{(j)}}{\sum_{j=1}^{k_o} n_o^{(j)}} = x_o,$$

where x_e and x_o are the overall experimental and observational mean values, then once *all* of the data have been observed, the joint posterior density of M and B given the individual sample results will be equivalent to what one would have seen if the two overall means x_e and x_o were revealed simultaneously.

The proof follows directly from the sufficiency of the sample mean as an estimator for the population mean for normal distributions. Straightforward application of Bayes' rule shows that, conditional on observing $x_e^{(1)}, x_e^{(2)}, \ldots x_e^{(k_e)}$ and $x_o^{(1)}, x_o^{(2)}, \ldots x_o^{(k_o)}$ the joint density of M *and* B is given by

$$f_{M,B|x_e^{(1)},x_e^{(2)},\ldots x_e^{(k_e)};x_o^{(1)},x_o^{(2)},\ldots x_o^{(k_o)}}(m,b)$$
$$= \frac{f_M(m) \times f_B(b) \times \prod_{j=1}^{k_e} f_{X_e^{(j)}|M=m}\left(x_e^{(j)}\right) \times \prod_{j=1}^{k_o} f_{X_o^{(j)}|M=m,B=b}\left(x_o^{(j)}\right)}{\int_{m'=-\infty}^{\infty} \int_{b'=-\infty}^{\infty} f_M(m') \times f_B(b') \times \prod_{j=1}^{k_e} f_{X_e^{(j)}|M=m'}\left(x_e^{(j)}\right) \times \prod_{j=1}^{k_o} f_{X_o^{(j)}|M=m',B=b'}\left(x_o^{(j)}\right) db'\,dm'}.$$

For normal distributions (see Freund 1971: 263) and constants C_e and C_o that do not depend on m and b, the likelihoods can be factored as

$$\prod_{j=1}^{k_e} f_{X_e^{(j)}|M=m}\left(x_e^{(j)}\right) = C_e \times f_{X_e|M=m}(x_e) \text{ and}$$

$$\prod_{j=1}^{k_o} f_{X_o|M=m,B=b}\left(x_o^{(j)}\right) = C_o \times f_{X_o|M=m,B=b}(x_o).$$

Consequently, upon substituting into the numerator and denominator of the posterior density shown above, the constants C_e and C_o cancel, leaving the expression below:

$$f_{M,B|X_e=x_e,X_o=x_o}(m,b) = \frac{f_M(m) \times f_{X_e|M=m}(x_e) \times f_B(b) \times f_{X_o|M=m,B=b}(x_o)}{\int_{m'=-\infty}^{\infty}\int_{b'=-\infty}^{\infty} f_M(m') \times f_{X_e|M=m'}(x_e) \times f_B(b') \times f_{X_o|M=m',B=b'}(x_o)db'dm'}.$$

This expression is the same as the one presented in the text, which shows that the order in which we observe the empirical results does not matter.

The implication of this proof is that one cannot squeeze additional information out of observational research by comparing initial observational estimates to experimental estimates, calibrating the bias of the observational studies, and then gathering additional bias-corrected observational data. This procedure, it turns out, produces the same estimates as simply aggregating all of the observational and experimental evidence and using the estimation methods described in the text.

REFERENCES

Achen, Christopher H. 1986. *The Statistical Analysis Of Quasi-Experiments*. Berkeley: University of California Press.

Angrist, Joshua A. 1990. "Lifetime Earnings and the Vietnam Era Draft Lottery: Evidence from Social Security Administrative Records." *American Economic Review* 80(3): 313–36.

Angrist, Joshua D., Guido W. Imbens, and Donald B. Rubin. 1996. "Identification of Casual Effects Using Instrumental Variables." *Journal of the American Statistical Association* 91(June): 444–55.

Berube, Danton. 2002. "Random Variations in Federal Sentencing as an Instrumental Variable." Unpublished ms, Yale University.

Box, George E. P. and G. C. Tiao. 1973. *Baysian Inference in Statistical Analysis*. Reading, MA: Addison-Wesley.

Cook, Thomas D. and Donald T. Campbell. 1979. *Quasi-Experimentation: Design and Analysis Issues for Field Settings*. Boston: Houghton Mifflin.

Cook, Thomas D. and Monique R. Payne. 2002. "Objecting to the Objections to Using Random Assignment in Educational Research," in Frederick Mosteller and Robert Boruch (eds.), *Evidence Matters: Randomized Trials in Education Research*. Washington, DC: Brookings Institution Press.

Eldersveld, Samuel J. 1956. "Experimental Propaganda Techniques and Voting Behavior." *American Political Science Review* 50 (March): 154–65.

Freund, John E. 1971. *Mathematical Statistics*, 2nd ed. New York: Prentice-Hall.

Gerber, Alan S. and Donald P. Green. 2000. "The Effects of Canvassing, Direct Mail, and Telephone Contact on Voter Turnout: A Field Experiment." *American Political Science Review* 94: 653–63.

Green, Donald P. and Alan S. Gerber. 2003. "Reclaiming the Experimental Tradition in Political Science," in *The State of the Discipline III*, Helen Milner and Ira Katznelson (eds.), Washington, DC: American Political Science Association.

Gosnell, Harold F. 1927. *Getting-Out-The-Vote: An Experiment in the Stimulation of Voting*. Chicago: University of Chicago Press.

Heckman, James J. and Jeffrey A. Smith. 1995. "Assessing the Case for Social Experiments." *Journal of Economic Perspectives* 9 (Spring): 85–110.

Howell, William G. and Paul E. Peterson. 2002. *The Education Gap: Vouchers and Urban Schools*. Washington, DC: Brookings Institution Press.

Imbens, Guido W., Donald B. Rubin, and Bruce I. Sacerdote. 2001. "Estimating the Effect of Unearned Income on Labor Earnings, Savings, and Consumption: Evidence from a Survey of Lottery Winners." *American Economic Review* 91(4): 778–94.

Katz, Lawrence F., Jeffrey R. Kling, and Jeffrey B. Liebman. 2001. "Moving to Opportunity in Boston: Early Results of a Randomized Mobility Experiment." *Quarterly Journal of Economics* 116: 607–54.

King, Gary, Robert O. Keohane, and Sidney Verba. 1994. *Designing Social Inquiry: Scientific Inference in Qualitative Research*. Princeton: Princeton University Press.

Kramer, Gerald H. 1970. "The Effects of Precinct-Level Canvassing on Voting Behavior." *Public Opinion Quarterly* 34 (Winter): 560–72.

Levitt, Stephen D. 1997. "Using Electoral Cycles in police hiring to Estimate the Effect of Police on Crime." *American Economic Review* 87(3): 270–90.

McConahay, John B. 1982. "Self-Interest versus Racial Attitudes as Correlates of Anti-Busing Attitudes in Louisville: Is it the Buses or the Blacks?" *Journal of Politics* 44(3): 2–720.

Morton, Rebecca. 2002. "EITM: Experimental Implications of Theoretical Models." *The Political Methodologist* 10(2): 14–16.

Rosenstone, Steven J. and John Mark Hansen. 1993. *Mobilization, Participation, and Democracy in America*. New York: Macmillan Publishing Company.

Sherman, Lawrence W. and Dennis P. Rogan. 1995. "Deterrent Effects of Police Raids on Crack Houses: A Randomized, Controlled Experiment." *Justice Quarterly* 12(4): 755–81.

Slemrod, J., M. Blumenthal, and C. Christian. 2001. "Taxpayer Response to an Increased Probability of Audit: Evidence from a Controlled Experiment in Minnesota." *Journal of Public Economics* 79(3): 455–83.

Weiss, Carol H. 2002. "What to Do until the Random Assigner Comes," in Frederick Mosteller and Robert Boruch (eds.), *Evidence Matters: Randomized Trials in Education Research*. Washington, DC: Brookings Institution Press.

13 Concepts and commitments in the study of democracy

Lisa Wedeen

Introduction

This chapter imagines a conversation among three different communities of scholarship on democracy. One approach, what we might call scientific studies of political economy, tends to explore the relationship between economic development and democracy (or sometimes more broadly, regime type); it uses formal theoretical models combined with statistical ones for empirical testing in an effort to come up with law-like rules or patterns governing political behavior. Another approach, framed by an interpretive, philosophical commitment to understanding the relationship between words and politics, examines the conceptual conundrums and meanings associated with words such as democracy. Still another approach, also termed "interpretive," investigates the substantive activities undertaken by individuals and/or groups comprising the political order – the everyday practices and systems of signification associated with democracy in particular places. Each is dedicated to solving problems of abiding relevance to empirical politics, and each employs one or more methods to do so. Yet the kinds of questions, the terms of debate, the definition of democracy, the importance of science, the role methods play, and the underlying epistemological commitments animating each differ in critical and recognizable ways. This chapter considers the questions that are enabled or foreclosed in opting for one approach over the other. It asks: what are the scholarly and political stakes involved in thinking about democracy in ways that emphasize scientific methods or that tackle long-standing theoretical confusions? How can the example of democracy help us consider the ramifications of embarking on a large-N

I owe thanks to Michael Dawson, Patchen Markell, Don Reneau, and Anna Wuerth for careful readings of all or parts of this chapter. An earlier version of this chapter was presented at Yale University in December 2002. At that conference, I benefited from suggestions made by Arjun Appadurai, José Antonio Cheibub, William E. Connolly, John Ferejohn, Anne Norton, Lloyd Rudolph, Susanne Rudolph, James C. Scott, and Ian Shapiro. I dedicate this essay to the memory of Jar Allah 'Umar and Rick Hooper.

analysis, on the one hand, or case study analyses, on the other? What combinations are possible and fruitful?

The chapter is divided into three parts. Part I examines an exemplary large-N, methodologically motivated analysis in which a minimalist definition of democracy as contested elections is operationalized. Part II discusses how a Wittgensteinian approach complicates the picture of what democracy means, offering us potentially valuable lessons about how we think about politics. Part III demonstrates how interpretive methods can be used to grasp substantive everyday political practices of contestation outside of electoral channels. Underlying each approach, I argue, are an important ideological commitment and scholarly understanding of the discipline, which have consequences for what politics means and how political science positions itself in relation to actual political conundrums.

I. Scientific studies of political economy

In a recent and influential book, *Democracy and Development: Political Institutions and Well-Being in the World, 1950–1990*, Adam Przeworski, Michael E. Alvarez, José Antonio Cheibub, and Fernando Limongi (2000) attempt to assess the impact of regime type on "economic well-being" or "economic performance." Przeworski *et al.* thus define the terms and develop the classificatory rules that are then applied to historical observations of 141 countries between 1950 (or the year of the country's independence) and 1990, as well as to a smaller data set used in the rest of the book. They find that transitions to democracy can occur in poor and wealthy countries, but the probability that democracies will survive increases as per capita incomes get bigger. Conversely, poor countries might make a transition to democracy, but they are less likely to sustain an electoral system over time (273). Moreover, regimes that successfully consolidate the electoral process generate more wealth and wealth brings benefits, such as marked improvements in public health and medical technology (277). Poverty is conducive to dictatorship, which makes lives miserable (277).

The authors rely on a Schumpeterian "minimalist" and formalistic definition of democracy as "a regime in which those who govern are selected through contested elections" (Przeworski *et al.* 2000: 15; see also Schumpeter 1942). Operationalizing this definition entails the specification of classificatory rules: the chief executive and legislature must both be elected in a multiparty system in which alternation in office via elections is observable. Contestation exists when there is an opposition that has "some chance of winning office as a consequence of elections" (16). Alternation in office is "*prima facie* evidence" of such contestation, which entails "*ex-ante* uncertainty," "*ex-post* irreversibility," and "repeatability" (16). The authors argue that uncertainty does not imply unpredictability because "the probability distribution of electoral chances is typically known. All that is necessary for outcomes to be uncertain is that it be possible for some incumbent party to lose" (17). Outcomes are uncertain because even though Mayor Daley of Chicago never loses, for example, the rules and procedures are institutionalized so that he could, in principle, lose.

The formalistic, minimalist definition has much to recommend it if the goal is to pursue a large-N, transhistorical study. Yet despite its attractions and value, the definition also has limitations for scholarly thinking about democracy. One problem is that the definition rules out degree-oriented notions of democracy that better describe the dynamics of lived political experience in many cases. Relatedly, by relying on a procedural notion of democracy that is itself derivative of the methods used to measure it, the authors also make troublesome claims that do not hold up under scrutiny. One such contention is that alternative definitions of democracy give rise to "almost identical classifications of the actual observations." The authors' classification system, "though more extensive than most . . . differs little from all others with which we could compare it" (Przeworski *et al.* 2000: 10).

But that assertion *cannot* be true, if those definitions are really different or "alternative." Rather, the authors' approach seems to require a minimalist conceptualization, which, in turn, generates particular kinds of classificatory rules that then necessitate a binary or nominal classification system. Or put differently, a binary classification system – in which a regime is either a democracy or a dictatorship – results from a shared reliance on a formalistic, procedural notion of what democracy *is*. Other definitions *could* conceivably yield different classification systems, but the authors' conceptualization forecloses such possibilities. The authors are thus able to analogize a ratio scale to "the proverbial pregnancy. . . . Democracy can be more or less advanced, [but] one cannot be half-democratic: There is no natural zero point" (Przeworski *et al.* 2000: 57). In their view, scholars who attempt to argue that democracy is a continuous feature over all regimes are simply wrong (e.g., Bollen and Jackman 1989: 612). Although some democracies are more advanced than others, dictatorships cannot be partially democratic. This binary formulation of democracy is true only if one accepts that the nominal classification system Przeworski *et al.* provide is what democracy *means*. This conceptualization seems too narrow to capture democracy's substantive connotations – the fact of its grammar and the conventions of its use.

There *are*, of course, definitions of democracy that might generate alternative systems of classification. As Ian Shapiro notes:

> Democracy means different things to different people. Sometimes it is identified with a particular decision rule; at other times it conjures up the spirit of an age. Democracy can be defined by reference to lists of criteria (such as regular elections, competitive parties, and a universal franchise), yet sometimes it is a comparative idea: the Athenian polis exemplified few characteristics on which most contemporary democrats would insist, but it was relatively democratic by comparison with other ancient Greek city-states. Many people conceive of democratic government in procedural terms; others insist that it requires substantive – usually egalitarian – distributive arrangements. In some circumstances democracy connotes little more than an oppositional ethic; in others it is taken to require robust republican self-government. And whereas some commentators insist that collective deliberation is the high point of democratic politics, for others deliberation is an occupational hazard of democracy. (Shapiro 1999: 17)

Imagine adopting a definition of democracy that stresses aspects of substantive representation rather than simply the existence of contested elections. Substantive representation implies that citizens have control over what a government does, and that governments are continually responsive and accountable to them. Iris Marion Young, for example, suggests a working definition of democracy in which citizens influence and have a direct political impact on the choices and actions of those who govern. For Young, there are regimes in which such connections are strong and those in which they are weak (Young 2000: 173; see also Cunningham 1987, 2001). In this view, citizens might have more or less control over rulers, and governments might be more or less responsive to citizens – or responsive in some contexts but not in others. Young's definition would presumably yield a system of classification based on a continuum (or arguably, on an ordinal system) rather than on the dichotomous system favored by Przeworski *et al.* And although contested elections might be one way to ensure responsive and accountable government, there is no necessary connection between contested elections and responsiveness and accountability.

Using a formalistic definition thus has an additional limitation – it tends to obscure concerns of central importance to substantive representation, such as how democratic rulers should act once elected, and what their duties and obligations as rulers are, or should be (Pitkin 1967; Nikolova n.d). As Hanna Pitkin notes, a formalistic view of what she calls "representative government" operates like a "black box shaped by the initial giving of authority, within which the representative can do whatever he pleases" (Pitkin 1967: 39). This view, what Pitkin calls the "authorization" position, makes elections the key criterion of representative democracy; representation is "seen as a grant of authority by the voters to the elected officials" (Pitkin 1967: 43). The criterion of political success becomes, rather than governance as such, the ability to capture elections. The problem with the formalistic approach from this perspective is that "if representing means merely acting with special rights, or acting with someone else bearing the consequences, then there can be no such thing as representing well or badly" (Pitkin 1967: 43). In other words, there can be no such thing as continual accountability or immediate responsiveness, no account of how those elected actually govern, and no standards for assessing whether their policies work for or against the citizens who have elected them.

In fact, the treatment of accountability in Przeworski *et al.* (1967) is inconsistent. In the first chapter, they state: "governmental responsibility either directly to voters or to a parliament elected by them is a defining feature of democracy" (15). Yet subsequently they also make the opposite claim that "accountability," "responsiveness," or "representation" should not be treated as definitional features of democracy (33).[1] The contradiction reveals some of the tensions inherent in formalistic understandings. On the one hand, there are good reasons for not including accountability in a definition designed to facilitate large-N coding – the term accountability is itself muddled, and it may be hard to measure. On the other hand, formalists should have no difficulty accepting a view of accountability and responsiveness in which holding politicians accountable simply means reelecting them or removing them from office (see Pitkin 1967). Both the authorization and the accountability view are "formalistic in the sense that their defining criterion for representation lies outside the activity of representing itself, before it begins or after it ends" (Pitkin 1967: 59).

A substantive view of accountability is more like Young's, in which citizens have an impact on what government officials actually do *while they are doing it*. Przeworski *et al.* may be sacrificing a key feature of what distinguishes democracy from other forms of governance, namely, that the relationship between ruler and ruled in a democracy is one that is representative in the substantive, active sense of the term. As Pitkin points out, "we show a government to be representative not by demonstrating its control over its subjects, but just the reverse, by demonstrating that its subjects have control over what it does" (Pitkin 1967: 232). Citizen control, in a substantive representative sense, would have to be exercised continuously through institutional and practical mechanisms, which may include elections, but cannot be reduced to them.

[1] I am grateful to Boriana Nikolova for bringing this contradiction to my attention.

Another problem with Przeworski *et al.*'s notion of electoral contestation is that it does not require widespread political participation. This absence may facilitate coding transhistorically, but it denies the context-specific ways in which we customarily understand democratic relations between rulers and ruled. Would a country in which contestation meant competition among a small, restricted elite be a democracy in any meaningful sense of the term today? Surely the formal (and perhaps *de facto*) disenfranchisement of large numbers of people disqualifies a political regime from being democratic in most of our ordinary language uses of the word. If highly restricted suffrage can nevertheless count as democracy, then the reasons for insisting on a multiparty system seem arbitrary and unhelpful. It is unclear why a one-party system could not be democratic if divergent interests might be represented. Indeed, as Boriana Nikolova points out, it is not even clear why, according to this minimalist view, elections should be held at all if divergent interests can be represented without citizens voting (Nikolova n.d.).

Przeworski *et al.*'s adoption of a thin or minimalist definition is, in part, a way of facilitating coding in the interests of scientific objectivity. Yet ideological biases trouble claims of impartiality, and the authors' belief in the inherent value of science as a method of producing objective truth about the real world coincides with a defense of US liberalism. For example, Przeworski *et al.* criticize Dahl's use of participation in his definition of democracy and justify their own exclusion of the phenomenon because Dahl sets the threshold of participation "too high," thereby disqualifying the United States "as a democracy until the 1950s" (Przeworski *et al.* 2000: 34). It is unclear, however, why excluding the United States is a problem *per se*. Indeed, Przeworski *et al.*'s *Democracy and Development* occasionally reads as if it were an advertisement for the Anglo-American system – "the yearnings of an ideology seeking repose" – to borrow from Sheldon Wolin's description of John Rawls' theory of liberalism, rather than the disinterested study it purports to be (Wolin 1996: 117; cited in Honig 2001: 798).

To put it bluntly: these problems suggest that Przeworski *et al.* might be better off avoiding the term "democracy" altogether. The overall objective of this large-N study would then be to explain the relationship between contested elections and economic development, without producing a general account of democracy that would seem to foreclose thinking about political participation and accountability within and outside of electoral confines. Yet even the notion of contested elections may require some account of the conditions that enable its exercise, such as the possibilities for participation in the electoral system, access to information and political office, the range of choices among candidates, and the lived political experiences of apathy, despair, impotence, joy, and accomplishment that determine whether elections are actually competitive in any meaningful sense of that word.

By identifying the general relationship between regime type and economic well-being, *Democracy and Development* and projects similar to it evacuate politics of the messy stuff of political contestation – of initiative, spontaneity, self-fashioning, revelation, ingenuity, action, creativity – which occurs *outside the domain of electoral outcomes*. It is possible to view democracy in terms of elections and recognize other forms of vibrant political life, but what is at issue is whether those other forms must be studied or can safely be disregarded in the interests of science. Election-oriented approaches make measuring democracy easier than it otherwise would be by identifying observable rules, thereby allowing the classification and comparison of a large number of cases (Schaffer 1998: 3). Large-N studies of democracy contribute to our knowledge about the effects of contested elections on economic performance, and about how regime types or particular electoral arrangements determine political outcomes (Schaffer 1998: 4; on the latter, see Lijphart's oft-cited discussion of PR systems, 1984). They are incapable, however, of capturing the substantive activities that define people's everyday political experiences. We need additional methods and a reconsideration of our concepts if we are to engage in the study of politics. As Wittgenstein shows us in his examination of how we learn and use language, the point is not that a methodologically driven definition of democracy as contested elections is wrong, but that the assumptions underlying such analysis miss something important about what we call democracy.

II. Wittgenstein and ordinary language use analysis

In the *Philosophical Investigations*, Ludwig Wittgenstein opens with a passage from Augustine's *Confessions* and demonstrates how Augustine's account of how humans learn language relies on a number of unjustifiable and incorrect assumptions, ones that Wittgenstein shared in his earlier work (Pitkin [1972], 1993: 31). Augustine presents us with a "picture of language," in which words are essentially the names of objects in the world, so that "naming something is like attaching a label to a thing" (Wittgenstein [1958], 1998: par. 15, 7e). "The individual words in language name objects. . . . Every word has a meaning. This meaning is correlated with the word. It is the object for which the word stands" (Wittgenstein [1958], 1998: par. 1, 2e). According to this common and seductive view of language, words signify or refer; they stand for an object

(Pitkin [1972]: 1993: 31). We might, for example, point to a doll and pronounce the word, self-consciously attempting to teach a child the relationship between the word and the thing itself.

Certain words, however, cannot be taught by "ostensive definition" (Pitkin [1972], 1993: 31–36). Children acquire words like "the" or "whether" or "despite" without being taught by pointing to objects or associating labels with a visible sign. Moreover, although one can point to a spouse one cannot teach the meaning of spouse by pointing to one (Pitkin, 1972: 30–50). The meanings of abstract concepts such as "democracy," "justice," "the nation," or "allegiance" are also not learned by pointing to an object, but rather by speaking the words and responding to speech in the context of their use. According to Wittgenstein, "What determining the length means is not learned by learning what length and determining are; the meaning of the word 'length' is learnt by learning, among other things, what it is to determine length" (Wittgenstein [1958], 1998: 225).

A concept deserves Wittgensteinian attention not because theorists have got it right and have used it to explain reality, but because the confusions and arguments the concept generates reveal to us important tensions in how we think and act in the world. Knowing how a word is used is crucial in knowing what the phenomenon in question is (Pitkin 1967: 9–11). The point, then, is not to arrive at an authoritative definition of democracy, but rather to explore how our different uses imply various understandings of what a democratic politics entails.

Wittgenstein advises us to ask: on what occasions and for what purposes do we say a particular word? What kinds of actions accompany these words? Imagine applying these questions to a concept, such as "democracy." We might begin by tracing the etymological roots of the word, because a word rarely shrugs off its etymology completely (Austin 1961: 149; cited in Pitkin 1972: 10). We might chart the ways in which the word has been used over time in ordinary language, in scholarly texts, or both. We might then think of its "family resemblance" to other words, the contexts within which we might say "equality," "freedom," "participation," "elections," or "contestation" rather than "democracy." What are the synonyms and antonyms associated with the concept? When do we use the concept in its verb forms and when do we use it as a noun? What do these uses tell us about the word?

Wittgenstein liked to say that such analyses can help "clear the fog," to understand the source of scholarly confusions in order to eliminate them. An ordinary language use analysis might also be employed to examine the meaning of a concept such as "democracy" in other countries. For example, in *Democracy in Translation* Frederick Schaffer uses the case of Senegal to explore the differences between elite and ordinary citizens' understandings of what democracy means. Although Schaffer problematically attributes these differences to "culture," the fieldwork evidence suggests that class affiliation may be of paramount importance: Senegalese elites tend to invoke the word "democracy" (or its French equivalent *democratie*) in ways similar to Schumpeter and Przeworski. A democratic system is one in which elections are contested and outcomes are uncertain. Lower-class, less-educated Senegalese use the Wolof equivalent, *demokaraasi*, to mean "equality" or equal distribution of material resources, however. Schaffer shows that the meanings of democracy may have consequences for how elections are implemented and what sorts of reactions they evince. He also suggests that attending to lived experience requires thinking through what democracy means in class contexts. Whereas Przeworski *et al.* have offered us a scientifically rigorous study of whether contested elections, in their definition, do or do not correlate positively with wealth, this is only partly what democracy means or what citizens understand it to mean.

Imagine a Wolof speaker who understands *demokaraasi* to mean equality interacting with a Przeworskian who wants to ask the question of whether democracy generates economic well-being or wealth. The question would hardly make sense because both the independent and dependent variables are too similar. Indeed, the Wolof speaker might understand Przeworski *et al.* to be asking whether economic well-being generates economic well-being. (In fairness, the authors' formulations of economic well-being are much less minimalist than their definitions of democracy, and therefore, the indicators are more varied than those selected for democracy.) Or consider Malcolm Bradbury's *Doctor Criminale* (1992). Gertla Riviero, the Hungarian ex-wife of a postmodern philosopher ruminates on the recent transitions to democracy and free market economics: "'Democracy, the free market,' she muses, 'do you really think they can save us? . . . Marxism [was] a great idea, democracy [is] just a small idea. It promises hope, and it gives you [Kentucky] Fried Chicken'" (cited in Comaroff and Comaroff 1997: 124; Bradbury 1992: 276). As anthropologists Jean Comaroff and John Comaroff argue:

> Gertla Riviero's question carries an obvious, ominous punch, precisely because it calls into doubt our taken-for-granted narrative of democratization, a heroic liberal myth which links the conventional practices of modernist politics to the prospect of material and social salvation. So, too, if in a different way, does the image of a patient, passive people standing in millennial lines to choose either cheap food or political candidates. (Comaroff and Comaroff 1997: 124)

One may disagree with the connection of fatty fast food preferences with preferences for political candidates. Indeed, one can imagine Przeworski *et al.*'s study being used to demonstrate how wrong the fictional Ms. Riviero is. But one can also imagine a fictional Ms. Riviero questioning a data set on democracy that ends in 1990 (a year after the onset of arguably the greatest global transformation of the last half century), or wondering how a study on "democracy" can paper over contemporary experiences of alienation, passion, outrage, and despair. Moreover, both scientifically minded scholars of politics and those who have experienced transitions to electoral regimes and a free market economy should demand a fuller account or theory of the mechanisms that could make elections generate economic well-being when they do. The absence of a theoretical story from the account is by no means a product of the methods used, and a Wittgensteinian ordinary language analysis is ill-equipped to provide a story of mechanisms, whereas other scientific large-N analyses might (see, e.g., Boix 2003; Stokes and Boix 2003).

What a Wittgensteinian ordinary language use analysis helps us do is to question our taken-for-granted assumptions when conceptual puzzlement prevents or "distracts us from our work" (Pitkin 20: 1972). Such an analysis might be used, as some political theorists use it, to achieve conceptual clarity, and by extension, to teach us how we think. By making these assumptions explicit, they can be evaluated against alternative perceptions of the world in various geographical locations, not with the aim of resolving conflicting phenomena, but with the objective of doing justice to them (Pitkin 1972: 23). In this sense, Wittgenstein allows us to think dialectically (23). By "dialectical," I mean a relationship in which two terms are defined and generated with reference to each other, yet can come into conflict, both conceptually in their meanings and causally in the world. The only way of handling such material is by synthesis – i.e., by maintaining an overview that includes both sides without stifling the conflict or denying their logical incompatibilities (Wedeen 2002).[2] Applied to democracy, a dialectical approach allows us to see how people are *both* authors and addressees of the law (Habermas 2001; Honig 2001), or how democracy can entail both rule by the people and the rule of law – both popular sovereignty and representative government. We can thus begin to understand how concepts reflect tensions between ideal standards and their practical realization, and how opposing definitions can have a plausibility that permits us to identify separately with the arguments of each (see Pitkin 1967).

[2] I am indebted to Hanna Pitkin for encouraging me to foreground this part of my argument in a different article.

III. Interpretive methods two: semiotic practices

(A) *About interpretation (briefly)*

Wittgensteinian ordinary language uses analysis is one way to practice "interpretive" social science, but there are others. Indeed, the label may be so elastic as to refer to everything and nothing at the same time.[3] Yet there *are* at least four characteristics uniting interpretivists, despite their differences. First, interpretivists, to borrow from Michel Foucault, question the "power that is presumed to accompany . . . science" (Foucault 1980: 84). Second, interpretivists are interested in language and other symbolic systems, what is sometimes termed "culture" in the literature. Third, interpretivists are also "constructivists" in the sense that they see the world as socially made, so that the categories, presuppositions, and classifications referring to particular phenomena are manufactured rather than natural. There is no such thing as ethnicity or race, for example, outside of the social conditions that make such classifications meaningful. The task of an interpretivist may be, then, to see the sort of work these categories do, while accounting for how they come to seem natural when they do. Fourth and relatedly, interpretivists tend to eschew the individualist orientation characterizing rational choice and behaviorist literatures. Although some interpretivists stress the importance of agentive individuals, they do not assume a maximizing, cost-benefit calculator who is unproblematically divorced from actual historical processes (e.g., Bourdieu 1978). Ideas, beliefs, values, and "preferences" are always embedded in a social world, which is constituted through humans' linguistic, institutional, and practical relations with others (Wedeen 2002). Wittgensteinian ordinary language philosophy is one kind of interpretative endeavor that tries to get at how people think by analyzing the ways in which they use concepts in context. A semiotic-practical approach, however, investigates what language and other symbolic systems *do* – how they are inscribed in practices that operate to produce observable political effects (Wedeen 2002). Thus, whereas a Wittgensteinian might explore how the concept of democracy is used and the conceptual puzzlements the term brings to the fore, other interpretivists might be more interested in the ways in which electoral or other participatory practices work in the world.

Interpretive methods may provide "thick descriptions" of lived political experience. They can also be used to produce explanatory arguments

[3] For a helpful discussion of interpretive social science, see Rabinow and Sullivan, eds., 1987.

and causal stories. Interpretivists can explain why some political ideologies, policies, and self-policing strategies work better than others; and why particular material and status interests are taken for granted, viewed as valuable, or become available idioms for dissemination and collective action. Interpretive methods can help us explain the mechanisms by which political identifications with particular forms of government get established, shift, or lose their importance. They can show how rhetoric and symbols, such as campaign advertisements, political speeches, and iconography, not only exemplify but also produce political support, or simply compliance (Wedeen 2002). Most importantly for our purposes here, they can identify and analyze participatory practices through which people experience their lives as political when they do.

In relation to the study of democracy, an interpretive approach that emphasizes semiotic practices allows us to shift our attention away from the minimalist, formalist notion of electoral procedures to animated worlds of peaceful contestation and political argumentation outside of electoral arrangements, to other dimensions of what might be construed as "democratic" practice. In what follows, I explore the reproduction of what Barrington Moore and Ian Shapiro have both termed "loyal" opposition (Shapiro 1999: 39; Moore cited in Shapiro 1999: 39) through everyday qat chew practices in Yemen. The example of Yemen demonstrates how a formalistic account fails to capture lived political experience in which animated public sphere activities operate to produce important forms of political engagement and participation. I end by suggesting why such vibrant forms of public sphere activity do not lead to contested elections, and thus return to the large-N work that best supplies possible answers to that question.

(B) The case of Yemeni qat chews as public sphere activities

Yemen cannot be considered "democratic" in the formalistic or "minimalist" sense of the term. The president is not selected through contested elections, and there has been no alternation in the executive since 1978. Viewed from a perspective of electoral contestation allowing for assessments of more or less democracy, Yemen is more democratic than most countries in the Middle East by virtue of a weak, but functioning, opposition made manifest in political parties that compete for votes in local and parliamentary elections. Political parties make efforts to satisfy constituent demands and to encourage the electorate to vote, but serious limitations exist. Opposition candidates compete under conditions that favor the ruling party, although victories in certain districts do occur, and Parliament's powers remain severely circumscribed by the executive.

Viewed from a perspective of citizen influence and impact, moreover, the fact of an armed population, of a courageous, albeit occasionally harassed press, and of civic associations operating independently of state control makes Yemen one of the most democratic countries in the Middle East. Human rights associations and related groups pressing for women's rights, a free press, fair treatment of prisoners, and a thorough and transparent judicial process hold conferences, sponsor intellectual debates, and publish articles in local newspapers. Mosque sermons often address social inequalities and criticize the government openly, paying particular attention to practices of political corruption and moral laxity. No visitor to Yemen can help but notice vigorous forms of non-electoral contestation that animate daily life (Carapico 1998).

The presence of political activism is not synonymous with democracy, however. This chapter does not intend to suggest, because Yemen is more democratic in some ways than other countries in the Middle East, that it *is* a democracy in either the sense of possessing fair, contested elections or adequate experiences of substantive representation. Rather, I am arguing that there are different sites of the enactment of democracy, and a strong democracy needs them all. The example of Yemen demonstrates that any political analysis that fails to take into account participation and the formation of "public spheres" as activities of political expression in their own right falls short of what a democratic politics might reasonably be taken to include. Thus, a consideration of democracy also requires theorizing about aspects of substantive representation that are evident in Yemen, namely, the widespread, inclusive mobilization of critical, practical discourses in which people articulate and think through their moral and material demands in public.

By investigating qat chew conversations as instances of what Habermasians call "public sphere" practice, we can specify the role everyday activities of political discussion play in Yemen, as well as perhaps more generally. Qat chews fulfill the two requirements Habermas sees as central to a "public sphere": that citizens engage in rational, critical discussion, and that mini publics, in tandem with other mini publics, play a mediated, reflexive role, helping to produce the impersonal, audience-oriented broader public of anonymous citizens.[4] An interpretive engagement with the example of qat chews in Yemen can also draw our attention to some of the *problems* with Habermas' formulation of the "public sphere," problems that have tended to be neglected by Habermas' critics and by deliberative democrats alike. I end by suggesting why such vibrant public sphere

[4] Thanks are owed to Iris Marion Young for pressing me to emphasize this aspect of Habermas' concept.

activities do not lead to contested elections, and thus return to the large-N work that best supplies a possible answer to that question.

Qat chews are political in at least three senses: first, they are a forum for negotiating power relationships between elites and constituencies, in which elites are not only held responsible, but also must be responsive to the needs of participants by guaranteeing goods and services – or by advocating on behalf of the village, electoral district, and/or local group. Second, during some qat chews, actual policy decisions get made: government officials, opposition parties, and networks of activists meet separately and with one another to decide how elections should be run, which alliances will be forged, what political positions taken, statements drafted, and grievances voiced and heeded. Third, people make avid use of qat chews to share information about political events and to discuss their significance publicly.

Although approaching these practices through Habermas encourages an understanding of them in terms of the structural historical conditions under which they have arisen, these conditions need not be the same for formal, everyday practices of deliberation to emerge. Or put differently, there is no one path to critical, rational debate in public, or if there is, then Habermas misidentified the circumstances under which the "subjectivity" – or conscious presentations of self – conducive to such deliberation appears. Different historical conditions of possibility from the ones Habermas specified in Europe have allowed for the emergence of vibrant, politicized, critical communities of argument in Yemen.

For Habermas, the term "public sphere" connotes a set of places (coffee houses and salons) where bourgeois citizens historically met to argue about literary and political matters; the substantive activity of private persons coming together as a public for purposes of rational-critical debate; *and* the mediated, reflexive ways in which these critical conversations in public places referred to, and actually influenced, events and arguments appearing in print.[5] Public debate, according to Habermas, "*was supposed to transform* voluntas *into a* ratio *that in the public competition of private arguments came into being as the consensus about what was practically necessary in the interest of all*" (emphasis in the original, 83). As substantive activity, the public sphere held out the promise of transcending the narrow confines of particular class interests because conversations tended to allow private bourgeois persons to come together as a public and to be "nothing more than human" in both their interactions with

[5] Habermas locates these new institutions in Great Britain and France: coffee houses emerged as popular gathering places in around 1680 and 1730, respectively, and the salons became important sites of critical debate in the period between regency and revolution.

others and in their "self-interpretations" (48). Or, put somewhat differently, the "fully developed public sphere" was a fiction, according to Habermas, that entailed the historical identification of "the public" of property owners with the more generalized notion of a public of "human beings pure and simple" (56). Thus the specifically bourgeois character of the actual public sphere could, *in principle*, be broadened to include everyone by generating the vocabulary, attitudes of equality, and activities of critical debate congenial to the "political freedom of the individual in general" (56).

Habermas's analysis of the public sphere also explores the ways in which institutions produce certain "subjectivities" that then become crucial for everyday substantive participation in political life. Scholars have wondered whether and how subjectivity matters for democracy. Some argue that democracy requires democrats, whereas others suggest that transitions to electoral regimes and their sustainability over time do not necessitate people committed to elections (Waterbury 1994; Kalyvas 1998). According to the latter view, institutional dynamics take on their own momentum so that the theologically minded Belgian Catholic movement, for example, ended up abiding by the rules of the democratic game despite members' anti-democratic inclinations (Kalyvas 1998). The consolidation of Belgian democracy did not result from a process of ideological or civic learning, according to this argument, but from "purely self-interested actions" (Kalyvas 1998: 317). In this view, people's ideas, ideologies, and identities matter less than the institutions within which they operate. Still, what counts as self-interest, how and why citizens become invested in particular institutional arrangements and not in others, and how and why institutions themselves produce the everyday self-enforcing practices conducive to electoral consolidation – these are issues central to interpretive social science. Why do people come to see their self-interests as best served by a state? Why don't citizens always view democracy as in their self-interest? What makes citizens risk tumult when they do? What does self-interest mean in different contexts? And what kinds of conversations or political arguments among members of non-democratic organizations or movements, such as the Belgian Catholic one, help produce circumstances under which non-democrats can imagine contested elections as being in their "self-interest"?

For Habermas, the institution of the family in eighteenth-century Western Europe played a key role in the formation of individuals whose audience-oriented sense of themselves made a public sphere possible. The substantive activities of rational debate within the public sphere, in turn, seem to have provided the felicitous conditions for popular sovereignty and the rule of law to flourish. Habermas argues that "the public understanding of the public use of reason was guided specifically by such private experiences as grew out of the audience-oriented subjectivity of the conjugal family's intimate domain" (28). The family was thus the seat of the production of a specifically bourgeois subjectivity that helped make the public sphere possible.

Habermas concedes that the notion of the family as a private world was based on a denial of the economic relations supporting the family's very existence. But the fiction was useful in establishing the conditions of subjectivity that would enable private men to join together in public, rational debates, to be audience-oriented, empathic, self-knowing, and capable of what Arendt would call "representative" thinking – seeing the world from others' points of view (Arendt 1961 [1954]: 241). The autonomy that property owners enjoyed in the relatively free liberal market "corresponded" to a "self-presentation of human beings in the family" (Habermas, 46). Individuals thus came to experiment with their subjectivities by writing letters, as well as diaries, autobiographies, and domestic novels. In short, the conditions of intimacy within the "patriarchal conjugal family" and of property-owning autonomy in the market permitted the individual to "unfold himself in his subjectivity," which was both an intensely private expression of an individual's interiority and also already oriented to an audience (*Publikum*) (49).

Critics of Habermas have pointed out that he conceives of the public sphere in the singular, thereby failing to consider the possibility of a multiplicity of public spheres, many of which are "subaltern" or "counter" discursive spaces that challenge the prevailing ways of thinking in the critical mainstream one.[6] Many scholars have also argued that Habermas exaggerates the possibilities for equal treatment in public and neglects to theorize the ways in which each participant is enmeshed in particular power relationships affecting his/her ability to speak and to be heard.[7] Social and political inequalities, even among people from the same economic class, help determine who speaks, when, and how convincing arguments are.

More importantly for our purposes here, I take issue with Habermas' location of the sources of modern, public-oriented subjectivity in the

[6] See, for example, Nancy Fraser's oft-cited piece, "Rethinking the Public Sphere: A Contribution to the Critique of Actually Existing Democracy," 1992.

[7] The idealized account of argumentative dialogue is especially apparent in Habermas' subsequent work, *Theory of Communicative Action* (1984, 1987). For criticism of this development, see Lee 1993 and Warner 2002. In addition, there are a number of essays that challenge the fiction of equal access to, and equal conditions within, the deliberative process. See, for example, Young 1996; Mansbridge 1991; and Sanders 1997.

bourgeois family. In order for public spheres to exist, in his sense, individuals need to have the reflexive capacity of critical thinking. Not everyone has such a capacity, and not all societies encourage it. To the extent that there is a public sphere in which critical discursive activities thrive, a significant segment of Yemenis experience it, but its source cannot be the bourgeois family. (Indeed, it is possible that Habermas is wrong about the source of the public sphere for modern Europe.) Whereas Habermas attributes the orientation toward an audience to circumstances within the conjugal family and to transformations within market relations that took centuries of capitalist development to unfold in Western Europe, experiences of subjectivity in Yemen operate in a context where persons most often self-identify and are classified as members of a large extended family or "*bayt*" (see Meneley 1996: ch. 3). The bourgeois individual is hardly the unit of analysis or identification for most people, nor are sources of agency and autonomy necessarily attributable to everyday practices associated with the commodity form. Approximately 75 percent of Yemenis live in the countryside and participate in some form of agriculture, and there exists no large or coherent middle class. In other words, this orientation toward an audience evident in rural and urban qat chew conversations, as well as in intellectual gatherings, the press, and civic associations exists independently of developments in market and family relations central to Habermas' account.

Second, this vibrant public sphere activity has led neither to contested elections nor to effective governance. Although public spheres may be a condition of possibility, in the short run they do not seem to offer one sufficient to prompt Przeworski *et al.*'s contested elections. Or put differently, although qat chews and other public sphere activities in which people discuss their political worlds represent sites of important political vitality, the lively and obvious presence of such public arenas in the absence of fair and free contested elections suggests that such everyday practices may not be as instrumentally related to contestatory electoral arrangements as some political scientists would claim – or at least not in the short run (see, e.g., Putnam 1993).[8] Indeed, one might argue that they operate as a "safety valve," allowing people to vent frustrations and

[8] For an important critique of Putnam's idea of civil society, see Cohen 1998. In that paper, Cohen argues that Putnam's notion of civil society is theoretically impoverished because it ignores the concept of the public sphere, thereby failing to "articulate the complex relation between social and political institutions." For Cohen, the concept of the public sphere is at the core of any conception of democracy. Modern constitutional democracies enjoy "legitimacy" to the extent that "action-orienting norms, practices, policies, and claims to authority can be contested by citizens and . . . affirmed or redeemed in public discourse" (all from p. 2).

displace tensions that otherwise might find expression in political action. Some Yemenis describe qat gatherings in just this way – as a mode of *tanfis*, a way of literally "letting out air." Scholars too assert that tolerated or authorized critical practices function to preserve a regime's dominance rather than undermine it (Guha 1983: 18–76; Adas 1992: 301; see also a discussion of *tanfis* in Wedeen 1999).

There are, however, at least two key problems with safety-valve arguments. First, it is impossible to demonstrate that were participants in qat chews not engaged in such activities then they would be organizing collectively against the regime. Second, and more importantly, political rallies and other forms of collective resistance *are* often organized during qat chews. The ideas and organization for oppositional political action arise out of qat chew discussions in which people cultivate shared concerns and act upon them.

Qat, a leafy stimulant drug with similarities in effect to caffeine, is often chewed in the context of structured conversations occurring in public or semi-public places. In large rooms in people's houses or in a civic association's offices, people, some of whom are strangers to one another, meet to debate critically about literary matters, political life, and social problems. Qat chews, like Habermas' depiction of salon and coffee house gatherings, vary in the size and composition of their publics, the mode of their proceedings, and the topics open to debate. They, like their historical Western European counterparts, often entail formal, on-going discussions, which might be prompted by international current events, by local domestic problems, or by a special guest in attendance. In addition, marriages are arranged and business deals solidified, village shaykhs (chiefs) chosen, and local troubles resolved over qat. Qat chews also enable the circulation of important information. The anthropologist Shelagh Weir likens these gatherings to a "kind of institutionalized grapevine" in which information about local and national affairs gets exchanged (1985: 125; see also Sheila Carapico who analogizes qat gatherings to American focus groups, 1998: xi). Qat chews occasion the generation and verification of rumors (including ones about electoral results), the trading of knowledge by peasants about farming techniques and current agricultural programs, the negotiation by religious experts of prevailing theological problems, and the bargaining among elites about policy, political positions, and alliances. Sometimes one issue is discussed in depth; at other times a number of issues may animate the afternoon. Most people, even those who do not chew, are organized in one way or another around the social world of afternoon qat consumption. Although not all qat chew gatherings occasion the discussion of substantive political issues in public, many, particularly those held among both educated and uneducated, rural and urban *men*, do.[9]

Patronage networks are reproduced during some qat chews in which villagers and self-identified tribesmen visit the home of a nationally powerful political leader or local notable in order to secure promises of goods and services. Politicians may use the qat chew to meet with constituents, and legal experts to promulgate legal judgments. Important officials take part in qat gatherings most afternoons and "everyone knows or can easily discover where they can be found. . . . Everyone has the right of access to officials" (Weir 1985: 125). Qat chews, which used to be exclusive parties for aristocrats, have become a generalized, institutionalized form of social life, diminishing the importance of other ways of socializing or securing patronage, and also endowing the qat chew with a formality these gatherings did not previously possess (Weir 1985: 144). Although supplicants may attend chews for the sole purpose of securing a favor, many are nonetheless treated (or subjected) to a sustained conversation about prevailing political matters.

Ordinary, everyday qat gatherings are public, then, in the sense that they are "open to anyone who wants to take part. Men simply choose which party they want to attend, walk in and take a seat; no one questions their presence" (Weir 1985: 124). This openness is due partially to the fact that guests are not a financial burden on a host. Yemenis generally bring their own qat to gatherings. During important ceremonies an affluent host might supply the qat, but occasions when qat is supplied in the absence of a ceremonial event are rare. When they do occur, they are an important way of influencing people by generating prestige for a host whose financial resources and generosity are thereby made manifest (Weir 1985: 123).[10]

9 Like the Bengali practice of *adda* – long, informal conversations – qat chews are considered by many to be a "flawed social practice" because they are segregated and predominantly male activities that are "oblivious of the materiality of labor" (see Chakrabarty's social history of *adda* in *Provincializing Europe* 2000: 181). Unlike *adda*, however, qat chewing is not a dying social practice, a vestige from a class-specific, urban modernist past, but rather an increasingly widespread, contemporary activity.

10 Weir's book, the most comprehensive discussion of qat and its social implications to date, makes observations about the "qat parties" which largely jibe with my own fieldwork experiences. Weir also notes that the host of a large chew may invite some guests to lunch beforehand, or lunch may be hosted independently before a qat chew. Lunches are also ways to manifest generosity and enhance prestige, and they prompt reciprocal invitations or favors from beneficiaries, which an everyday qat chew does not. Guests at lunch are present because of their relationship with the host, and they may not know (or have an interest in knowing) each other. Conversations during lunch are brief and interaction limited. Food is consumed quickly and without extended socializing, so that guests can move on to the day's central activity, the qat conversation. Other studies on qat include a survey by Maxime Rodinson, "Esquisse d'une monographie du qat, *Journal Asiatique*, 265 (1977), 71–96; R. B. Serjeant, "The Market, Business Life, Occupations,

Everyday qat chews, by contrast, are what Weir characterizes as "*free associations of people who have chosen to attend*, not because they are invited, but for a variety of personal reasons" (1985: 124).

The accessibility of officials and the openness of everyday qat sessions are qualities suggestive of the horizontal or equality-inducing aspects of Yemeni public life. Yet qat chews are also occasions for reproducing hierarchy and reaffirming the variations in social status among participants (Gerholm 1977; Weir 1985). The seating positions at a qat chew roughly correspond to each person's status relative to others in the room. Paradoxically, it is the very equality-inducing openness of many qat chews that also reproduces status distinctions by generating occasions in which social classes mix, but hierarchically. In public places accessible to intimates and strangers alike, policies and political practices can be challenged or reinforced through critical discussion, but people's seating positions nevertheless indicate their social position, and perhaps, the seriousness with which their argument is likely to be received.

One might argue that the proliferation of qat chew gatherings as a particular instance of public sphere activity corresponds to identifiable economic and political factors. By all accounts, the number of men and women participating in qat chews has increased over time.[11] In the former People's Democratic Republic of Yemen (PDRY), qat chewing was prohibited on working days, which meant that gatherings were restricted in most places to Thursdays and Fridays. In the former northern Yemen Arab Republic, the influx of labor remittance income in the 1970s resulted in a marked increase in the consumption of qat and of both men and women's qat gatherings (Weir 1985: 85; see also Mundy 1981: 61). Researchers reported the proliferation of formal and informal contexts in which a majority of men and a large minority of women spent their afternoons chewing qat in sex-segregated public sessions, the increasing number of workers and small shopkeepers chewing qat on the job, the expansion in size and number of markets where qat was sold, and

the Legality and Sale of Stimulants," in R. B. Serjeant and G. R. Lewcock, eds., San'a': *An Arabian Islamic City* (London 1983), 159–178; Varisco 1986: 1–13. In Arabic, see the articles in *Al-Qat fi Hayat al-Yaman wa al-Yamaniyin* (San'a' 1981) and 'Abbas Fadil al-Sa'di, *Al-Qat fi al-Yaman: Dirasa Jughrafiya* (Kuwait 1983).

11 There are specifiable groups of Yemeni men who do not chew, however. Self-identified Salafis (people who adhere to a puritanical version of Islam inspired by Saudi Wahhabi traditions) tend not to chew, although there are those who do. Some Western-educated Yemenis also identify qat with "backwardness" and do not chew; some have even organized campaigns to dissuade people from consuming qat. See, for example, the pamphlets published by al-'Afif Cultural Center. In the regions of the Hadramawt and al-Mahra chewing is still considered crass, although more inhabitants are engaging in the activity. One educated Hadrami thought that about 30 percent of men now chew in al-Mukalla', for example.

the growing amount of land devoted to qat cultivation (Gerholm 1977: 53; Tutwiler and Carapico 1981: 49–59; Weir 1985).[12] Despite the dire straits of the bust economy of the 1980s and the subsequent return of an estimated 800,000 to 1.5 million labor migrants from Saudi Arabia in 1991, qat consumption has continued to rise – and thus so too have the occasions for critical debates in public. Even with the constriction of institutionalized electoral possibilities in the aftermath of the brief 1994 civil war, qat chew conversations have flourished as a key enclave of publicity through which frank discussions among politicians and ordinary citizens continue to take place.[13]

Qat chews are democratic in the sense that they are occasions for inclusive participation in which "loyal" opposition can crystallize through critical rational discussions that are refracted through communications media and formal organizations; impersonal media, in turn, also generate topics that become the center of qat chew conversations. During my fieldwork, qat chew conversations afforded opposition politicians the occasion to ask how they should protest against constitutional amendments curtailing democracy (September 2000), and regime officials wondered aloud how to respond to US foreign policy in the aftermath of September 11. Participants from various political parties, ideological persuasions, and regional backgrounds raised questions such as "who will benefit from the September 11 attacks?" (September 2001). "Are ordinary Afghanis better off under the Taliban or under American occupation?" (September 2002). "What sorts of positions should we take on Iraq?" (September 2002). Activists also met to discuss how to organize resistance to proposed amendments (September 2000) and garner support for political parties (September 2002).

[12] Weir estimates that qat production at least doubled during the 1970s (1985: 86). She cites the World Bank Country Report on the Yemen Arab Republic, which notes that there is "ample evidence that qat growing has increased rapidly in recent years" (1979: 93; Weir 1985: 176). A USAID report (10), also cited by Weir, estimated that qat production increased "two to three fold" during the period of the Yemen government's first Five Year Plan (1976/77–1980/81). According to the International Monetary Fund's 2001 Country Report (No. 02/61), qat chewing is a "widespread practice" (104). This IMF report cites a study from the 1960s that "found that in the city of Taiz about 60 percent of males and 35 percent of females were 'habitual chewers.'" The report notes that there is insufficient quantified data on contemporary qat production and use, in part because of "official uneasiness with the importance of qat in Yemeni society," but according to available information, "consumption is in fact widely thought to have increased in recent years"; qat cultivation is estimated to have increased by 10 percent between "1995 and 1990" [*sic*], redistributing wealth from the urban areas to the countryside (104). Approximately 170,000 families (i.e., 1 million people) benefit directly from the sale of qat. The IMF's information on qat is derived from interviews with officials and from an unpublished World Bank paper (Ward *et al.* 1998).

[13] For detailed descriptions of conversations from qat chew gatherings, see my *Peripheral Visions: Political Identifications in Unified Yemen*, in preparation.

Participants engage in dialogue with others, thereby sharpening some views and abandoning others. In the course of a qat chew on "culture and identity," for example, one educated male from a pan-Arab political party began to change his mind about the ways in which people differed from one another. Whereas before, he had thought about identities as ascriptive or "vertical" – "imposed on people so that they do not have a choice, because they do not choose their families, tribes, or nationalities" – in the course of the conversation he began to consider some identities as more chosen or "horizontal." He argued that the former vision was a "traditional" one, whereas a sense of horizontal identities was "modern." Contrary to a pan-nationalist vision held by his colleagues, in which Arab-ness is often construed in primordial or essentialist terms, he began to think aloud about the ways in which distinctions considered to be essential could be provisional and strategic rather than always already there: "juxtaposed to distinctions that are vertical are those that are horizontal, whose bases are moral and cultural . . . and what importantly distinguishes them is that we choose them by our will and with awareness." The critical point here is not that his statements were true or false, but rather that they were revelatory for his politics:

> My words were an important announcement for me; they announced the liberation of my thinking. I had thought that the significance of such discussions about identity was symbolic for me, but I found that the conversation was extremely important as a basis for my discussions about a war against Iraq, about the issue of Palestine – that the example of a horizontal, humanitarian distinction was more important and powerful support for my positions than an appeal to a vertical, given identity [such as Arab origins]. (January 2002; reiterated in a letter, March 8, 2003)

Qat chews occasion reflection and revelation. They do so by providing a home for politicized debate, for entertaining competing perspectives, for discussing the question of "what shall we do?," and for thinking collectively about issues of power and responsibility. They are, in some ways, like Arendt's notion of the political, in which people join together and cultivate the human capacity for action (see especially Arendt's *The Human Condition*). The "public" of the qat chew is thus not simply a social category, but also an existential one (Dietz 1995), a particular context for enacting "personhood" through which participants stage scenes of "self-disclosure" and make worlds in common (Warner 2002: 59). Put differently, qat chews in which politics come to the fore are also occasions for personal performances of courage and eloquence. They are moments of self-fashioning, interpretation, and revelation in which participants stake out positions, define themselves for others, and take friends and foes, leaders and peers to task.

Whereas for Habermas the fiction of free and equal individuals defines the public sphere and distinguishes it from other spaces and forms of activity, in Yemen it is the multiclass, peaceful nature of political debate that distinguishes public spheres from other worlds where violence is the ultimate arbiter and force is privately held. Yemeni political subjectivity remains hierarchical in the sense that people are self-conscious about socially stratified familial relationships, occupations, and reputations – all of which may find expression in the seating arrangements at qat chew gatherings. But political subjectivity within the qat chew also entails an expectation of openness to a variety of political views and people. Political subjectivity in qat chews emerges out of the commitment to verbal disagreement and the (temporary) forswearing of the use of violence to accomplish political ends. In Habermas' idealized version, the European public sphere appeared against the backdrop of states that had achieved a more or less secure monopoly over violence so that the "force" of social position was supplanted by the force of the better argument, rather than the force of weapons. By contrast, qat chews are not predicated on the existence of robust state institutions, but actually thrive in their absence. Qat chews defer or replace political violence with discursive contestation, producing expectations among participants that they will be seen and heard by others whose political positions will be different from their own. Qat chew gatherings are thus a mode of personal accountability and a forum for challenging conventional wisdoms. If not a guarantee of consensus, they provide an awareness of the diverse ways in which any particular political problem might be seen and solved imaginatively.

To summarize thus far: in Yemen, qat chews are sites of active political argument where issues of accountability, citizenship, and contemporary affairs can be negotiated.[14] Although the proliferation of such meetings can be historically situated in changes of capital accumulation in the oil boom period of the 1970s, neither the bust period of the 1980s nor the constriction of formal electoral politics since 1994 has diminished the frequency or diluted the content of these gatherings. The persistence of qat chew gatherings as sites for animated political debate suggests that they do not depend strictly on those historical conditions to exist; that people, having discovered such public sphere activities, find them valuable and therefore persist in reproducing them; and that tangible problems are handled by means of them. It is in the repetition of the ways in which problems can be discussed and sometimes resolved that the robustness

[14] The vast majority of the 250 qat chews I have attended entailed political discussion. Although I realize that these gatherings do not constitute a random sample, my fieldwork does suggest that political debate at chews, far from being a rare event, is an integral part of many gatherings and a choice available to both urban and rural Yemeni men.

of these activities lies. In the North, there is a long-standing history of consultative, participatory politics in which public meetings were also occasions for the authority of "tribal" leaders to be evaluated (Dresch 1990, 1994, 2000; Caton 1990; see Comaroff and Comaroff 1997 for a similar phenomenon in Botswana). The presence of cooperative local development projects in the 1970s similarly testified to the workings of "civil society" in the absence of state institutions capable of delivering goods and services (Carapico 1998). The conditions for substantive discourses about governance have existed both inside and outside of the regime for years. Unsatisfactory conditions encourage radical critiques, and qat chews are one particularly widespread site where such criticisms are expressed, challenged, and refined.

Since the brief civil war of 1994, these everyday practices of political participation operate in a context in which the electoral process has come to be seen by many ordinary citizens as a way of containing populist politics rather than enabling its expression (Wedeen 2003). Yet qat chews are also a key place where people are able to exchange conceptions of fair and free elections, while also deliberating about how to respond to the rigging. That exchange and deliberation are the very substance of both the development and practice of democracy. Topics of debate include registration infractions (September 2002), the trials and tribulations of running candidates (March 2001; September 2002), the efficacy of distributing goods to constituencies (March 2001; September 2002), the moral obligations of voters who receive goods in expected return for their support (September 2000), and the nature of what counts as a compelling political message or a political program worth endorsing (September 1999; January 2001; September 2002). Przeworski *et al.* would recognize such conversations as political, but they would be irrelevant to how democracy gets defined. In contrast, a less minimalist understanding of democracy would think of these debates as themselves constitutive of democracy, as an instantiation of the "deliberative genesis and justification of public policy in political and civil public spaces" (Cohen 1998: 2).

It may also be true that insofar as groups or individuals that could be potentially in conflict take part in qat chew conversations with one another, they are helping to avoid violence: talking in qat gatherings may function as diplomacy does in international relations – helping people to work out disputes, to keep aware of threats, and register changes in status.[15] In this view, qat chews help alleviate uncertainty about intentions that might otherwise overwhelm good will and cause violence, especially in situations where the population is as armed as Yemen's is. This may be

[15] Thanks are owed to Matthew Kocher for this point.

an important social function of qat chewing, but it is not a democratic one.

The democratic nature of qat chews has to do with the kind of political subject formation that takes place through the practice of discussion and deliberation in public. In other words, the political activity of discussing and deliberating is part of what a democrat does. Such conversations are themselves predicated on the expectation of argument and disagreement, on what Arendt called a politics of "unique distinctness" (1958: 176) in which a person's speech can be a revelatory form of action in its own right. In the Yemeni context, qat chews are an instantiation of an Arendtian politics of collective action and revelatory discussion in which Habermasian norms of "egalitarian reciprocity" (Benhabib 1992) find expression in each agent's ability to speak and be heard. At the same time, participants are aware of themselves and others as beings whose identities are assumed (by participants) to be coherent, related to others hierarchically, and "always already there." Or to put it differently, the possibilities for performative action and egalitarian reciprocity exist within socially sedimented classifications and identifications, which may be experienced not as natural states, qualities, or properties, but rather as provisionally settled forms of categorization and affiliation – themselves the subject of qat chew debates.

The regime tolerates the discursive activity generated through qat chews if only because it is unable to suppress it, and meanwhile it takes advantage of what it can. The regime, for example, may benefit from the information-rich environments that qat chews afford. The practice makes it easy for the regime to keep tabs on who might be interested in challenging the regime violently. President ʿAli ʿAbd Allah Salih himself often calls members of the political elite from a variety of political tendencies after a qat chew ends to find out what the topics of conversations were, who held what positions, and what, if anything, was potentially politically subversive. Indeed, members of the political opposition tend to interpret his subsequent political speeches as testimony both to his knowledge of current conversations and his relative approval or disapproval of positions taken by participants in a qat chew gathering. For example, on the occasion commemorating the fortieth anniversary of the North's "revolution" of September 26, 1962, President Salih castigated some of the opposition's political parties for "loitering in front of foreign embassies."[16] Some members of the opposition interpreted this attack as a veiled reference to a well-attended qat chew conversation the day before,

16 *Al-Thawra*, September 27, 2002; see also *al-Thawra*, last page, September 28, 2002, which featured various Arabic media sources' excerpts from, and brief commentaries on, the speech. The editors noted that media throughout the Arab world were particularly

September 25, 2002, in which several politicians had voiced their qualified support for American intervention in Iraq. Arguing on the grounds of Saddam Husayn's brutality and the possibilities for democratic transformation in the region, members of the opposition raised the question of whether Iraq would be better under American occupation or under Husayn. The point is not that Salih was, in fact, referring to this qat chew, but that participants could plausibly expect him to, and could even name those who would potentially inform.

In general, qat chews as an institutionalized everyday practice can serve both ruler and ruled, providing an authoritarian regime with crucial information while giving ordinary individuals and an organized opposition more freedom than they would enjoy in other authoritarian circumstances. The "off-equilibrium" dimension of this information richness is that it should, and does, promote collective action against the regime under certain conditions. More importantly, collective action for or against a regime obviously requires prior discussions and deliberation. Qat chews are sites where that discussion takes place, independent of more or less closed party councils. Forms of political inclusiveness and temporary equality based on the commonality of shared concerns, disclosed in interaction with others, would seem to be part of what democracy means. Politically functional approximations of public spheres exist independently of Habermas' historical derivation of "the public sphere" from bourgeois private experience. Public sphere activities, moreover, exist without inventing a universal subject (brought about by expressing those private experiences in and to a public).

Large-N studies such as Przeworski *et al.*'s leave out, for the sake of methodology, lived experiences that interpretivist accounts have shown to be vital to any adequate concept of democracy. It nevertheless remains worthwhile to consider why everyday practices of political debate and oppositional consciousness do not seem to lead to contested elections in which outcomes are procedurally uncertain and the ruling party cedes power. Here large-N studies can help us explain why certain states are less likely to experience enduring contested electoral arrangements (although it is important to keep in mind that because statistical studies are concerned with the median, they have little to say about variants or outliers, such as, say India or Singapore). Przeworski and Limongi, for example, find that although it is impossible to explain what triggers a transition

interested in Salih's statement about the opposition parties, which was reproduced in the following newspapers: *al-Ra'y* (Qatar), *al-Bayan* (Dubai), *al-Ittihad* (UAE), *al-Ra'y al-'Amm* (Kuwait).

to elections, wealthy countries are more likely than poor ones to maintain such a system once it is put in place (Przeworski and Limongi 1997). Carles Boix's *Democracy and Redistribution* (2003; see also 2000) provides a theory of why this is so. Boix considers variables such as initial levels of inequality, the distribution of assets and the demands for redistribution, and the types of capital (mobile or not so mobile) that help to determine whether leaders will or will not favor competitive elections. Boix finds that "at low levels of capital mobility capitalists have a direct and strong interest in tutoring the state. Since capital is hardly sensitive to taxes, voters have a high incentive to impose heavy taxes." As a result, holders of fixed capital, such as oil wells, agricultural products, and mines will invest considerable effort in blocking democracy, since the costs of not doing so are so high. In other words, the lack of alternative uses to the holders of capital make them particularly interested in shaping policy: "It is better for capital to incur a certain cost to block democracy than suffer quasi-confiscatory taxes" (Boix 2000: 16; see an elaboration of this argument in Boix 2003).

Thus high per capita income is related to democracy only to the extent that the former resides in mobile kinds of capital. High-income countries that base their prosperity on fixed natural resources, such as oil or diamonds or sugar, are likely to remain authoritarian (Boix 2000: 20; 2003). In this view, dictatorships are the "direct consequence of a strong concentration of fixed natural resources" (Boix 2000: 21). With regard to types of capital and to poverty indices, Yemen's chances of sustaining electoral arrangements prove low. Yemen is one of the poorest countries in the world, with average yearly per capita income estimated at anywhere from $270 to $444 and a population growth rate of 3.4 percent per year. Illiteracy rates among men reach 50 percent and among women approximate 70 percent – among the highest in the world. A modest level of oil production is the most important contributor to GDP, economic growth, fiscal revenues, exports, and foreign exchange earnings in the 1990s (World Bank report 2002).

Yemen did undergo a brief transition to fair and free elections upon unification in 1990. Prior to 1991, when labor migrants were expelled *en masse* from Saudi Arabia, Yemen's chief source of revenues derived from labor remittances (Chaudhry 1997). Although poverty was acute, the gap between rich and poor was not nearly as pronounced or as apparent as it has come to be since the civil war. It is possible, then, to suggest that the conditions of inequality and a reliance on fixed capital that obtain in Yemen today were not as characteristic of the immediate unification period. The impetus for unification came from reform-minded politicians in the former socialist South who pressed for a unified democratic polity, in part as a way of saving the party from additional civil wars. The "transition" might be explained by the combinatorial power of these immediate democratic pact-like concerns and the structural conditions of possibility that made elections seem desirable. Boix's argument helps us understand why fair and free elections were difficult to sustain, given the prevalence of variables such as poverty, inequality, and a reliance on fixed capital, which tends to undermine these prospects.

Large-N studies, then, aid us in explaining the presence or absence of competitive elections, but they may be less helpful in thinking through why a regime like Yemen is as democratic as it is (Wedeen 2003), or how participation is enacted when it is. For analyzing what public sphere practices do, how they work as performances of citizenship by facilitating collective deliberation, we may have to adopt an interpretive approach – studying the meanings and workings of substantive political activity, in which contestation refers less to procedural elections and more to the participatory experiences characteristic of contemporary political life in many parts of the world.

Such an approach need not romanticize these forms of political practice or imply their durability. Technological innovations such as the introduction and widespread accessibility of cell phones may end up unsettling the structure of conversations, making them less sustained and coherently deliberative than they have typically been. Television too has been gradually introduced into some qat chew gatherings, with consequences for the kinds of interactions and depth of conversations pursued among people who used to gather without its seductions. The quality and intensity of political experience within a qat chew gathering is also dependent, in part, on the individuals who run them. To the extent that a generation of politicized (primarily) men saw it as part of their duty as citizens to address what Char, in the context of the French Resistance called the real "affairs of the country" (cited in Pitkin 1998), the death of these men – by assassination, illness, and war – may make the imaginings of political action and the concerted deliberations about self-rule all the more elusive.

The recent interventions of the United States in domestic Yemeni affairs, dramatized by the coordinated Yemeni-US incineration of an alleged al-Qa'ida leader, "Abu 'Ali" al-Harithi, may also undermine the vibrancy of these gatherings – if not directly through the shoring up of state capacities to control populations, then indirectly through the short-term violence such interventions unleash. The state's inability to monopolize violence may have its own chilling effects, as its failure to protect citizens makes them particularly vulnerable for what they say. In Yemen, qat chew conversations invite accountability, and such conversations are reiterated publicly in newspaper accounts and mosque sermons, as well as in private conversations among friends and government informants. Sometimes views expressed in intellectuals' conferences or through civil society organizations are initially "tried out" in qat chew contexts; and often issues dealt with formally in a conference are pursued further in afternoon qat sessions, thereby making citizens accountable (and vulnerable) to multiple publics. If political assassinations – of "loyal" opposition members (such as Jar Allah 'Umar), of insurrectionary figures (such as al-Harithi), and of regime officials (such as perhaps is the case in the 2003 car accident of the ruling party's Yahya al-Mutawakkil) – become a preferred mode of political expression, then the instances of public sphere activity are likely to atrophy, as the openness and accessibility of deliberative political life retreat in the shadows of death and uncertainty.

Concluding remarks

This chapter does not judge whether one approach or the other is right or wrong, but rather, how different epistemological commitments sometimes yield divergent questions, sources of data, and explanations of political life. An approach that works with continua and gradations of political phenomena identifies evidence that tends to dodge the purview of many large-N studies; large-N studies provide us with insights about a number of cases that are sometimes lacking in interpretive projects. Large data sets often rely on various simplifications for the sake of the model. An interpretivist is likely to find such simplifications inadequate for the purpose of understanding lived political experience. A large-N practitioner is likely to find the messiness and complications to which interpretivists attend unnecessary and distracting for the tasks of generating law-like patterns. All three approaches discussed above can be used to solve problems of abiding concern to political scientists. And all three approaches can be theoretical – providing generalizable accounts of how and why the world is as it is. Large-N work does this by specifying law-like patterns of human action. Wittgensteinian ordinary language philosophy does this by clarifying concepts that tell us how various communities – scholars, "informants," and ordinary speakers – think about the world. Other interpretivists engaged in an analysis of meaning-making do this by specifying what language and symbols do, and how they are themselves the product of institutional arrangements, structures of domination, and strategic interests. Fruitful combinations of the three approaches may generate the richest, most persuasive accounts of politics. This essay has been devoted to elucidating the strengths and weaknesses of each by putting them into conversation with one another. Problem-driven work will always be, it seems to me, more intellectually compelling than projects driven primarily by methodological predilections, but problems can be solved in multiple ways. What a conversation among these approaches allows us to envisage is not only the use of multiple methods, but also the ways in which methodological choices affect our understanding of what the problem is. It is in this sense that knowing what the puzzle is, is essential to knowing how we might go about solving it, but applying conceptual methods may also help us deepen our knowledge of what, in fact, the problem is – or to twist a memorable phrase of Geertz's, what the devil we scholars are up to.

REFERENCES

Adas, Michael. 1992. "South Asian Resistance in Comparative Perspective." *Contesting Power: Resistance and Everyday Social Relations in South Asia.* Douglas Haynes and Gyan Prakash (eds.). Berkeley and Los Angeles: University of California Press.

Arendt, Hannah. 1961 [1954]. *Between Past and Future: Six Exercises in Political Thought.* New York: The Viking Press.

1974 [1958] *The Human Condition.* Chicago: University of Chicago Press.

Austin, J. L. 1961. *Philosophical Papers.* Oxford: Clarendon Press.

Benhabib, Seyla 1992. *Situating the Self: Gender, Community, and Postmodernism in Contemporary Ethics.* London: Routledge.

Benhabib, Seyla (ed.). 1996. *Democracy and Difference: Contesting the Boundaries of the Political.* Princeton: Princeton University Press.

Boix, Carles. 2003. *Democracy and Redistribution.* Cambridge: Cambridge University Press.

2000. "Democracy and Inequality." Unpublished manuscript. University of Chicago.

Bollen, Kenneth A. and Robert W. Jackman. (1989) "Democracy, Stability, and Dichotomies." *American Sociological Review* 54: 438–57.

Bourdieu, Pierre. 1978. *Outline of a Theory of Practice.* Cambridge: Cambridge University Press.

Bradbury, Malcolm. 1992. *Doctor Criminale.* New York: Viking Penguin.

Carapico, Sheila. 1998. *Civil Society in Yemen: The Political Economy of Activism in Modern Arabia.* Cambridge: Cambridge University Press.

Chakrabarty, Dipesh. 2000. *Provincializing Europe: Postcolonial Thought and Historical Difference.* Princeton: Princeton University Press.

Cohen, Jean L. 1998. "American Civil Society Talk," in *Report from the Institute for Philosophy and Public Policy.* University of Maryland Working Papers, summer.

Chaudhry, Kiren Aziz. 1997. *The Price of Wealth: Economics and Institutions in the Middle East.* Ithaca: Cornell University Press.

Comaroff, John L. and Jean Comaroff. 1997. "Postcolonial Politics and Discourses of Democracy in Southern Africa: An Anthropological Reflection on African Political Modernities." *Journal of Anthropological Research* 53(2) (Summer).

Dahl, Robert A. *Polyarchy*. 1971. New Haven: Yale University Press.

Dietz, Mary. 1995. "Feminist Receptions of Hannah Arendt," in *Feminist Interpretations of Hannah Arendt*. Bonnie Honig (ed.). University Park: Pennsylvania State University Press.

Dresch, Paul. 1990. "Imams and Tribes: The Writing and Acting of History in Upper Yemen," in *Tribes and State Formation in the Middle East*. Philip S. Khoury and Joseph Kostiner (eds.). Berkeley and Los Angeles: University of California. 252–87.

1994. *Tribes, Government, and History in Yemen*. Oxford: Oxford University Press.

2000. *A History of Modern Yemen*. Cambridge: Cambridge University Press.

Dreyfus, Hubert L. and Paul Rabinow. 1983 [1982]. *Michel Foucault: Beyond Structuralism and Hermeneutics* (2nd edn). Chicago: University of Chicago Press.

Foucault, Michel. 1980 *Power/Knowledge: Selected Interviews and Other Writings*. Colin Gordon (ed.). New York: Pantheon Books.

Fraser, Nancy. 1992. "Rethinking the Public Sphere: A Contribution to the Critique of Actually Existing Democracy." *Habermas and the Public Sphere*. Craig Calhoun (ed.). Cambridge, MA: MIT Press.

Geertz, Clifford. 2000. *The Interpretation of Cultures*. New York: Basic Books.

Gerholm, Tomas. 1977. *Market, Mosque and Mafraj: Social Inequality in a Yemeni Town*. Stockholm: Stockholm Studies in Social Anthropology.

Guha, Ranajit. 1983. *Elementary Aspects of Peasant Insurgency*. Delhi: Oxford University Press.

Guinier, Lani and Stephen L. Carter. 1995. *The Tyranny of the Majority: Fundamental Fairness in Representative Democracy*. New York: Free Press.

Habermas, Jürgen. 1984 [1987]. *Theory of Communicative Action*. Vols. I and II. Boston: Beacon Press.

1996 [1962]. *The Structural Transformation of the Public Sphere: An Inquiry into a Category of Bourgeois Society*. Thomas Burger (trans.). Cambridge, MA: MIT Press.

2001. "Constitutional Democracy: A Paradoxical Union of Contradictory Principles?" *Political Theory* 29(6) (December): 766–81.

Honig, Bonnie. 2001. "Dead Rights, Live Futures: A Reply to Habermas's 'Constitutional Democracy.'" *Political Theory* 29(6) (December): 792–805.

Kalyvas, Stathis N. 1998. "Democracy and Religious Politics: Evidence from Belgium." *Comparative Political Studies* 31(3) (June): 292–320.

Lee, Benjamin. 1993. "Textuality, Mediation, and Public Discourse," in *Habermas and the Public Sphere*. Craig Calhoun (ed.), Cambridge, MA: MIT Press.

Mansbridge, Jane. 1991. "Feminism and Democratic Community." *Democratic Community. Nomos*, no. 35. John W. Chapman and Ian Shapiro (eds.). New York: New York University Press.

Meneley, Anne. 1996. *Tournaments of Value: Sociability and Hierarchy in a Yemeni Town*. Toronto: University of Toronto Press.
Nikolova, Boriana. Exam for "Rethinking Democratic Practice." N.d.
Pitkin, Hanna Fenichel. 1967 *The Concept of Representation*. Berkeley and Los Angeles: University of California Press.
1993 [1972]. *Wittgenstein and Justice*. Berkeley and Los Angeles: University of California Press.
1998. *The Attack of the Blob: Hannah Arendt's Concept of the Social*. Chicago: University of Chicago Press.
Przeworski, Adam and Fernando Limongi. 1997. "Modernization: Theories and Facts." *World Politics* 49(2) (January): 155–83.
Przeworski, Adam, Michael E. Alvarez, Jose Antonio Cheibub, and Fernando Limongi. 2000. *Democracy and Development: Political Institutions and Well-Being in the World, 1950–1990*. Cambridge: Cambridge University Press.
Putnam, Robert D. 1993. *Making Democracy Work: Civic Traditions in Modern Italy*. Princeton: Princeton University Press.
Rabinow, Paul and William M. Sullivan. 1987. *Interpretive Social Science: A Second Look*. Berkeley, Los Angeles, London: University of California Press.
Sanders, Lynn. 1997. "Against Deliberation." *Political Theory* 25(3) (June): 347–76.
Schaffer, Frederick Charles. 1998. *Democracy in Translation: Understanding Politics in an Unfamiliar Culture*. Ithaca, NY: Cornell University Press.
Schumpeter, Joseph A. 1942. *Capitalism, Socialism, and Democracy*. London: George Allen and Unwin.
Shapiro, Ian. 1999. *Democratic Justice*. New Haven: Yale University Press.
Stokes, Susan and Carles Boix. 2003. "Endogenous Democratization." *World Politics*, July.
Tutwiler, Richard and Sheila Carapico. 1981. *Yemeni Agriculture and Economic Change: Case Studies of Two Highland Regions*. Sanaa: American Institute for Yemeni Studies.
Varisco, Daniel Martin. 1986. "On the Meaning of Chewing: The Significance of Qat (*Catha edulis*) in the Yemen Arab Republic." *International Journal of Middle East Studies* 18(1) (February): 1–13.
Warner, Michael. 2002. *Publics and Counterpublics*. Cambridge, MA: MIT Press.
Waterbury, John. 1994. "Democracy without Democrats? The Potential for Political Liberalization in the Middle East." *Democracy Without Democrats? The Renewal of Politics in the Muslim World*. Ghassan Salame (ed.). London and New York: I.B. Tauris.
Wedeen, Lisa. 1999. *Ambiguities of Domination: Politics, Rhetoric, and Symbols in Contemporary Syria*. Chicago: University of Chicago Press.
2002. "Conceptualizing Culture: Possibilities for Political Science." *American Political Science Review* 96(4) (December).
2003. "Seeing Like a Citizen, Acting Like a State: Exemplary Events in Unified Yemen." *Comparative Studies in Society and History* 45(4), October.
Weir, Shelagh. 1985. *Qat in Yemen: Consumption and Social Change*. London: British Museum Publications.

Wittgenstein, Ludwig. 1958. *Philosophical Investigations* (3rd edn). G. E. M. Anscombe (trans.). New York: Macmillan.
Wolin, Sheldon. 1996. "The Liberal/Democratic Divide: On Rawls' Political Liberalism." *Political Theory* 24 (1): 97–142.
Young, Iris Marion. 1996. "Communicating and the Other: Beyond Deliberative Democracy." *Democracy and Difference: Contesting the Boundaries of the Political*. Seyla Benhabib (ed.). Princeton: Princeton University Press.
Young, Iris Marion. 2000. *Democracy and Inclusion*. Oxford: Oxford University Press.

14 Problems chasing methods or methods chasing problems? Research communities, constrained pluralism, and the role of eclecticism

Rudra Sil

> The idea of a method that contains firm, unchanging, and absolutely binding principles for conducting the business of science meets considerable difficulty when confronted with the results of historical research. We find, then, that there is not a single rule, however plausible, and however firmly grounded in epistemology, that is not violated at some time or another. It becomes evident that such violations are not accidental events . . . [D]evelopments, such as . . . the Copernican Revolution . . . occurred only because some thinkers either *decided* not to be bound by certain "obvious" methodological rules, or because they *unwittingly broke* them.
>
> (Feyerabend 1993: 14)

> [I]f the notion of a theory-neutral observation language had been viable, and if theory changes were in fact cumulative, and if all scientists subscribed to the same methodological standards, and if there were mechanical algorithms for theory evaluation . . . then the positivists might well have been able to show wherein scientific rationality and objectivity consist. But that was not to come to pass . . . These days, social constructionists, epistemological anarchists, biblical inerrantists, political conservatives, and cultural relativists all find in the surviving traces of positivism grist for their mills; for what they find there appears to sustain their conviction that science has no particular claim on us, either as a source of beliefs or as a model of progressive, objective knowledge.
>
> (Laudan 1996: 25)

Feyerabend and Laudan have been represented, by Laudan himself among others (e.g., Sanderson 1987), as competing perspectives in the philosophy of science. Feyerabend's skeptical treatment of

I am grateful to the volume's editors and to participants at the Conference on "Problems and Methods in the Study of Politics" (Yale University, December 6–8, 2002), especially William Connolly, Keith Darden, Margaret Levi, John Mearsheimer, Lloyd Rudolph, and Susanne Rudolph, for valuable comments and suggestions. The discussion on eclecticism partly builds upon Katzenstein and Sil (forthcoming 2004) and Sil and Katzenstein (2003).

method-driven social science reflects a position of epistemological anarchism that, although not a defense of relativism, encourages the unfettered proliferation of incommensurable theories. Laudan rejects such an anarchistic view of science and puts his faith in a view of "progress" that depends on competing research traditions demonstrating their relative value by identifying and solving problems. A flexible reading of their positions in the context of the question addressed in this volume suggests that the two positions may not be as starkly opposed as they appear. As the above epithets imply, Feyerabend and Laudan begin to converge on the point that rigidity in the application of uniform methodological tenets has neither formed the basis for the most significant breakthroughs in the history of science, nor constituted a scientific ideal that natural and social science should aspire to realize. For both, when something resembling progress can be identified in the history of science, it is more than likely to be the result of scholars either eschewing rigorous adherence to uniform methodological principles or transgressing the boundaries between competing research traditions. In keeping with the spirit of this projected common ground, this chapter offers three related arguments, developed in the three sections that follow, on how we might fruitfully think about the relationship between problems and methods in the social sciences.

The first section considers what the choice between a problem-driven or method-driven approach means for the social sciences as a whole, for individual researchers, and for research communities. While the collective output of social scientific research tends to be evaluated outside the academe primarily in terms of its utility in relation to concrete problems, individual scholars do face tradeoffs between acquiring greater mastery of a given method and seeking a deeper understanding of a given substantive problem. Any principled choice in this regard, however, is not terribly meaningful since, in the end, individual scholarly projects usually need to offer claims about both methodological rigor and substantive utility in order to be taken seriously in the eyes of a particular *research community*. I use this term as an ideal-typical category to refer to emergent clusters of scholars who communicate more regularly and more fluently with each other than with others because they share a common academic interest in a particular substantive problem as well as some minimum familiarity with a particular methodological approach and theoretical vocabulary.

Given that research communities in the social sciences embrace often conflicting understandings of what problems are significant and what methods are appropriate for tackling them, the second section focuses on the vexing question of how to establish some common basis for the unity of social scientific research while preserving a spirit of intellectual pluralism. Some resolve this tension by designating a uniform set of methodological principles and translating them into rules to guide a wide range of approaches, and others see no alternative to an anarchistic social science consisting of varied research products that can never be integrated or evaluated relative to each other. In contrast to both perspectives, I articulate a position of "constrained pluralism," reflecting a flexible conception of methodological rigor that takes into account the assumptions, traditions, objectives and expectations associated with research communities. Extending an analogy developed elsewhere (Sil 2000a), I suggest that intellectual pluralism and a unified conception of social science – to the extent that both are desirable ideals – can be realized simultaneously only by thinking of the intellectual endeavors associated with different research communities as performing distinct yet interdependent "roles" within an increasingly complex "division of labor" (Durkheim 1984). This analogy between a social and scholarly division of labor is not intended to suggest a unified purpose for all social scientists or a rigid system of intellectual specialization but rather to reveal the need to negotiate a common understanding of the different goals, payoffs, tradeoffs and evaluative standards implied in a research community's choice of problem and method. Such an understanding could provide a modest but feasible basis for a unified view of social science while enabling a meaningful intellectual pluralism marked by wider appreciation of the distinctive contributions of the intellectual products generated by different research communities.

The third section takes this logic one step further in examining the potential contributions of self-consciously *eclectic* research endeavors that fall outside the domains of identifiable research communities. Such eclectic approaches neither reject the scholarship produced in research communities, nor offer substitutes for it; they simply depart from the established practices of any one research community and explore the payoffs from different permutations of problems, methods, theories, and empirics in generating new conceptual schemes and substantive interpretations. To be sure, an eclectic stance entails risks, costs, and limits that may not consistently justify the value added to work being done within existing research communities. But, for the social sciences as a whole, eclecticism serves a distinctive and valuable function by expanding the scope of communication across a wider range of research communities, and by experimenting with permutations of components of varied research products that are formulated in different theoretical languages and cast at different levels of abstraction. This would help not only in the negotiation of appropriate expectations for different kinds of research products, but also in reducing the gap between messy concrete problems confronting actors in everyday life and the more neatly sliced analytic problems investigated by members of a particular research community. The conclusion of this chapter relates these propositions to the aforementioned middle ground projected between Feyerabend's anarchistic view of method and Laudan's effort to seek progress through "problem-solving."

Problem and method in action: research communities in social science

As methodological choices ultimately rest on unverifiable assumptions about the nature of the social world and about our efforts to make sense of this world, any attempt at establishing uniform rules of scientific analysis and uniform standards of methodological rigor is also an attempt at privileging a particular set of ontological and epistemological assumptions. For better or for worse, this has not come to pass. If anything, as Laudan's epithet above anticipates, there has been more divergence than convergence among scholars on issues pertaining to method. However, therein lies a serious challenge for a method-driven social science. The fact that scholars have perpetually disagreed on methodological rules and criteria for "rigor" – as evident in the differences among many of the contributors to this volume – implies that judgments about research products based solely on methodological sophistication will be regularly contested, potentially weakening the standing of the entire enterprise of social science.

For this reason alone, social scientific research as a collective enterprise cannot but be problem oriented, at least when it comes to defending its legitimacy in the eyes of others. It is a good deal easier to identify agreements on substantive conclusions resulting from the application of different methodological tools than to defend a particular method in the face of inconsistent conclusions generated by it. In international relations, for example, if realists and constructivists both conclude that a particular country does not pose a threat to its neighbors, the fact that they use different logics and different kinds of data to make this case should not detract our attention from the agreement on the conclusion itself. Similarly, a game-theoretic model and a historical-institutionalist argument, while founded on quite different epistemological premises and employing quite different methodological tools, can yield similar substantive conclusions about the likelihood of revolution or the viability of particular electoral systems in a given social context. Thus, as a practical matter, if the enterprise of social science is to continue to receive validation (and funding) from outside the academe, this is likely to be for perceived substantive contributions that bear on problems in public life rather than for the demonstration of sophistication in the application of a particular set of methodological tools. It is in this sense that I contend that the social sciences must necessarily be problem oriented.

However, this is little more than a pragmatic posture in light of the lack of unity among social scientists on epistemic norms and methodological principles and standards. As such, it does not constitute an answer to the question of whether social scientists, as *individuals,* ought to privilege problem or method in pursuing their research projects and intellectual agendas. Given that most of us cannot easily grasp the empirics of a wide range of concrete problems while also mastering a wide range of methods, it is quite reasonable to ask whether problem-oriented expertise or methodological skill takes precedence from the point of view of both the process of research and the evolution of a professional career in the academe. Some will definitively adopt a problem-driven understanding of "good" social science, exploring similar kinds of problems in the course of a career, and rely on their intimate familiarity with these problems in assessing the value-added of particular research products; but those following such a strategy will tend to be susceptible on the question of how exactly social scientists trained in different methods can come to any consensus about how much value is actually being added by a particular contribution on the subject. Others will definitively adopt a method-driven strategy, addressing a wider range of concrete problems in the course of their careers as they seek meaningful applications of the methodological techniques they have mastered; but those following such a strategy will tend to be susceptible on the question of how to authoritatively demonstrate the relative merits of particular contributions on a given problem to others who have been exploring this problem using different methods. The benefits and costs of each of these strategies suggest a quite real tradeoff for individual social scientists deciding how to invest their time and energy over the course of a career.

When we turn to the question of how research products are presented and evaluated, however, a principled choice in favor of a problem- or method-driven approach does not mean very much. In everyday research practice, most social scientists typically make reference to the proficiency with which a particular method has been applied (whatever the method and the criteria for rigor) as well as to the value-added in our understanding of some substantive problem (however that problem is framed and investigated). For the most part, these dual claims are not difficult to sustain given the simple fact that problems can be made to chase methods, and vice versa, with relative ease. That is, a "problem" can be selected and posed in ways that make it more easy to apply a given method; since many socially important "real world" problems are too complex for any one individual to tackle, researchers are in a position to constitute their own analytic problems in ways that take advantage of the strengths of a particular methodological approach. Similarly, the study of certain problems, particularly those connected to particular times and places, can require such an investment of time and effort (for example, in mastering a language and becoming familiar with the relevant historiographic debates and social contexts) that it is more worthwhile to subsequently invest in expanding one's methodological toolkit so as to build on the problem-specific expertise already acquired.[1] Whether problems are chasing methods or methods chasing problems, in practice, the reputations of scholars within particular scholarly circles generally depend on whether they appear to be both knowledgeable about a particular subject matter and skillful in the application of particular methods.

To explore the implications of this for the social sciences as a collective enterprise, I find it helpful to think of the organization of social scientific research not in terms of disciplines or subfields but rather in terms of *research communities.* I use this term as an ideal-typical category to capture a group of scholars who form more dense intellectual associations with each other than with other scholars because of their shared interest in a given problem and their common preference for a given method. As implied in the well-known opposition of "community" to "society" in classical sociology (e.g., Tönnies 1957), the notion of "research community" suggests the importance of the *similarity* in roles and skills among individual members (as opposed to the differentiation of roles and skills within a larger, more complex social organization). In an ideal-typical research community, a shared familiarity with a similar set of problems and methods is likely to be accompanied by implicit agreements on an appropriate theoretical vocabulary, criteria for evaluating works by members, and, above all, a set of foundational assumptions that validate the investigation of particular problems and the use of particular methods. For the purposes of this volume, however, research communities are distinguished not by *a priori* agreements on this entire range of metatheoretical issues, but rather in terms of emergent clusters of researchers who happen to form connections with one another as they become aware of their common interests and their common ability to talk about a given problem in a common analytic language that suits the application of a given method. Although research communities are likely to have fluid boundaries and overlapping membership, they can be

[1] This is evident, for example, in efforts to apply game theory in order to develop different sorts of interpretations in response to familiar problems that have previously been explored through qualitative methods (e.g., Bates *et al.* 1998).

approximated reasonably well in everyday research practice by reference to the group of scholars for whom one is primarily writing, from whom one expects to receive informed evaluations, and with whom one expects to engage in ongoing scholarly conversations or collaborations.[2] Examples might include historical institutionalists focusing on social policies in advanced industrial countries, those using statistical methods to examine relationships between levels of trade and conflict, those employing game-theoretic models to make sense of electoral politics in multiethnic societies, or those doing archival work to produce a social history of labor in a given locale.

Thinking about the social sciences in terms of research communities begs two related questions addressed in the following two sections. First, given the differences in methodological principles and standards employed across emergent research communities, what, if anything, might constitute a reasonable common basis for the "unity" of social science as a bounded collective enterprise? Second, given the differences in the identification and framing of substantive problems, what, if anything, might be done to integrate components of scholarship produced in different research communities in the hope of generating something resembling "progress"?

Beyond unified method: constrained pluralism in a "division of labor"

One common answer to the first of these questions is articulated in a popular text on methods of social inquiry (King, Keohane, and Verba 1994): the unity of the social sciences lies in the essentially uniform logic undergirding all genuinely social scientific efforts at recording and interpreting observations. That is, since "the difference between the amount of complexity in the world and that in the thickest of descriptions is still vastly larger than the difference between the thickest of descriptions and

[2] In this sense, research communities are less expansive and less rigid than a "scientific paradigm" (Kuhn 1962) or "research programme" (Lakatos 1970) and are closer in character to "research traditions" (Laudan 1977, 1996), which are characterized as more flexible in allowing varied substantive propositions, some of which may be commensurable with propositions articulated in other traditions. A research community may be understood as a subset of a research tradition to the extent that it exhibits common "beliefs about what sorts of entities and processes make up the domain of inquiry" and agreement on "epistemic and methodological norms about how the domain is to be investigated, how theories are to be tested, how data are to be collected, and the like" (Laudan 1996: 83). At the same time, what makes a research community a significantly narrower category is that it also reflects shared stocks of empirical knowledge on a substantive problem *within* the domain of inquiry, and by shared competence in applying a methodological approach and theoretical vocabulary in articulating and exploring that problem.

the most abstract quantitative or formal analysis" (King, Keohane, and Verba 1994: 43), the same basic methodological principles and standards can and should be applied to the process through which any kind of inference, however "thick" or "thin," is generated from empirical observations. In effect, this means that the rules, norms, and standards employed in performing and evaluating statistical analysis, including measures to enforce the norm of replication (King 1995), can be directly translated into propositions for increasing and evaluating "rigor" across a wide range of methods, including qualitative and interpretive ones. Thus, the unity of the social sciences would be reflected in the acknowledgment of a *uniform* set of rules for designing and evaluating empirical research (King, Keohane, and Verba 1994; Goldthorpe 1996), while the diversity of the social sciences would be limited only to the types of data employed and the degrees of "thickness" or "thinness" in the interpretations offered.

Unfortunately, this effort to ground the unity of the social sciences upon a uniform view of methodological rigor requires the authors to deliberately bypass even a cursory discussion of the foundational assumptions on which their understandings of science and method are founded. Although adopting a seemingly pluralist posture in translating the methodological principles, norms, and standards used in quantitative analysis for a wider range of efforts at "discovering truth" (King, Keohane, and Verba 1994: 38), the authors also end up ignoring the difference between the quite varied objectives scholars may pursue depending on their beliefs about the extent to which different "truths" are accessible to human observers, the level of abstraction at which "truths" are to be formulated, and the extent to which these "truths" can be generalized across contexts. Even if the amount of complexity in the world is indeed vastly larger than the difference between thick descriptions and formal models, and even if the process of abstracting from this complexity involves moving in the same *direction* whether we are engaged in statistical or interpretive analysis, *how far* one chooses to travel in this direction, and *for what purpose*, remain important questions that have enormous implications for just how meaningfully and fruitfully methodological principles and standards can be translated across different kinds of research products. In the end, this effort at establishing a common basis for the unity of the social sciences, while laudable for stimulating conversations about method among a wider range of scholars, offers standards and principles for scientific research that are only meaningful and legitimate among those who share the authors' empiricist epistemological perspective – a perspective that discounts ontology, emphasizes causal and descriptive inferences without reference to causal mechanisms, and equates social knowledge with falsifiable law-like propositions concerning observable regularities in social life.[3]

An alternative perspective on what unifies the social sciences emerges from a "realist" philosophical position (Bhaskar 1986; Devitt 1991; Rouse 1987) that, in contrast to empiricism, views the social world and the laws that govern it as having an independent "real" existence apart from what the individuals in it experience or observe. While realist assumptions may inform approaches ranging from game theory and neorealist international relations (e.g., Waltz 1979) to some types of constructivist theory (e.g., Wendt 1999) and historical sociology (e.g., Somers 1998; Steinmetz 1998), the strand of realism that most clearly embraces a unified science of society marked by uniform principles and standards may be traced back to Hempel's (1966) naturalistic elevation of the logic driving concept formation and scientific inquiry. This "theoretical realism" (Kiser and Hechter 1998) not only privileges ontology over epistemology but also emphasizes the centrality of inferences about *unobserved* causal mechanisms so long as the inferences are logically consistent, and so long as the unobserved mechanisms generate some set of observable effects. While issues related to selection bias or sampling may be important in the development of empirical tests for propositions, these tests neither serve as the most significant basis for evaluating theoretical statements nor have the same methodological standing as the inner logic driving concept formation and formal theorizing in social science (Lalman, Oppehneimer, and Swistak 1993). As Bueno de Mesquita puts it (1985: 128): "internal logical consistency is a fundamental requirement of all hypotheses . . . [and] formal, explicit theorizing takes intellectual, if not temporal, precedence over empiricism." While his contribution in this volume adopts a more pluralist tone and acknowledges the multiplicity of methods, the unity of social scientific research is still presumed to depend first and foremost on establishing the "rigorous logical foundation of propositions," something that is supposedly achieved more reliably through mathematical modeling than through ordinary language.

[3] Interestingly, even among empiricists, enthusiasm for the methodological canons outlined by King *et al.* (1994) has been qualified at best. Some point out that these canons of research design, while appropriate for formulating certain kinds of claims about observed data, do not take into account the different priorities that scholars may be pursuing at different stages of research and may actually discourage a range of scholarly activities that may otherwise serve to illuminate spurious inferences or generate new typologies (Caporaso 1995; Collier 1995). Moreover, some qualitative research – for example, fieldwork based on archival research or interviews in different areas of the world – require overcoming distinctive challenges (involving contested historiography, data availability and reliability, and adaptation to new languages and social environments) through shifting strategies that do not lend themselves to simple rules for avoiding selection bias and constructing replicable data archives (Ames 1996; Lustick 1996).

A minimalist version of such a claim is not problematic. After all, even a Derrida or a Lyotard would shudder at the thought that their exercises in textual analysis or deconstruction might contain logical contradictions. But, the claim that inner logical consistency is greatly facilitated by the use of mathematical symbols in place of ordinary language depends on many contestable assumptions about the nature of the social world and the character of social knowledge. There is, for example, the basic assumption that mathematical modeling can, in fact, be adapted to all kinds of investigations and can replace ordinary language in most instances. As some game theorists themselves recognize, there are a myriad factors that simply cannot be represented in a meaningful way within a formal mathematical model (Myerson 1992: 66). In particular, the translation of human experiences communicated discursively by actors into mathematical symbols and functions is a complicated process into which many hidden assumptions (and even a logical fallacy or two) can be smuggled. The very process of coding processes, sequences, episodes, and patterns of social relations across time- and space-bound contexts entails having to make a long list of claims about the meaning of each act and event, claims that often are not themselves made explicit or subjected to critical scrutiny. While attributing meaning to actions and events is an intrinsically problematic challenge for social science, it is unclear why a mathematical representation of complex social realities should be less problematic or inherently superior. By the same token, ordinary language is not inherently more prone to miscommunication or logical fallacy; so long as conversations continue, misunderstandings can be repaired, errors in reasoning can be identified, and equivalency can be negotiated in the use of ordinary words (Schegloff 1992).

Thus, both the empiricist and theoretical realist are able to reconcile methodological pluralism and a unified conception of method only by avoiding any meaningful discussion of the foundational positions informing the projects of different research communities. One indication that this is fundamentally an unsustainable position is implied by the fact that these two perspectives on unified methodology differ significantly on what is most important for "good" social science. The problem becomes even more evident when we consider how meaningfully any one conception of "rigor" may be extended to scholarly activities that are founded on the increasingly wide range of non-positivist or postpositivist epistemologies that have proliferated over the past century (Laudan 1996).

Those who adopt a pragmatist epistemology view social reality as essentially intersubjective and consider the articulation of explanatory theory to be fundamentally intertwined with the search for understanding; while causal explanation is not ruled out, it is limited to the identification of concrete mechanisms and processes within distinctive historical sequences or social structures that are objects of interpretation in their own right.[4] Other research communities directly reject a positive science of society, and instead proceed from the view that "the web of meaning constitutes human existence to such an extent that it cannot ever be meaningfully reduced to . . . any predefined elements" (Rabinow and Sullivan 1987: 6). Research communities engaged in more phenomenologically oriented exercises such as "thick description" (Geertz 1973), hermeneutic interpretation (Gadamer 1976; Ricoeur 1971), or ethnomethodology (Garfinkel 1967) share the view that the interpretation of human behavior in any given context primarily aims to uncover the meaning behind ordinary social practices and involves a process akin to the *translation* of texts. In addition, there is the tradition of critical theory – frequently identified with the Frankfurt School – which proceeds from the premise that all knowledge claims have implications for politics, ethics, and aesthetics, and that it is neither possible nor desirable to enforce an artificial separation between the "objective" practices of scientific investigation and the articulation of normative critique or advocacy of political action (Habermas 1987). Even more starkly relativistic is the view of many postmodernists who reject the very idea of interpreting social reality on the grounds that the subjectivity inherent in each and every human act makes it impossible to establish the "truth" of even the most basic observational statement about the social world (Fish 1989).

The point is not that the methods associated with these varied perspectives are all equally valuable for analyzing all problems; nor is it that the incommensurability of their foundational postures necessarily precludes any basis for judgments across research communities. The point is simply that it is unlikely that different research communities will ever come to share an enduring agreement on the rules for designing or evaluating the research products they generate, in part because any such agreement would require a prior agreement on a range of ontological and epistemological issues on which there has been little evidence of convergence to date (Laudan 1996: 25). Different kinds of scholarly pursuits reflect preferences for dealing with problems at a particular level of generality through approaches that require different degrees of abstraction from the time- and space-bound contexts being analyzed. Such preferences will be shaped by quite different, long-enduring, understandings

4 "Pragmatism" has a long and complicated lineage in social theory (see Joas 1993), but for the purposes of this chapter, its significance lies in the effort to negotiate a flexible "middle ground" between objectivism and relativism, and between causal explanation and interpretive understanding, Such a position is consistent with Weber's (1978) quest for "explanatory understanding."

of the extent to which aspects of social reality can or should be interpreted, and of the extent to which such an interpretation constitutes a replicable process in which the effects of an investigator's subjectivity can or should be negated by deploying universal canons of research. Under these circumstances, attempts to effect a meaningful reconciliation between a unified social science and intellectual pluralism end up privileging one or the other ideal. The translation of a uniform set of methodological principles and standards across different strategies of data analysis ultimately requires acceptance of a related set of ontological and epistemological assumptions, thereby excluding intellectual activities that proceed from a different foundation. By the same token, an unqualified defense of intellectual pluralism effectively renders meaningless even a modest attempt to bound the social sciences, ultimately leading to the kind of anarchism reflected in Feyerabend's encouragement of the proliferation of incommensurable theories. A more practical alternative in view of the present intellectual environment may be a *constrained* pluralism based on cultivating an understanding of what definitions and standards of "rigor" are employed by research communities themselves given the particular objectives, constraints, opportunities, and expectations their members take on when they choose to explore a given problem at a certain level of abstraction using a particular set of concepts and methods.

To identify the distinctive logic behind this notion of constrained pluralism, it may be helpful to think about the relations within and among research communities in light of an analogy I have previously explored between social scientific research and the evolution of a "social division of labor" as articulated in the work of Emile Durkheim (Sil 2000a). For Durkheim, whereas a simple division of labor features a "mechanical solidarity" arising from a similarity of roles among individual members, a more complex division of labor is sustained by an "organic solidarity" among individuals who focus on different activities but also develop a shared, abstract consciousness of the interdependence among their roles and tasks (Durkheim 1984: 230). In the context of the questions addressed in this volume, if we regard an ideal-typical research community as characterized by a rough *similarity* in the kinds of approaches adopted by members, then it is possible to conceptualize the relationship of research communities to one another and to the wider arena of social science in terms analogous to the differentiated roles evident in a more complex division of labor. According to this logic, the insistence on a uniform set of methodological principles and standards corresponds to a view of the social sciences as a simple division of labor sustained by mechanical forms of solidarity and a likeness of tasks and roles.

A position of constrained pluralism, by contrast, proceeds from a view of social science as a *complex* division of labor within which discrete research communities may be regarded as performing distinct, interdependent roles in the process of applying different methods to generate different kinds of intellectual insights into different sets of problems.

The notion of treating clusters of research in terms of their distinct roles in a scholarly "division of labor" is not meant to suggest that there is some consensus on what the collective product of social scientific research ought to be, or on how individual scholars should develop and deploy their skills and expertise. In fact, the one significant limitation of the Durkheimian analogy is that the notion of specialization in a social division of labor presupposes a view of production that is *not* applicable to a social scientific research: research products do not lend themselves to aggregation in the way that the roles and tasks of individuals would in a complex organization designed to achieve pregiven objectives. Nevertheless, since methodological debates have such a powerful and divisive impact on the work and lives of social scientists (in terms of hiring, tenure, grants, and basic civility), the analogy seems useful at least for the purpose of articulating a flexible view of social science in which the ideals of unity and pluralism are reconciled not through essentially similar methodological principles but through a shared awareness of the distinctive payoffs and limitations associated with different kinds of endeavors.

Cultivating such an awareness requires translating commonplace statements supporting, criticizing, or defending the typical products of different research communities into a description of the distinctive roles each might be regarded as playing within the collective process of making sense of the social world through words, numbers, and symbols. Elsewhere, I have made an initial attempt at articulating the tradeoffs accompanying the decision to adopt different kinds of methods in relation to problems cast at different levels of abstraction (Sil 2000a). Here, Figure 14.1 provides a sense of what these tradeoffs might be by juxtaposing a series of ideal-typical research products along a continuum that captures a range of two-dimensional choices between extreme forms of nomothetic and ideographic analysis. The first dimension, represented by the vertical axis, is intended to capture variations in the degree of parsimony or complexity in the interpretation of some set of empirical observations. This axis captures the substantial variation between abstract models characterized by a minimum number of symbols and operations, on one extreme, and detailed narratives presenting a large number of observations about concrete social contexts. The second dimension, represented by the horizontal axis, captures a range of foundational perspectives concerning the possibility and desirability of uncovering generalized propositions in relation

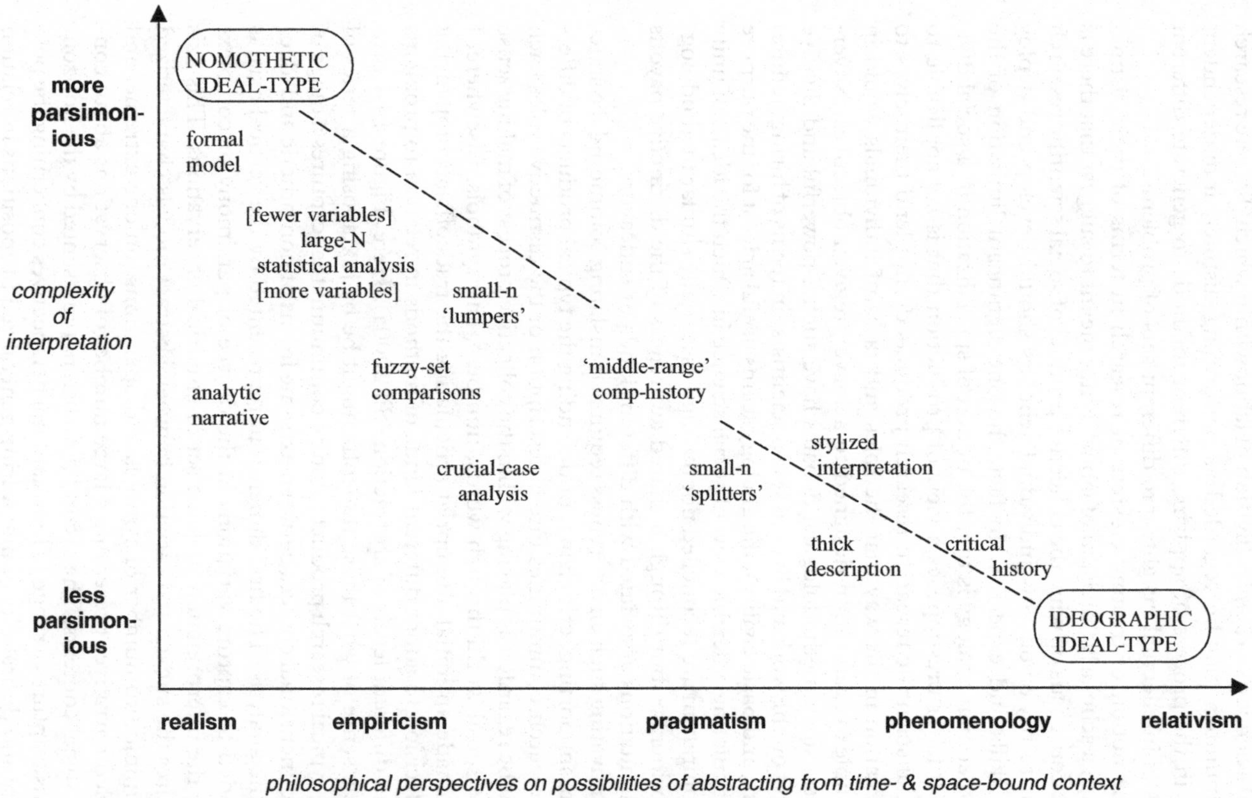

Figure 14.1 Varieties of Research Products in the Social Sciences.
Adapted from Sil (2000a, 519).

to processes identified in specific time- and space-bound contexts. This axis stretches between the aforementioned positions of theoretical realists and empiricists on one side to phenomenological and relativist perspectives on the other, with pragmatism representing an ideal-typical center along this axis. Combining the extremes along the two axes yields ideal-typical versions of nomothetic and ideographic research, but as the putative locations of typical research products in Figure 14.1 suggest, there is no reason to assume a one-to-one correspondence between a given epistemological position and a particular level of parsimony.

For the purposes of this volume, the significant implication of Figure 14.1 is this: the decision to pursue an interpretation at a certain level of complexity or parsimony, together with the epistemological perspective reflected in one's express choice of goals and methods, results in distinctive orientations toward the subject matter being studied as well as different combinations of intellectual payoffs and limitations. This holds true even for diversity-oriented approaches that recognize the tradeoffs inherent in alternative methods and seek to combine particular elements of certain methods to split the difference between complexity and parsimony.[5] Figure 14.1 essentially suggests that whatever unity may be possible across the social sciences depends on first establishing a common understanding of the correspondence between varied methodological approaches, the epistemological postures these reflect, the kinds of narratives these generate, and the specific tradeoffs each faces in its attempt to shed light on a given problem.

[5] This is evident in the case of both the "middle-range" comparative-historical approach I myself have sought to define and pursue (Sil 2002: 47–50, 295–300) as well as the distinctive brand of qualitative comparative analysis (QCA) developed by Ragin (2000). Both approaches explicitly acknowledge the diversity of social scientific methods and seek to take advantage of this diversity by seeking to combine features of methods aiming to produce analyses with different degrees of complexity or parsimony. However, what I call "middle-range" comparison proceeds from a pragmatist foundation, seeking to negotiate not only the difference between complexity and parsimony, but also the wide epistemological chasm between realist and relativist perspectives. For this purpose, an appropriate strategy continues to rely on small-N comparison, but aims to link problem-specific analytic constructs to case-specific interpretations that are attentive to the problems created by multiple historiographic traditions. Ragin, by contrast, is squarely within the empiricist tradition and is concerned primarily with bridging the gap between large-N and small-N studies that also follow from this tradition. His most recent articulation of QCA aims to limit the respective shortcomings of, and combine the advantages of, large-N and case-study approaches by employing an intermediate number of cases and comparing them with regard to variables each with measures reflecting degrees of membership in "fuzzy sets." Such efforts are fruitful in that they enable new strategies for bounding and linking the analytics and empirics of a given problem, but this advantage comes with its own tradeoffs (for example, the opportunity cost of sacrificing both the elegance of a formal model and the nuance of a context-sensitive historical interpretation).

This implies that principles and standards normally used in evaluating the work of those belonging to one research community can be fairly translated for the purpose of evaluating work produced by another only where the typical research products of both communities can be located roughly around the same position along the graph above. In all other cases, the principle of constrained pluralism requires not the *translation* of a uniform set of methodological postulates for each and every location on the graph but rather the *extrapolation* of different sets of rules and standards, each reflecting a distinct logic related to the specific objectives, constraints, and expectations attached to a given scholarly endeavor. This process of extrapolation must necessarily be a negotiated, diaological, and incomplete one, and, for this very reason, it can provide the focal point for a more inclusive conversation among members of discrete research communities about what *kinds* of insights are associated with what *types* of methods. Figure 14.1 is only one way of mapping how the work done by different research communities may be understood in relation to each other within a wider arena of social scientific research. Similar undertakings by others and, more importantly, a more inclusive discussion of the issues involved might go a long way toward providing a common understanding of what kinds of expectations are appropriate for the scholarly endeavors of diverse research communities in view of the distinctive payoffs, limitations, and tradeoffs inherent in these endeavors.

The role of eclecticism: communication and experimentation

In advancing this process, a particularly useful function is served by self-consciously eclectic approaches – approaches that are not readily identifiable with existing research communities although they incorporate analytic vocabularies, interpretive logics, and empirical data from several of them. Such approaches cannot produce elegant theories and cannot be evaluated against existing approaches in terms of methodological sophistication or expertise on a problem as posed by members of a given research community. What they can do is articulate potentially significant complementarities between problems, analyses, interpretations, and observational statements that may otherwise remain hidden because of the different analytic vocabularies through which socially important problems are posed and investigated by bounded research communities. Without dismissing the contributions of those working within an established research community, it is important to recognize not only the distinctive substantive insights eclectic approaches can generate but also their role in facilitating greater communication and experimentation across a wider range of research communities across the social sciences.

By eclecticism, I do not mean synthesis. Given the metatheoretical, substantive and methodological positions demarcating research communities, even "coalitions of the willing" will find that there are analytic boundaries that preclude synthesis (Ruggie 1998: 37). This may be why recent efforts to explore the interface between rational choice theory or game theory, on the one hand, and interpretive or comparative-historical work, on the other, have so far generated not syntheses but a juxtaposition of alternative approaches aimed at demonstrating the value of each in illuminating different dimensions of a given problem (e.g., Bates *et al.* 1998; Mahoney 2000; Munck 2001). While such exercises reveal an impressive array of skills and foster an atmosphere conducive to the sort of intellectual diversity being embraced here, it is important to recognize that genuine theoretical synthesis requires something very rare and extraordinary: a marked departure from the distinctive core metatheoretical assumptions, conceptual formulations, and intellectual objectives characterizing each of several approaches in order to generate a new array of problems, concepts, methods, and interpretive logics. This would result in the formation of a larger research community whose members can claim to subsume the intellectual products of the preexisting research communities. In the absence of such a wholesale transformation, we are usually left with a juxtaposition of separate research products accompanied by rhetorical gestures in the direction of pluralism (Sil 2000b: 369–76).

What I refer to as eclecticism is less ambitious than theoretical synthesis, but more significant than the juxtaposition of different approaches. Eclectic modes of analysis are distinguished by the fact that features of analyses embedded in separate research communities are separated from their respective frameworks and recombined as part of an original permutation of concepts, methods, analytics, and empirics. Because research communities are not usually so rigid as to produce uniform research products that will predictably converge on substantive interpretations, explanations, or predictions, it is possible to suspend or adjust metatheoretical postulates to permit a more direct comparison between, and greater integration across, components of research products that deal with similar or related phenomena even if these are normally transformed into distinct analytic puzzles in different theoretical languages. Such an effort, however creative and innovative, is understood to be eclectic rather than synthetic in that it is tailored to a given problem and does not represent a common foundation for a new, unified tradition of scholarship around which a new research community can form. But, this is precisely what enhances the distinctive value of eclecticism: its role in enabling greater scope for communication and experimentation across research communities.

Whatever the specific mode of disaggregation and recombination, what ultimately makes eclecticism a worthwhile endeavor is not the greater likelihood of theory cumulation or of novel empirical discoveries, but the possibilities of posing and exploring more complex sets of intellectual problems that more closely approximate the actual problems faced by actors in the social world. Most intellectual products generated by research communities inherently focus on those aspects of the social world that can be more readily problematized using a preferred conceptual and methodological apparatus. There is nothing inherently wrong with paring down the complexity of social life into manageable problems that can be meaningfully addressed using the skills and expertise at the disposal of a given research community. Such analytic accentuation can be fruitful and is sometimes necessary in light of practical research constraints, enabling scholars with similar backgrounds and skills to more thoroughly examine selected aspects of a problem. But, following Shapiro (2002: 593), it is important to recognize the fundamental distinction between *analytic* problems as framed in terms of the theoretical priors shared by members of a research community and *concrete* problems that can be posed in ordinary language and are of interest to scholars as well as other actors. In organizational terms, we may say that privileging the former at the expense of the latter is sometimes necessary and helpful as it enables more fluid communication and more facile evaluation within research communities, but this also intensifies the compartmentalization of research communities within the larger arena of social scientific research.

The significance of eclecticism lies not in its ability to overcome this tradeoff but rather its ability to reverse it. The benefits of engaging in scholarship mediated by research communities – the cultivation of a recognizable professional identity, efficient communication based on shared stocks of knowledge and methodological skills, a common set of evaluative standards linked to explicit methodological assumptions, and the psychological and institutional support provided by fellow members – are sacrificed for the purpose of posing a problem in terms that more closely approximate the complexity and messiness of the social world and that can be explored through different permutations of concepts, data, and interpretive logics originally constructed within separate pre-existing research communities. The case for analytic eclecticism then is dependent not on its ability to solve specific problems already identified by one or another research community, but on the possibility of *expanding the scope of research problems beyond that of each of the competing research communities.*

For the purposes of recasting the social sciences as a more unified (though not uniform) social activity akin to an evolving division of labor, the expanded scope of problems serves to open further channels for communication across more a wide-ranging set of research communities, providing opportunities for members of separate research communities to increase their familiarity with the range of metatheoretical perspectives, theoretical languages, and methodological traditions available to probe substantive problems while also becoming more conscious of the intellectual payoffs, tradeoffs, and expectations that characterize other research communities. Extrapolating from Mead's (1964) pragmatist treatment of the basis for social integration, we might say that eclecticism enables a more closely interrelated set of research communities by permitting one to take on the roles and perspectives of members embedded in varied research communities.

This logic also points to the possibilities for lowering other barriers to intellectual communication. One is the barrier between public administration and traditional academic disciplines which, as Lindblom (1959) has pointed out, is partially a function of research communities' unquestioning elevation of particular replication methods, none of which bear much resemblance to the more incremental process of "muddling through" evident in the efforts of administrators or policymakers to grapple with complex social problems. Another barrier is that between research communities embedded in different national or organizational settings which can have the effect of privileging intellectual concerns and interpretive styles that have historically emerged in some regions while foreclosing certain lines of interpretation that have more resonance in other regions (as is evident in European debates over the effects of American scholars' concerns in international relations theory (Makinda 2000)). By "recomplexifying" analytic problems favored by research communities in particular settings, eclecticism increases our chances of translating these problems in ways that make them more recognizable and relevant to administrators, policymakers, and scholars in varied social settings.

A commitment to eclectic approaches does entail costs and risks. First, eclectic approaches lack the "protective belt" (Lakatos 1970) that can shield substantive analyses from questions about core epistemological and methodological assumptions; they also lack the standards and norms that enable research communities to reward individual contributions and proclaim some degree of internal progress. Second, eclectic approaches are likely to leave themselves open to a wider range of criticism motivated by standards and practices developed within the various research communities whose products are being disaggregated and selectively redeployed in a different type of analysis. Third, eclectic analyses also share with attempts at theoretical synthesis the dangers of conceptual muddiness as they work with multiple analytic languages in conjunction with a wider range of empirics cast at different levels of generality (Johnson 2002). Finally, given the investment of time and energy required to meaningfully engage members of different research communities, it may be tempting for eclectic researchers to ignore potentially relevant theories and studies or to settle for translating existing research into different analytic languages; in this case, there would be neither much value-added in terms of substantive insights, nor much of a contribution to communication and experimentation across research communities.

Indeed, eclecticism that is unresponsive to existing discourses in a field and inattentive to the dangers of conceptual fuzziness can lead to uncompelling, structurally flawed, arguments that add little value to the work being done in a more systematic fashion within research communities. But, the unconstrained proliferation of eclecticism at the expense of more disciplined scholarship in research communities is not what is being advocated here. What is being advocated here is the *accommodation* of scholars who consciously, selectively frame their problems in new ways precisely because of a suspicion that existing modes of posing and addressing problems in different research communities are obscuring connections and complexities that need to be revealed, investigated, and at least partially, theorized. In this sense, the meaning and value of eclecticism depends entirely on the work being done in separate research communities, for without this work, the main benefit of eclecticism – enhancing communication between research communities and revealing connections and complementarities across clusters of problems and interpretations – would be lost.

There is still the question of whether the limited resources available for social scientific research should be expended on eclectic projects in the absence of prior indications of their utility and of the standards to which they may be held. As Sanderson (1987: 321) puts it in his vigorous critique of eclecticism: "one cannot follow up on every conceivable lead. Because the vast majority of these leads will not produce anything of value, to follow this strategy is to waste enormous amounts of time and energy. A better alternative . . . is to adopt the theoretical tradition that seems at the time most useful and to follow it as intensively as possible." For better or for worse, the track record of research traditions and communities does not justify such a conclusion except in the eyes of their members. The fundamental problem of devising appropriate indicators of "progress" for the social sciences as a whole remains as elusive as before, and devoting resources only to scholarship that is disciplined by separate research communities only serves to perpetuate this problem. While eclecticism does not by any means resolve the issue, eclectic researchers are in a position to recognize and evaluate what is going on in different research communities concerned with different aspects of problems that may be, in empirical terms, very much intertwined. For this reason alone, it is possible and desirable to gamble at least *some* of our resources on the intuition that eclectic modes of analysis can advance our collective ability to communicate across the boundaries of research communities and, in the process, to create greater space for experimenting with different permutations of concepts, theories, and empirics in relation to complex problems encountered in the social world. Even if the payoffs from eclecticism are not immediately evident and even if issues of methodological rigor are left unsettled, the risk that eclectic research may lead to a waste of time and resources is not so great as the risk that we will simply miss opportunities for making sense of real-world problems because of the different lenses through which separate research communities choose to analytically view these problems.

Conclusion

This chapter does not represent a call for intellectual pluralism for its own sake. Rather, it is an effort to proactively tackle issues arising from social scientific research as presently organized and practiced, an effort that is only partially supported by greater pluralism. The more significant challenge has been to outline a tentative framework for identifying and relating the different kinds of intellectual payoffs, limitations, and tradeoffs inherent in the scholarly activities of research communities differentiated by distinctive combinations of problems, methods, theoretical vocabularies, and related metatheoretical orientations. This is a challenge that is not likely to be met either by those who insist on the universality of a particular set of methodological principles and standards or by those who simply reject the very idea of discipline or rigor, although both positions may be seen as founded on reasonable philosophical positions. In fact, given the impossibility of asserting a common epistemic foundation for the unity of the social sciences, the best that can be achieved is an evolving common framework, however abstract, that enables members of one research community to recognize the distinctive challenges and insights associated with the work pursued by other research communities, and to appreciate the basis for the standards employed by the latter in light of the relevant assumptions, objectives, constraints, and expectations.

In this process, I have argued, the accommodation of eclectic approaches can help in two ways: providing more channels for communication between members of otherwise discretely bounded research communities, and creating space for experimenting with different permutations of concepts, methods, data, and interpretations in the process of teasing out complementarities across problems that are analytically sliced in different ways in separate research communities. Such eclecticism entails costs and risks that can only be justified insofar as eclectic research is responsive to the more disciplined work done in research communities. Where this is the case, eclecticism increases the chances that the social sciences as a whole might hit upon agreements about socially important problems and solutions that would otherwise elude adherents of particular research communities working with separate understandings of "rigor" and "progress."

This is, in fact, the basis for the middle ground between Feyerabend and Laudan anticipated above. The former's epistemological anarchism can be moderated to suggest that problems of incommensurability are not absolute given the possibilities for translation by those less invested in particular research communities; and the latter's conception of "problem-solving progress" can be refined to emphasize that "amalgamation" across research traditions is often critical to identifying new problems and opening up of new lines of research (Laudan 1977: 104). That is, Feyerabend's position can be viewed as moving toward Laudan's if it is interpreted as a defense of the proliferation of different kinds of research practices – some eclectic, some within research communities – rather than an injunction to individual scholars to embrace anarchism; and Laudan's position can be viewed as moving towards Feyerabend's if the focus on problem-solving is interpreted as a statement challenging the use of uniform methodological principles as the main basis for evaluating work produced by different research traditions. Thus, the simultaneous proliferation of research communities and eclectic approaches does not suggest either an unconstrained intellectual pluralism or a sweeping rejection of method but rather an intellectually fruitful component of a still more complex division of labor. The diverse products generated by research communities may be expected to satisfy different standards depending on the character of the problems investigated and the opportunities and constraints inherent in the methods employed; eclectic approaches would be evaluated in terms of the opportunities they provide for disaggregating and recombining constituent elements of varied research products in order to shed light on complex social phenomena that are only partially problematized within existing research communities. Such a tentative arrangement may not constitute a comprehensive answer to the question of the relationship between problem and method in social scientific research, but in light of the current state of affairs, less may be more.

REFERENCES

Ames, Barry. 1996. "Comparative Politics and the Replication Controversy." *APSA-CP: Newsletter of the APSA Organized Section in Comparative Politics* 7(1) (Winter).

Bates, Robert, Avner Greif, Margaret Levi, Jean-Laurent Rosenthal, and Barry Weingast. 1998. *Analytic Narratives*. Princeton: Princeton University Press.

Bhaskar, Roy. 1986. *Scientific Realism and Human Emancipation*. London: Verso.

Bueno de Mesquita, Bruce. 1985. "Toward a Scientific Understanding of International Conflict." *International Studies Quarterly* 29: 121–36.

Caporaso, James. 1995. "Research Design, Falsification, and the Qualitative-Quantitative Divide." *American Political Science Review* 89(2) (June): 457–60.

Collier, David. 1995. "Translating Quantitative Methods for Qualitative Researchers: The Case of Selection Bias." *American Political Science Review* 89(2) (June): 461–6.

Devitt, Michael. 1991. "Aberrations of the Realism Debate." *Philosophical Studies* 61: 43–63.

Durkheim, Emile. 1984 [1933]. *The Division of Labor in Society*. New York: Free Press.

Feyerabend, Paul. 1993. *Against Method* (3rd edn). London: Verso.

Fish, Stanley. 1989. *Doing What Comes Naturally*. Durham: Duke University Press.

Gadamer, Hans-Gorg. 1976. *Philosophical Hermeneutics*. Berkeley: University of California Press.

Garfinkel, Harold. 1967. *Studies in Ethnomethodology*. Englewood Cliffs, NJ: Prentice-Hall.

Geertz, Clifford. 1973. *The Interpretation of Cultures*. New York: Basic Books.

Goldthorpe, John. 1996. "Current Issues in Comparative Methodology." *Comparative Social Research* 16: 1–26.

Habermas, Jurgen. 1987. "Modernity – an Incomplete Project," in Rabinow and Sullivan (eds.).

Hempel, Carl G. 1966. *Philosophy of Natural Science*. Englewood Cliffs: Prentice-Hall.

Joas, Hans. 1993. *Pragmatism and Social Theory*. Chicago: University of Chicago Press.

Johnson, James. 2002. "How Conceptual Problems Migrate: Rational Choice, Interpretation and the Hazards of Pluralism." *Annual Review of Political Science* 5: 223–48.

Katzenstein, Peter and Rudra Sil. 2004. "Rethinking Security in East Asia: A Case for Analytical Eclecticism," in Jae-Jung Suh, Peter Katzenstein, and Allen Carlson (eds.), *Rethinking Security in Asia: Identity, Power and Efficiency*. Stanford: Stanford University Press.

King, Gary. 1995. "Replication, Replication." *Political Science* 28(3) (September): 444–52.

King, Gary, Robert Keohane, and Sidney Verba. 1994. *Designing Social Inquiry: Scientific Inference in Qualitative Research*. Princeton: Princeton University Press.

Kiser, Edgar and Michael Hechter. 1998. "The Debate on Historical Sociology: Rational Choice Theory and its Critics." *American Journal of Sociology* 104(3) (November): 785–816.

Kuhn Thomas. 1962. *The Structure of Scientific Revolutions*. Chicago: University of Chicago Press.

Lakatos, Imre. 1970. "Falsification, and the Methodology of Scientific Research Programmes," in Imre Lakatos and Alan Musgrave (eds.), *Criticism and the Growth of Knowledge*. New York: Cambridge University Press. 91–196

Lalman, David, Joe Oppehneimer, and Piotr Swistak. 1993. "Formal Rational Choice Theory: A Cumulative Science of Politics," in Ada Finifter (ed.), *Political Science: The State of the Discipline II*. Washington, DC: American Political Science Association.

Laudan, Larry. 1977. *Progress and its Problems: Toward a Theory of Scientific Growth*. Berkeley: University of California Press.

1996. *Beyond Postivism and Relativism*. Boulder: Westview.

Lindblom, Charles. 1959. "The Science of 'Muddling Through'." *Public Administration Review* 19 (Spring).

Lustick, Ian. 1996. "History, Historiography, and Political Science: Multiple Historical Records and the Problem of Selection Bias." *American Political Science Review* 90(3) (September): 605–18.

Mahoney, James. 2000. "Rational Choice Theory and the Comparative Method: An Emerging Synthesis?" *Studies in Comparative International Development* 35(2) (Summer).

Makinda, Samuel. 2000. "International Society and Eclecticism in International Relations Theory." *Cooperation and Conflict* 35(2) 205–16.

Mead, George Herbert. 1964 [1956]. *On Social Psychology*. Chicago: University of Chicago Press.

Munck, Gerardo. 2001. "Game Theory and Comparative Politics: New Perspectives and Old Concerns." *World Politics* 53(2) 173–204.

Myerson Roger. 1992. "On the Value of Game Theory in Social Science." *Rationality and Society* 4: 62–73.

Rabinow, Paul and William Sullivan (eds.). 1987. *Interpretive Social Science: A Second Look*. Berkeley: University of California Press.

Ragin, Charles. 2000. *Fuzzy-Set Social Science*. Chicago: University of Chicago Press.

Ricoeur, Paul. 1971. "The Model of the Text: Meaningful Action Considered as Text." *Social Research* 38: 529–55.

Rouse, Joseph. 1987. *Knowledge and Power*. Ithaca: Cornell University Press.

Ruggie, John G. 1998. *Constructing the World Polity*. New York: Routledge.

Sanderson, Stephen K. 1987. "Eclecticism and its Alternatives," in John Wilson (ed.), *Current Perspectives in Social Theory*. Greenwich, CT: JAI Press.

Schegloff, Emanuel. 1992. "Repair after Next Turn: The Last Structurally Provided Defense of Intersubjectivity in Conversation." *American Journal of Sociology* 97(5) (March).

Shapiro, Ian. 2002. "Problems, Methods, and Theories in the Study of Politics, or What's Wrong with Political Science and What to Do About It?" *Political Theory* 30(4) (August): 588–611.

Sil, Rudra. 2000a. "The Division of Labor in Social Science Research: Unified Methodology or 'Organic Solidarity'?" *Polity* 32(4) (Summer): 499–531.

2000b. "The Foundations of Eclecticism: Agency, Culture and Structure in Social Theory." *Journal of Theoretical Politics* 2(3) (July): 353–87.

2002. *Managing "Modernity": Work, Community, and Authority in Late-Industrializing Japan and Russia*. Ann Arbor: University of Michigan Press.

Sil, Rudra and Peter Katzenstein. 2003. "Analytic Eclecticism in the Study of Asian Security." Presented at the Annual Meeting of the International Studies Association, Portland, Oregon, February 26–March 1.

Skocpol, Theda, and Margaret Somers. 1980. "The Uses of Comparative History in Macrosocial Inquiry." *Comparative Studies in Society and History* 2(2) (April): 174–97.

Somers, Margaret. 1998. "We're No Angels: Realism, Rational Choice, and Relationality in Social Science." *American Journal of Sociology* 104(3) (November): 722–84.

Steinmetz, George. 1998. "Critical Realism and Historical Sociology." *Comparative Studies in Society and History* 40: 170–86.

Tönnies, Ferdinand. 1957. *Community and Society [Gemeinschaft und Gesellschaft]*. Charles P. Loomis (trans. and ed.), New York: Harper and Row.

Waltz, Kenneth. 1979. *The Theory of International Politics*. Reading, MA: Addison-Wesley.

Weber, Max. 1978. *Economy and Society, Volume 1*. Berkeley: University of California Press.

Wendt, Alexander. 1999. *Social Theory of International Politics*. New York: Cambridge University Press.

15 Method, problem, faith

William E. Connolly

The shape of contending problematics

The question we are invited to address, as I understand it, is whether it is better to be governed in research by the imperative of a method or the pressure of a political problem. I suppose those who give primacy to method would say that they focus on issues most congenial to the favored method to build a reliable fund of knowledge that can later be brought to bear on problems that now seem intractable. Those who give primacy to the problem might say that critical issues – such as the sources of terrorism, the sources of urban crime, the rise of cross-state religious fundamentalism, or the effects of global capital – need to be engaged now, even if popular methodological approaches are not currently well equipped to respond to them. Stated in this way, I go with those who give priority to the problem.

My sense, however, is that the question reflects a presupposition in need of examination: that those who focus on problems and those who focus on method can be separated neatly. To me, an intervening "variable" compromises the question. Let me put it this way: a particular orientation to method is apt to express in some way or other a set of metaphysical commitments to which the methodist is deeply attached; and a close definition of a political problem is apt to be infiltrated by similar attachments. I do not mean anything technical by "metaphysical commitment" in this context. It is merely the most profound image of being or the world in which your thinking is set. Differences in world image between Plato, Sophocles, Augustine, Epicurus, Spinoza, Kant, Nietzsche, Freud, James, Charles Taylor, Bertrand Russell, Jon Elster, and Merleau-Ponty might all be taken to be metaphysical in this sense. Freud, James, and Merleau-Ponty diverge not only in the method of dream analysis each pursues, but also in their faiths about the shape of the world in which dreaming occurs. If, however, you flinch at the heavy term "metaphysical," I will drop it in favor of another phrase.[1] Indeed, "existential faith" highlights something that the cold term "metaphysic" downplays. An existential faith is a hot, committed view of the world layered into the affective dispositions, habits and institutional priorities of its confessors. The intensity of commitment to it typically exceeds the power of the arguments and evidence advanced. On my reading, then, each thinker listed above is a carrier of a distinctive existential faith. The faith in which each is invested has not yet been established in a way that rules out of court every perspective except it. It is a contestable faith. This is not to deny that impressive, comparative considerations might be offered on its behalf, or that it might be subjected to critical interrogations that press its advocates to adjust this or that aspect of it. An existential faith is not immune to new argument and evidence, as I will try to show; commitment to it, rather, is seldom exhausted by them.

Existential faith is ubiquitous to life. Not only Jews, Christians, and Islamists are invested in it. To be human is to be invested by and in existential faith before you reach the age of critical reflection, even if these investments are often laden with ambivalence, periodically marked by crisis, and potentially susceptible to reconfiguration. A shift in faith becomes lodged on the visceral register of being as well as the more refined intellectual register to which the first is connected. Moreover, on my reading, an existential faith is not fully insulated from your public views and judgments. The private/public division is not as sharp as some pretend. There is a degree of separability, but the line is more like a porous membrane than an iron curtain.[2]

Some existential faiths are theistic, others non-theistic. Some discern an intrinsic purpose in being; others project a world covered by rigorous laws; others yet project protean forces into an undesigned world not entirely susceptible to law-like explanation. Yet other faiths – such as that of many Buddhists and American devotees of William James – are lodged in a fugitive space between the options of theism and non-theism. On another register, some faiths appear to many at one time to be defeated by the rise of new evidence or new argument, as certain

[1] The excellent study by Stephen White (2002) is pertinent here. White examines comparatively the work of George Kateb, Charles Taylor, Judith Butler, and myself. He does so, first, to bring out the "weak ontology" of each and the role it plays in their different theories, and, second, to assess them to see how effectively and honestly they come to terms with the element of contestability in the distinctive ontology that informs the work of each.

[2] Hence I am unable to endorse the view of a variety of academic secularists, Habermasians, Arendtians, and postmodernists who proclaim modernity (and themselves) to be "post-metaphysical."

proponents of the Enlightenment thought Christianity was by the Newtonian revolution. But a philosophical response to that apparent defeat might emerge, as for instance, when Kant sought heroically to make Newtonianism and Christianity safe for each other again. Or consider how the Epicurean idea of a swerve in nature was put to rest in early natural science because of the discrepancy between its unruliness and the new, law-like image of the world, only to return in revised form later. Or how the Catholic Church eventually rendered its doctrine roughly compatible with the theory of Galileo, after condemning it for centuries, and with a version of evolutionary theory, after resisting it for a century.

An existential faith, again, is not immune to argument, evidence, and theoretical revolution. Its most reflective proponents are immersed in them and haunted by them. And some faiths bite the dust. But a persistent, affectively imbued existential faith can marshal impressive resources to protect, amend, or revise itself modestly in the face of new and unexpected developments. It can protect core elements through new "clarifications." The reemergence of theories of "intelligent design" – presented as an alternative to the law-like model of evolution through natural selection – represents a recent example. It will not be surprising to see the theory of intelligent design replace Creationism and become more sophisticated in the years ahead.

What, more closely, are the relations between an existential faith, the method appropriate to it, and the political problems congenial to it? My sense is that these dimensions are connected, but that you can seldom simply read off the connections by describing a position globally as "Aristotelian," "Kantian," or "Nietzschean." The devil always resides in the details. It is thus necessary to work with specific cases.

A few formulations about method and faith from another time and place might help to illuminate this issue. *On Christian Doctrine*, by Augustine, was a problem-driven book, set in the fraught political context of the Church and the Roman Empire in the fifth century "AD." I pursue Augustine on this front to become more thoughtful about the relations between faith, problem, and method. The comparisons I make to contemporary traditions, however, are merely suggestive; more thick delineations would be required to disclose the linkages in these cases.

Augustine sought to resolve problems that arise when holy Scripture, taken to reveal the true faith, contains formulations that appear to transgress that very faith. To resolve this thorny problem he devised an authoritative method for reading the sacred text. The first rule is a "rule of faith":

This faith maintains, and it must be believed: neither the soul nor the human body may suffer complete annihilation, but the impious shall rise again into everlasting punishment, and the just into life everlasting. (Augustine: 30)

When you consult Scripture you must follow this rule of faith. Otherwise you cannot participate in the spirituality that permeates it; and your interpretations will be weak, sinful, or incoherent. Such a rule might be compared to an operational rule of faith (or "regulative ideal") in rational choice theory, that preferences are individual, orderable, and transitive.[3] To *say* that the latter is merely a convenient assumption of inquiry may not suffice. The question is whether it plays a more fundamental role in the thinking of those governed by it, and whether elaborate strategems are deployed to protect it as an assumption of inquiry. Augustine's rule of faith might be compared, as well, to the operational faith of most conventional empiricists and rational choice theorists that the analytic-synthetic dichotomy is an elemental distinction that everyone must adopt, and that the world is susceptible in principle to rigorous explanation through general laws.

[3] As Donald Green and Ian Shapiro (1994: 14–15) summarize rational choice theory, rational behavior is "typically identified with 'maximization of some sort' . . . it must be possible for all of an agent's options to be rank ordered . . . " and "reference orderings are transitive" so that if "A is preferred to B, and B is preferred to C then the consistency rule requires that A be preferred to C." As they also say, such a theory treats the preferences of individuals to be basic data, rather than exploring how identities and interests are formed through historically specific conjunctions of child rearing, church attendance, educational experience, searing personal experiences, surprising public events, etc. Green and Shapiro, however, test the evidentiary success of rational choice theory against a model of empiricism not itself put up for critical scrutiny. I am also fond of an earlier challenge to empiricism, perhaps comparable to this critique of rational choice theory (Connolly 1967).

Our review of the areas of contention between Dahl and Mills and of the organization of their respective studies reveals, I believe, that their resulting analyses are in large part self-fulfilling. In each case the focal points of analysis (that is, the political arena vs. the interrelations of the identified elite) and the conceptual decisions made seem to foster the interpretation rather than to provide a test of it. An analysis of the testing operations actually employed by pluralists and elitists will solidify this appraisal . . . moreover, each reinforces his own position by making favorable assumptions about key (and for practical purposes, untestable) aspects of the system under investigation. (p. 30)

What if neither rational choice theory nor empiricism lives up in practice to the standards to which they hold others in theory? The term "ideology" deployed in that book shares something with the notion "existential faith" in this chapter. But it is difficult to divest the former of the idea that only others participate in the endeavor, and it is also not commonly extended to those whose interpretations are strongly invested with theistic faith of some sort or other. The Dahl referred to there is unlike the Dahl who made a closing address at the Yale conference at which this paper was presented. That Robert Dahl emphasized how the object of politics is more complex than that of physics and concentrated on the role of contingency in political events. He and William James are now kindred spirits.

The Augustinian rule of faith, in turn, operates in tandem with a second methodological rule, one that calls upon the reader to place each concept in the larger context in which it operates, with the most fundamental context being the rule of faith itself. This second rule comes into play when a phrase or paragraph seems to contradict the basic faith:

When investigation reveals an uncertainty as to how a locution should be pointed or construed, the rule of faith should be consulted . . . But if both meanings, or all of them, in the event that there are several, remain ambiguous after the faith has been consulted, then it is necessary to examine the context of the preceding and following parts of the ambiguous places, so that we may determine which of the meanings which suggest themselves it would allow to be consistent. (79)

This methodological stricture bears a resemblance to those articulated today in hermeneutical, interpretive, and phenomenological approaches to political inquiry. Both he and they play up the interwoven character of language, as well as the role of context in determining meaning and developing interpretations. They would also converge in saying that contemporary empiricist and rationalist images of being, in adopting the analytical-synthetic dichotomy and designative philosophies of language, underplay the open texture and complex web of connections constituting the intersubjective life of linguistic beings. Many practitioners of the hermeneutical tradition seek contact with a *trace* or *whisper* of transcendence in the world. Augustine shares a lot with them, though he invests transcendence with more definitive presence and authority than contemporaries such as, say, Charles Taylor, Paul Ricoeur, and Merleau-Ponty do. Other contemporary interpretationists, however, do not hear the whisper of God. This, too, affects the way they proceed. We are apt to adopt a double-entry orientation to interpretation, oscillating as a matter of principle between genealogical critiques of already consolidated interpretations and the positive elaboration of revised interpretations appropriate to ever changing conditions. We are less apt to pursue "deep" interpretation than hermeneuticists and less apt to open ourselves to the soft breeze of mystical experience. The virtual register replaces mystical experience in immanent naturalism (Deleuze and Guattari 1987).

Back to Augustine. He concedes that the commands of the Church can sometimes appear to be countermanded by Scripture, even after a devout reader has adhered to the first two rules of method. In these instances a supplemental guideline is needed. That which "appears in the divine Word that does not literally pertain to virtuous behavior or to the true faith you must take to be figurative" (87–8).

Now, some sort of distinction between the literal and the figurative seems unavoidable. But once Augustine forges the distinction, he also places at risk the closure of Christian faith amidst the dangerous heresies he sees around him. So he must confine its use closely. If he does not, interpretations of Jesus as a Jew, pagan interpretations of the Book of Job (probably a pagan text initially), and a tragic reading of the Book of *Genesis* would all become highly possible (Connolly 1993; Vermes 1993). Thus the serpent in Genesis must be a figure that stands for Lucifer; it cannot embody the wisdom many pagan faiths invested in it. So Augustine insists that everything in Scripture that corresponds to the first rule of faith is literal, that only those formulations that appear to be at odds with that rule are figurative, and that the figures must each be read in an authoritative way. Now you have the makings of a true, rigorous reading of the holy text.

Augustine's division between the literal and the figurative might be compared to the distinction in contemporary empiricism between the designative dimension of language and its expressive, connotative, or performative dimensions. The two sets of distinctions do not correspond to each other, but they do function in the same way. In the empirical tradition, the designative dimension is given priority over the performative and expressive dimensions; and valiant efforts are made to demarcate the former sharply from the latter. For to admit that these two dimensions are interlocked would be to take a giant step toward an interpretive orientation to political inquiry. Both rational choice theorists and empiricists thus keep their eyes focused anxiously on this flank, to ensure that the adjustments they make in method to resolve problems in one area do not transport them too close to the dreaded territory of interpretation. As with Augustine in his relation to pagans and heretics, empiricists and rational choice theorists need such distinctions. But they also need to monitor them. This combination is accomplished less by positive demonstration of the purity in language commended, more by endless repetition of the *accusation* that advocates of other methods use language in imprecise, vague, normative, impressionistic, or ideological ways, while "we" refine our terms to be precise, rigorous, free of value connotations and susceptible to incorporation into testable propositions. Indeed, you might – against the probable wishes of the proponents – label that as a key method of self-protection adopted by empiricists and rational choice theorists. The perpetual task of language cleansing is as critical to protection of these faiths as Augustine's anxious management of the distinction between literal and figurative meanings is to his.

Still, Augustine realizes that uncertainties persist even after the reader has applied the rule of faith, attended to context, and managed the literal/figurative distinction. So he introduces a final rule:

But just as it is a servile infirmity to follow the letter and to take signs for the things that signify, in the same way it is an evil of wandering error to interpret signs in a useless way. However, he who does not know what a sign means but does know that it is a sign, is not in servitude. Thus it is better to be burdened by unknown but useful signs than to interpret signs in a useless way so that one is led from the yoke of servitude only to thrust his neck in the snares of error. (87)

Let us set aside the horrendous toll in Europe and the New World historically paid by those "idolatrous pagans" accused of confusing signs with the things they signify. We focus now on the importance of the "unknown" in the Augustinian problematic. The introduction of what might be called *the rule of mystery* does considerable work for Augustine. He invokes it when pagans ask him what his omnipotent God was doing "before" It invented time, and on other strategic occasions as well. It must not, however, be invoked *before* Augustine has revealed his God to be omnipotent, omniscient, benevolent, and salvational. It is wheeled out *after* such confessions, to protect those articles of faith from challenge or doubt.

Muted invocations of selective mystery emerge in the hermeneutic and interpretative traditions today. Is there a functional equivalent in rational choice theory and classical empiricism? Not exactly. But they do propose another rule at the same strategic point. The disdain advocates of both sometimes display toward old and "new mysterians" seems to express an underlying faith that the world must be fully explicable in the last instance. That faith is then articulated as a "regulative ideal" everyone "must" adopt if inquiry itself is to proceed. Some proponents, no doubt, deny that they harbor such a faith. But if and when it is *really* rendered problematical as a universal imperative of inquiry, the result will be displayed in a modesty of explanatory purpose, real hesitancy about the universality of the specific method adopted, more ready appreciation of a plurality of methods, and habitual forbearance toward theistic *and* non-theistic faiths that do project an element of mystery into the world.

What, then, is the relation between an existential faith and a confession of method? The review of Augustine suggests to me that they are loosely interwoven. Sometimes, as with Augustine, the basic faith secretes a method that helps to secure it and to insulate the faith from other problematics in circulation. At other times induction into a method and the successes it occasions recoils back upon the faith with which you started, encouraging adjustments in it. These two dimensions can then work back and forth upon one another until an equilibrium point emerges. I want to suggest, then, that faith and method are interconnected without asserting a general formula to cover the shape of that connection in every case. You can't, for instance, say that those who seek a place for transcendence in the world must, for that reason alone, adopt one method over another. The generic idea of "transcendence" is way too vague for that; and there are multiple ways to respect or promote these variants. There is a world of difference between, say, William James – the hesitant protestant who expresses faith in a limited God in a pluralistic universe – and Augustine – the aggressive Christian who insists upon an omnipotent God and finds so many Christian heretics to condemn. Nonetheless, the generous faith of James is relevant to the method of "radical empiricism" he embraces; his critiques of the narrowness of intellectualism, rationalism, and conventional empiricism are designed in part to foster techniques that open the door to mystical experience.

If method is inflected by faith, and a faith – once its shape is specified sufficiently – is more consonant with some methods than others, what can be done to adjudicate disputes between method/faith/problem complexes, or – as I call them – problematics? This is the triumphal question posed to people like me. My sense is that there are numerous considerations that might move you in this way or that, but no super-problematic above the fray is so closely defined and cleansed of contestable faith that it provides a sufficient, authoritative touchstone to resolve inter-method disputes. The demand for a definitive resolution is haunted by what Hegel (1977) called the "dilemma of epistemology."[4] The problem is not that there are not enough considerations to pursue. The problem is that there are so many, and that they so often point in different directions.

Moreover, a shift from one problematic to another typically involves a shift in thought-imbued sentiments layered into the visceral register of identity as well as the more refined intellectual register. A significant shift of problematic, then, is akin to a *conversion*. Here argument, evidence, identity issues, status considerations, and existential hopes jostle and mingle, sometimes fomenting a shift in one direction and sometimes triggering a reactive attitude of protection against assaults on your problematic. The intensity of methodological disputes in the social sciences flows, first, from the fact that our visceral identities and faiths are touched by these differences and, second, from the historically dubious expectation installed in this discipline in particular and secular culture in general

[4] Hegel says that the epistemologist treats the knowing faculty either as a "medium" through which knowledge is received or as an "instrument" by which to test knowledge claims. The problem is "that the use of the instrument on a thing does not let it be what it is for itself, but rather sets out to reshape and alter it." And a medium "does not receive the truth as it is in itself, but only as it exists in and through the medium" (para 73, p. 46). The dilemma is that each attempt to establish a medium or an instrument as final runs into the problem of either introducing another to justify it or repeating the first to justify itself.

that the question of method can be divorced entirely from existential faith. The sense of hubris veils the inner connection between faith and method, while the implication of visceral identities in these very issues encourages the consolidation of methodological dogmatism. Wherever there is dogmatism there is faith, but, as the example of James shows, with the right kind of tactical work on the self, there can also be faiths that cultivate a more generous orientation to other problematics.

One consideration in favor of this reading is that it helps us to understand the *intensity* accompanying methodological disputes in the human sciences, and elsewhere as well. Another is that it points toward reworking the global distinctions in secular culture between religion and philosophy, and between religious dogma and secular probabilism. Such rethinking has been going on for some time in sectors of popular culture, anthropology, and religious studies. It is perhaps timely to open up the issue in the human sciences. A third is that by paying attention to the ugly history of religious conflict we can pick up leads to follow in coping with struggles of method in the human sciences. For methodological disputes are imbued with a feeling of religiosity.

I know that all this is dangerous, and that underlying the argument against opening up the issue in the way I propose is an anxiety that the human sciences might become subject to too much outside political control. It is a worry. But doesn't the worry itself speak obliquely to the connections suggested here? My sense is that the political cover provided to the social sciences through the separation of faith from method has become more fragile and unconvincing to others than it once was. It has also helped to promote the hegemony of a couple of problematics that would otherwise not exercise so much hegemony. It is timely to confront the issue squarely, and to run the dangers that engagement with it carries.

Immanent naturalism and existential faith

So far I have proceeded as if I may float above the issues discussed. It is, for god's sake, unlikely that I embrace Augustinianism, rational choice theory, or empiricism in their classic forms. And I have recently become critical of some dimensions of interpretive theory. Partly to overturn the impression that I think the loosely textured relations between method, faith, and problem do not apply to me, I will say something about the perspective I adopt.

A quick way to distinguish my perspective from that of Augustine is to say that it is naturalistic and non-theistic. By naturalism I mean the faith that all human activities function without the aid (or obstruction) of a divine force. The specific form of naturalism I embrace, however, represents a minority confession within naturalism. It questions the *sufficiency* of the law-like model of nature endorsed in classical natural science. I concur with Charles Peirce when he says that "conformity to law exists only in a limited range of events and even there it is not perfect, for an element of pure spontaneity or lawless originality might, or at least must be supposed to mingle, with law everywhere" (quoted in Menand 1991). The view I embrace also emphasizes how nature and culture are differentially mixed into each other, depending upon the layer of embodiment at which the conjunction arises: there are crude, affectively imbued judgments and desires on the visceral register and more refined beliefs and judgments at work in the most complex human brain systems, with each layer connected to the others through a series of relays and feedback loops (Connolly 2002).

Let us define *eliminative naturalism* to be a metaphysical faith that reduces the experience of consciousness to non-conscious processes. Let us construe *lawlike naturalism* to be a view that denies any role to transcendence while finding both non-human nature and human culture to be amenable in principle to full explanation by precise laws. *Immanent naturalism*, by comparison, claims that the classical image of science lives off the remains of an old theology it purports to have left behind. According to immanent naturalism, if the world is not designed by a god it is apt to be more unruly in its mode of becoming or evolution than can be captured fully by any set of law-like explanations. On this reading, Stephen Jay Gould and Ilya Prigogine qualify as immanent naturalists in the fields of evolution and chemistry respectively. Gould (2002) doubts that if we had been in a position to know the volatile initial conditions from which biological evolution started we could have predicted the actual outcome of the evolutionary process to this point. Prigogine (1997) studies systems in "disequilibrium" which periodically issue in new formations that could not be explained or predicted in advance of their formation. This is because the capacity for self-organization within the system periodically becomes activated in new ways when it encounters novel factors from the external environment.

Perhaps the most basic projection (or existential faith) of immanent naturalism is vague essentialism. Such a projection finds differential presence in the work of Nietzsche, Prigogine, Gould, Peirce, and James. Here is an expression of it by Gilles Deleuze and Felix Guattari. They speak of immanent forces intensive enough to make a difference in an undesigned world of becoming when brought into conjunction with historically established entities, but not fixed enough to be identified as stable theoretical objects:

So how are we to define this matter-movement, this matter-energy, this matter-flow, this matter in variation that enters assemblages and leaves them? It is a destratified, deterritorialized matter. It seems to us that Husserl has brought thought a decisive step forward when he discovered the region of *vague and material essences* (in other words, essences that are vagabond, anexact and yet rigorous), distinguishing them from fixed, metric and formal essences . . . they constitute fuzzy aggregates. They relate to a *corporeality* (materiality) that is not to be confused either with an intelligible formal essentiality or a sensible, formed thinghood. (Deleuze and Guattari 1987: 407)

Thus, for them "uncertainty" exceeds the assumption of limited information marking conventional empiricist and rational choice theories; it also resides in the difference between human capacities of observation/conceptualization and the volatile character of the matter-energy flows under investigation. An example of an immanent field of matter-energy is the human body/brain system before it has become finely wired through a series of intensive exchanges with an established culture. Even after it is wired it is still traversed by surplus energies, unstable mixtures and static that might, given an unexpected shift in circumstances, issue in something new and surprising.

Perhaps the most distinctive move an immanent naturalist makes in cultural theory is the attempt to reconfigure established images of causality. The idea is to challenge the sufficiency of both efficient models of causality in social science and acausal images of mutual constitution in interpretive theory. It is not that efficient causality never operates; rather, this concept is inadequate for dealing with contexts of emergent causality. Emergent causality, when it occurs, is causal (rather than simply definitional) in that a movement at the immanent level has effects at another level. But it is emergent in that, first, the character of the immanent activity is not knowable in precise detail prior to effects that emerge at the second level, second, the new effects become *infused* into the very being or organization of the second level in such a way that the cause cannot be said to be fully different from the effect engendered and, third, a series of loops and feedback loops operate between first and second levels to generate the stabilized result.[5] The new emergent is shaped not only by external forces that become infused into it but *also by its own previously*

[5] The Deleuzian idea of change through spiral repetition is relevant here. As Jane Bennett (2002: 40) summarizes, the "point about spiral repetition is that sometimes that which repeats itself also transforms itself. Because each iteration occurs in a . . . unique context, each turn of the spiral enters into a new and distinctive assemblage." For other works that speak to emergent causation in the context of social and cultural theory see Massumi (2002) and De Landa (1997). De Landa is particularly effective in exploring the autocatalytic loops that help to generate new forms out of old molds in ecosystems and the composition of national languages.

under-tapped capacities for reception and self-organization. So the new emergent is the result of a spiraling movement back and forth between relatively open systems. If you now introduce looping entrant and re-entrant combinations between several relatively open systems, emergent causality becomes even more complex.[6]

Such relays result in the *emergence* of the new. Emergent causation issues in real effects without being susceptible to full explanation or precise prediction in advance, partly because what is produced could not be adequately conceptualized before its production. Of course, some domains taken now to exemplify emergent causation may be reduced to complex patterns of efficient causation later. But the wager-faith of the immanent naturalist is that this is not apt to happen fully or completely. The immanent naturalist neither places full explicability where Augustine located God nor denies a place for mystery as most post-Augustinian naturalists do. Rather, we *naturalize a place for mystery*, folding a modicum of it into emergent causality. Thinking, culture, and politics are sites *par excellence* of emergent causality.

In the sphere of ethics, immanent naturalism issues in an ethic of cultivation anchored in gratitude for being rather than a morality of command authorized by a God, a transcendental subject, or a fictive contractual agreement between already fixed agents. This is where the existential faith of its proponents, perhaps, becomes most visible. For its final source of ethical sustenance is drawn from attachment to the abundance of a world of becoming which, when you set each subsystem of the world into the appropriate time horizon, courses through us as well as circulating around us.[7]

6 Here, for instance, are some things Stephen Gould says about the need to rethink cause in evolutionary theory: in Darwin, "the organism supplies raw material in the form of 'random' variation, but does not 'push back' to direct the flow of its own alteration from the inside . . . By contrast the common themes in this book all follow from serious engagement with complexity, interaction, multiple levels of causation, multidirectional flows of influence, and pluralist approaches to explanation in general" (Gould: 31). Gould also emphasizes how Nietzsche's genealogical mode of analysis prefigures the approach to evolutionary theory he adopts (1214–1218).

7 Theorists in the neo-Kantian tradition are apt to say that an immanent naturalist such as me commits "the naturalistic fallacy" at this point in moving from "is to ought." But the move is only a "fallacy" if a set of prior assumptions by those who advance that charge are incontestable. The assumptions are (a) that morality intrinsically takes the form of law or obligation, and (b) that every reasonable person must somehow recognize this characteristic of morality or be deficient as an ethical agent. Kant, for instance, claimed that the fundamental character of morality as law was something every ordinary person recognizes "apodictically." It cannot, he said, be demonstrated by argument; it must simply be recognized without further fanfare. Immanent naturalists contest Kant at precisely this point. We construe the "recognition" that morality takes the form of law to be a cultural formation deeply inscribed in predominantly Christian cultures rather

Immanent naturalism challenges the hubris of total explicability "in principle" often pursued by empiricists and rational choice theorists; it resists the propensity to deep, authoritative interpretation advanced by some practitioners of interpretive theory; and it contests the theme of transcendental reason pushed by descendants of the Kantian revolution. Indeed, it replaces the transcendental field of Kant with an immanent field that, like it, is inscrutable in detail but, unlike it, neither generates the "necessary postulate" of a designing, legislative God nor projects subjective faith in life after death. Does immanent naturalism, then, contradict or subvert itself by seeking to explain a world from which it subtracts the expectation of full explicability? Some neo-Kantians doubtless will say so. I propose, rather, that it supports a double-entry orientation to the paradox of political interpretation. As a new inquiry is launched, the launchers may act *as if* complete explanation is possible. But as we step back from the course of inquiry we then, in a second gesture, contest the hubris invested in such a regulative ideal – doing so for ethicopolitical reasons. It is in negotiating the space between these two orientations that inquiry proceeds.

The critical task of a student of politics in the tradition of immanent naturalism is to occupy strategic junctures where significant possibilities of change are under way, *intervening* in ways that might help to move the complex in this way rather than that. *Explanation as surface intervention; intervention as engaged involvement; engaged involvement as ethicopolitical activity*. Intervention, that is, involves a mixture of explanation, participation, and evaluation in a compound not reducible to any of these concepts in its putatively pure form.

What about the *problems* immanent naturalists feel pressed to engage? The favored domains of intervention are those proponents other traditions often ignore or defer indefinitely into the future: social movements with a potential to *invent new rights* or *promote new identities* not heretofore registered on the cultural field; volatile modes of entrant and reentrant

than an apodictic recognition. I suspect that those who confess Buddhism or Hinduism share this view with us. We contend that a commitment to ethical life is first grounded in attachment to the abundance of being, and that obligations, moral laws, rights, and responsibilities are second-order formations growing out of this first level of affirmation. We thus support an ethic of cultivation over a morality of command. One advantage of our view is that we can participate in the politics by which new rights come into being without having to pretend *retrospectively* that those rights were "implicit" in a set of principles all along, even if nobody had explicitly "recognized" them. The charge of the naturalistic fallacy converts the debate between Buddha (perhaps), Jesus, Lucretius, Spinoza, Hume, Nietzsche, James, and Deleuze on one side, and Paul, Augustine, Kant, and Hegel on the other, as above and beyond the reach of contestable faith, as if the first set of practitioners simply made a mistake.

combinations between religious and secular practices that disturb previous lines of separation; unique electoral configurations that issue in realignment or a court takeover of the electoral process; protean forces that emerge as if from nowhere, such as the collapse of a regime or the effects of rapid climatic change on the political ecology of contemporary life; practices of humiliation that issue in a riot or civil war, or a new settlement previously unexpected by any party. We are drawn to such "events" – where event is defined as the eruption of the unexpected into the routinized – like moths to fire, partly because they signify to us chaotic forces already in play within regularized processes and partly because they spawn new actualizations that brim with promise or draw something dismal into the world.

I hope I have said enough about this existential *faith*, comparative considerations on its behalf, the open-ended *method* connected to it, the *problems* to which its practitioners are drawn, the concept of *emergent causality* it entertains, and the characteristic orientation it adopts to *ethics* to suggest comparable lines of connection between faith, method, problem, and ethos in other traditions as well. Perhaps, its delineation will help to loosen fixed lines of debate in the human sciences between "explanatory" theorists and "interpretive" theorists heretofore set in stone.

An ethos of academic engagement

Today some schools of religious studies appear more effective in negotiating differences between faiths within their field than many departments of political science are in negotiating differences of method. There are several reasons for this. But one factor seems highly pertinent. Faculty in religious studies now often conclude that differences of religious faith are apt to persist as long as human life remains on the face of the earth. The question therefore becomes how to respond to persistent diversity. Many now conclude that the methods Augustine adopted to sterilize the play of diversity within the Church are too ugly and destructive to emulate. His criticisms of Jews, his eagerness to mark alternative practices of Christian faith as heretical, and his tolerance of harsh methods to subdue "the pagans," while mild by comparison with the conquest of Amerinidians in America and the Holocaust in Europe, are nonetheless too severe for many students of religious studies to sanction.

The setting of political science is different. Many political scientists, feeding off the hubris of secularism and the Enlightenment, feel in their bones that if only their method were triumphant they would be more secure in their faith and politics would become more fully explicable.

My sense, however, is that as long as human beings seek to interpret the politics in which they participate, a significant diversity of problematics is apt to persist. I don't mean to say that the future shape of diversity will correspond to the set of options now before us. I doubt that this will be the case. Immanent naturalism, for instance, did not have an active presence in American political inquiry ten years ago. Its activation waited upon creative work in chemistry, evolutionary theory, and neuroscience, where unexpected problems of causality, explanation, prediction, and method were encountered (Connolly 2002). And its transfiguration into an approach appropriate to the human sciences required creative engagements with culturally oriented philosophers such as Nietzsche, Charles Peirce, Henri Bergson, William James, and Gilles Deleuze whose work already resonated with developments in the first three fields. Once that option is added to the field of alternatives, a new set of contrasts, comparisons, and contests becomes available. And new problematics, heretofore undeveloped, may emerge. Indeed, one problematic may come to command the respect of most practitioners for a time. But I suspect that methodological pluralism is a condition that will persist in some form into the indefinite future, unless one faction is successful in grabbing all the reins of institutional power. Even if that were to occur, new factions and schisms would arise out of the debris.

So the question arises: in a discipline where method, faith, and problems are loosely intertwined and where no single problematic has yet proven itself to be dispositive to all reasonable parties, what is the most admirable ethos to negotiate between contending problematics? And how is such an ethos to be fostered by those who themselves are also partisans to the disputes?

The first thing is to articulate comparatively the problematic you support with as much energy and skill as you can, pressing others to take it seriously as a possibility to respect, while attending to the relations between method, faith, and problems in it. For here the devil is in the details, more than in general glosses of global relations between generic dimensions. The second thing is to come to terms publicly with the persistent element of *contestability* in the eyes of others of the faith infusing your problematic. William James, who as a "radical empiricist" anticipates several themes adumbrated here under the heading of "immanent naturalism," is exemplary on both counts. After presenting a strong case in favor of treating the universe as "unfinished" and "pluralistic," after making it clear that this image of the world challenges the sufficiency of rationalism and empiricism in their traditional guises, James reflects upon the status of the existential faith he embraces:

> But if you consider the pluralistic horn to be intrinsically irrational, self-contradictory, and absurd, I can now say no more in its defense . . . Whatever I say, each of you will be sure to take pluralism or leave it, just as your own sense of rationality moves and inclines. The only thing I emphatically insist upon is that it is a fully co-ordinate hypothesis with monism [the alternative he considers]. This world *may*, in the last resort, be a block-universe; but on the other hand it *may* be a universe only strung along, not rounded in and closed. Reality may exist distributively, just as it sensibly seems to, after all. On that possibility I do insist.[8] (1996: 327–8)

I agree with James. I would love to convert you to my view. But I remain open to the probability that many will resist conversion. Besides, the elements involved in conversion are too multiple in type and layered into being to be under any individual's or group's easy control. Next, it is pertinent to work on my regret that you do not become a convert, so that this disappointment does not become transfigured into resentment against you or the world for being so recalcitrant. But, along with James, I insist that, unless you do a hell of a lot better job than you and your ilk have done so far, immanent naturalism remains a possibility to which it is reasonable for me and others to be attached. I invite you to enter into a relation of agonistic respect with us on this question, even while each tries within reasonable constraints to convert the other. Who knows, one of us may blink, as Augustine did when he converted from a flirtation with Epicureanism to Manicheanism and, later, from Manicheanism to Augustinianism. If only he had engaged the Greek tragic tradition sympathetically, he might have blinked again.

If the partisans involved admit that faith, method, and problem are loosely bound together in spiral patterns of emergent causation, if we reciprocally concede that it is unlikely in the near future that any single problematic will marshal conclusive support in its favor, if we acknowledge that some new problematic might yet emerge that moves many in a surprising direction, and if we work upon ourselves to overcome resentment against the world for harboring such uncertain tendencies and possibilities, the field of political science may find itself in a decent position to pursue an academic ethos of generosity and forbearance.

[8] James (1996) postulates a god as an actor in the pluralistic universe. His god is limited rather than omnipotent. It is also projected as a contestable act of faith on James's part. I concur with James on the latter point, except my contestable projection does not include a god. The universe is pluralistic in encompassing agents of multiple sorts and degrees of agency, including atoms, evolutionary forces, geological processes, tornadoes, volcanoes, snakes, individual human beings, and collective human agents. Again, an agent is a force whose behavior is not entirely reducible to law-like explanation; it makes a difference in the world without quite knowing what it is doing.

What kind of practitioner is well equipped to come to terms affirmatively and generously with this stubborn dimension of inquiry? Perhaps those who have cultivated modesty about our place in the world and our capacity to explain the cultural processes in which we participate. And perhaps those who extend the term agency – where an agent is a being who makes a difference in the world without knowing quite what it is doing – to a diverse range of actors stretching beyond humanity alone. The hubris often invested today in the twin projects of consummate explanation and deep interpretation may be linked to the subterranean desire to receive *compensation* for the unruliness of the world. But what, in turn, underlies the implicit sense of *entitlement* to compensation? And who or what *could* bestow it? We may never agree on the answer to these questions. To me, the demand for compensation resides in the insistence that if the world is not designed by a loving god then it must at least be susceptible to the ardent human desire for consummate knowledge and mastery. Either/Or. Either the world must exist for us in one way *or* in the other. But, again, whence comes *either* must?

Perhaps it is timely to contest both imperatives together. Nietzsche – an immanent naturalist who diagnosed the "advent of nihilism" accompanying the crisis in Christendom *and* the conceit invested in the new image of science – may have something to say about the kind of practitioner needed as the hubris of a universal method loses its power to enthrall:

> Who will prove to be the strongest in the course of this? The most moderate; those who do not require any extreme article of faith; those who not only concede but love a fair amount of accidents and nonsense; those who can think of man with a considerable reduction of his value without becoming small and weak on that account . . . (Nietzsche 1968: 38)

REFERENCES

Augustine. 1958. *On Christian Doctrine*. D. W. Robertson (trans.). New York: Macmillan.

Bennett, Jane. 2001. *The Enchantment of Modern Life*. Princeton: Princeton University Press.

Connolly, William E. 1967. *Political Science and Ideology.* New York: Atherton Press.

2002 [1993]. *The Augustinian Imperative*. Rowman and Littlefield.

2002. *Neuropolitics: Thinking, Culture, Speed.* Minneapolis: University of Minnesota Press.

De Landa, Manuel. 1997. *One Thousand Years of Nonlinear History*. New York: Zone Books.

Deleuze, Gilles and Guattari, Felix. 1987. *A Thousand Plateaus.* Brian Massumi (trans.). Minneapolis: University of Minnesota Press.

Gould, Stephen Jay. 2002. *The Structure of Evolutionary Theory*. Cambridge, MA: Harvard University Press.
Green, Donald and Shapiro, Ian. 1994. *Pathologies of Rational Choice Theory*. New Haven: Yale University Press.
Hegel, Friedrich. 1977. *Phenomenology of Spirit*. A. V. Miller (trans.). Oxford: Clarendon Press.
James, William. 1996. *A Pluralistic Universe*. Lincoln: University of Nebraska Press.
Massumi, Brian. 2002. *Parables of the Virtual*. Durham: Duke University Press, 2002.
Menand, Louis. 2001. *The Metaphysical Club: A Story of Ideas in America*. New York: Farrar, Straus, and Giroux.
Nietzsche, Friedrich. 1968. *The Will To Power*. Walter Kaufmann (trans.). New York: Vintage Books.
Prigogine, Ilya and Stengers, Isabelle. 1997. *The End of Certainty: Time Chaos and the New Laws of Nature*. New York: The Free Press.
Vermes, Geza. 1993. *The Religion of Jesus the Jew*. Minneapolis: Fortress Press.
White, Stephen. 2002. *Sustaining Affirmation: The Strength of Weak Ontology*. Princeton: Princeton University Press.

16 Provisionalism in the study of politics

Elisabeth Ellis

Introduction

At a panel on the state of the discipline at the 2001 meeting of the American Political Science Association, Professor David Laitin said that political theory plays an indispensable role in contemporary political science, because it provides the foundational questions that the rest of the field tries to answer (Laitin 2001a). For Laitin, however, political theory's continuing relevance is a sign of political science's immaturity (Laitin 2001b: 24). Political theory's function of "accounting for and justifying different public outcomes" should be taken up by political science generally. Such investigators, he argues, may do better than theorists have been doing lately, as they may avoid getting mired in "diversionary tacks inducing us to grapple with ontologies or epistemologies" (23).

Laitin's vision for a theory-free political science fails to account for the special nature of political science's object of inquiry: political actors are historically constituted, limited rational beings in the process of making their collective worlds. What differentiates a political science that includes theory from a "purified," "positive" political science is this: only the former can account for its object of inquiry *qua* constructor of value.[1] However, Laitin's point about diversionary tacks applies all too well, both

Early versions of some of the ideas in this chapter were presented at the 2002 meetings of the Western Political Science Association, the Southern Political Science Association, and the American Political Science Association. I learned a great deal in discussion of those early ideas with Mika LaVaque-Manty, Sung Ho Park, and Doug Dow. Thanks to my colleagues Cary Nederman and Ed Portis for providing the forum in which these ideas could be developed, and to visitors to the Texas A&M University Political Theory Convocation, especially Jim Johnson and Ian Shapiro, for their generous encouragement. Seyla Benhabib and Alan Ryan provided thoughtful and productive criticism of a draft of this chapter, for which I am grateful. Part of the research that contributed to this chapter was supported by a grant from the Texas A&M University Scholarly and Creative Enhancement Fund.

[1] Ian Shapiro's suggestion that political science fails to characterize its dependent variables adequately is relevant here, although my complaint covers inadequate understanding of variables on both sides of the equation (Shapiro, concluding remarks, Conference on Problems and Methods in the Study of Politics, Yale University, December 6–8, 2002).

to political theory and to the discipline of political science as a whole. In focusing too narrowly on subdisciplinary methodological issues, political scientists and political theorists alike have missed opportunities to address political questions of broad, substantive interest (cf. Isaac 1995).

In this chapter I shall argue for a new practice in political theory: one that takes a provisional, rather than a conclusive, perspective; and one that operates in the historical, rational, and empirical modes. I illustrate this practice with the extended example of a critique of contemporary democratic theory. I claim that a provisional rather than a conclusive perspective allows political theorists to focus away from attempts to influence decisions emanating from the traditional but relatively undemocratic locations of authoritative judgment, and toward critical analysis of the places where political actors may gain or lose capacities for substantive self-rule. Furthermore, I argue that most generally interesting political questions cannot be addressed by historical analysis, contemporary political philosophy, or empirical research alone. Take, for example, political analysis involving international human rights. First, ahistorical inquiry would miss key points, such as the fact of institutions for the defense of international human rights that precede the national state by hundreds of years, or the fact that claims for human rights have sometimes followed a sequence from civil through political to social. Second, contemporary political philosophy has clarified the consequences for international human rights of reasoning from various premises, such as the premise of individual rational agency or the premise of culturally distinct group memberships. Finally, without empirical inquiry we would have nothing but speculative responses to important questions, such as whether the concept of human right or the concept of human dignity has more international currency, or the question of whether advocacy groups mobilized on the basis of commitment to human rights have any advantages over or even dynamics different from other such groups.

Here I offer an example of provisional theory in the historical, rational, and empirical modes with an analysis of deliberative democratic theory. I argue that deliberative theory offers distinct advantages over its competitors, grouped roughly as participatory and minimalist democrats. However, deliberative theory is saddled with three basic implausibilities – psychological, philosophical, and institutional – and thus requires modification if it is to be a useful guide to the norms of democratic politics. First, historical analysis reveals deliberative theory's weak psychological basis, namely its reliance on individual-level rational enlightenment under deliberative conditions. A less psychologically implausible model of deliberative change, I argue, would focus on the longer-term historical dynamics of the public sphere.[2] Second, deliberative theory's main philosophical weakness is its continued reliance on conclusive argument with regard not only to procedure but also to policy outcomes. However noble the principles on which these policy conclusions are based, they undercut deliberative theory's bedrock democratic commitment to substantive self-rule by real citizenries (as opposed to formal self-rule by idealized citizenries), resulting in the kind of paternalism democratic theory is supposed to deny. Finally, empirical research on actual instances of deliberative polling demonstrates the weakness of current deliberative theory's institutional vision, while research on the concrete political power of widely held democratic ideals operating in the public sphere over the medium and long term provides a better alternative.

These three modes of political theory are not arbitrary paths to understanding public life; they are rooted in the nature of our object of inquiry: historically limited rational beings in the process of making their collective worlds. Only a provisional theory attentive to historical, rational, and empirical questions can do justice to the ineradicably normative and fundamentally provisional nature of politics.

Provisional theory

Provisional theory takes its starting point from the Kantian notion that the concept of provisional right applies to institutions which imperfectly mirror their own normative principles, and thus that practical politics must follow a rule of provisional rather than conclusive right. A focus on provisional rather than conclusive right makes it possible to analyze institutions that do not make absolute pronouncements on what counts as just. For example, in most people's experience of citizenship, what matters is their degree of access to the legal system rather than any one of the system's particular claims. A general formulation for provisional right in specifically Kantian language is: "always leave open the possibility of entering into a rightful condition" (Kant VI:347). Kant identified the distinction between provisional and conclusive right at the end of the eighteenth century, and applied it to various political topics, such as the determination of which violations of international right were immediate threats to the possibility of world peace, and which might be provisionally tolerated (Kant VIII:343–9). More importantly, provisional

[2] This criticism does not apply to deliberative theories of the public sphere (e.g., Habermas 1997, 1992, 1991; Benhabib 1996; Dryzek 1990). As I shall presently argue, the most defensible democratic theory would combine the provisionality of Gutmann and Thompson's theory and the historical perspective of Habermas's view, while jettisonning Gutmann and Thompson's historical naïvité and Habermas' conclusivist ambitions.

theory may, I argue, be applied with good results to current political problems.

Some areas of research in contemporary political theory have already moved beyond stale distinctions by means of the concept of provisional right. For example, Amy Gutmann and Dennis Thompson argue that a view of political principles as "morally provisional" allows us to avoid static debates among "first-order" theories like communitarianism and liberalism by focusing on a "second-order" theory for assessing the procedural justice of any deliberation (Gutmann and Thompson 2000). To take a different example, Jane Mansbridge has argued that an essential element of any democratic polity must be the maintenance of a variety of sources of opposition. Rather than attending strictly to problems of "political obligation and civil disobedience," democratic theory ought to recognize the "ongoing imperfection of democratic decision." She grounds this view on the premise that the principles that regulate democratic procedures must always remain "provisional" (Mansbridge 1996).

The provisional perspective allows political theory to address problems that have all too often been left to our colleagues in other subfields. Provisionalism allows the theorist to focus on the midrange problem of maintaining the possibility of progress, rather than on determining particular policy or even regime type outcomes and expecting the practical problems to be resolved separately. For example, traditional defenders of liberal rights tend to focus on the courts and their decisions. A provisional theory would focus instead on the institutions and processes that restrict or promote citizens' capacities for self-rule. The appointed panels making decisions for international trade regimes, devices like base-closure commissions used to get around representational failure, governmental and non-governmental agencies devoted to ensuring inclusive access to the levers of everyday power, electoral reforms designed to represent citizens on the basis of participation rather than inheritance, stakeholder schemes for resolving environmental disputes, micro-lending agencies that encourage citizen autonomy, and so forth; these locations ought to interest political theorists, because these are the places where citizens either gain or lose the capacity to determine their fates themselves.

An example: deliberative democratic theory

James Bohman and William Rehg usefully define deliberative democratic theory as referring to "the idea that legitimate lawmaking issues from the public deliberation of citizens" (Bohman and Rehg 1997: ix). For more than a decade now, under various headings such as "discursive" and "deliberative," democratic theorists have been revising previous paradigms in what John S. Dryzek calls "a strong deliberative turn" (Dryzek 2000: 1).

Roughly speaking, deliberative democratic theory falls between the neo-Rousseauian politics of public good and the neo-Lockean politics of liberal individualism; participatory democracy, communitarians, and civic republicans would fall into the former camp, while Schumpeterian minimalists, liberals of many stripes, and aggregative theories of democracy belong to the latter.[3] No matter what the image of theoretical terrain employed by a particular deliberative democrat, deliberative theory always seeks to provide an account of democratic politics that preserves the best of both ends of the spectrum, while avoiding the conundra associated with each. For example, a standardized debate between, say, a civic republican and a liberal individualist would result in a less-than-satisfactory set of alternatives on the divide between public and private. Many liberals would defend a broad concept of the private sphere in the name of non-tyranny. For them, the state exists to defend private interest. Civic republicans would point out that the narrow public sphere that remains in the liberal vision is arbitrarily beholden to a private status quo, and limits political freedoms to mere public formalities. Nearly all advocates of deliberative theory, on the other hand, regard some distinction between public and private as necessary to protect civil freedom, while insisting that protection of substantive (not merely formal) political liberties entails some blurring of the public/private divide. Gutmann and Thompson, for example, argue that no "matter how earnestly citizens carry on deliberation in the spirit of reciprocity, publicity, and accountability, they can realize these ideals only to the extent that each citizen has sufficient social and economic standing to meet his or her fellows on terms of equal respect" (1996: 349).

Most theorists of deliberative democracy argue that not only public policy, but also procedural standards themselves are included among potential items for public deliberation; they have provisional status (Gutmann and Thompson 1996: 351–354). Gutmann and Thompson, for example, explicitly identify deliberative theory as having "a provisional status" and say that it "expresses a dynamic conception of politics" (1999: 276). They add: "this provisional status does not merely express some general sense of fallibility: it is specific to resolutions of deliberative disagreements in politics and carries implications for political practices and institutions" (1999a: 277). The logic of deliberative democracy is hypothetical rather than categorical; if a system is to be legitimately democratic, then certain

[3] Not all neo-Schumpeterian democrats accept Lockean premises about the priority of society over the state. See, for example, Ian Shapiro's *Democratic Justice* (1999b).

deliberative conditions must be met. Deliberative theorists use provisional principles in the name of making good on the bedrock promises of democratic self-rule.

It follows that deliberative democratic theory may not foreclose any topic from the domain of public choice unless the choice might undermine democracy itself. As I have just mentioned, deliberative democracy's liberal constitutionalist competitors traditionally protect the private sphere from collective interference by categorically limiting the legitimate scope of state activity, while their more participatory counterparts complain that the resulting polity is too thin to satisfy the demands of positive freedom. Thus a significant advantage of deliberative theory over its alternatives is that it can broaden the scope of legitimate state activity to address the moral reality of political life without sacrificing liberal freedoms. "Moral argument already pervades democratic politics, and although it often fails to produce moral agreement, citizens evidently hope it will, or they would not engage in it" (Gutmann and Thompson 1996: 346). Indeed, much of the criticism of deliberative democratic theory has in fact run along the same lines as the classic modernist critique of the Old Regime's conflation of moral and temporal power. Such a modernist might argue, for example, that it took devastating wars and theorists of the caliber of Hobbes and Locke to free modern politics from religion as a permanent source of debilitating conflict. Along these lines, William Galston complains that deliberative theory underrates the achievements of "mere toleration" in the areas of peace and stability: "In most times and places, the avoidance of repression and bloody conflict is in itself a morally significant achievement . . ." (Galston 1999: 45). Gutmann's and Thompson's constitutionalist democratic foils make a similar point: the democratic state must be limited in scope – and excluded from enforcing moral or religious judgments – in order to keep the more fundamental promise of providing basic security to its citizens.

Given the force of these arguments, why bring morality back in? Because a system of political legitimacy that would exclude moral questions entirely from the activity of the state makes three fundamental errors. First, denial of the state's right to action in a particular realm does not make the state value-neutral; rather, such a denial privileges the status quo by throwing moral issues to non-state (and therefore non-accountable) actors to decide.[4] Whether the determining players are specific private sector interests or the more diffuse forces of culture and history, such a point of view arbitrarily limits the scope of democratic

[4] For an interesting argument about the myth of state neutrality from a different point of view, see Kymlicka 2000.

self-rule. Gutmann and Thompson argue similarly that denying moral reason a place in politics amounts to excluding all considerations but power and interest from the public sphere. "Why," they ask, "assume that a power-based conception of democracy is the default position? . . . At the foundational level, no one has yet provided a justification of a power-based conception that is any more adequate than a justification of a conception based on moral reason" (Gutmann and Thompson 1996: 353).

Second, democratic politics cannot legitimately be limited by theory in advance to a few supposedly neutral values, such as "security" or "economic prosperity." The very selection of these few values distorts the institution itself. To take a contemporary example: as currently constituted, the World Trade Organization (WTO) may legitimately sanction member nations only for actions deemed to be illegitimate restraints on free trade. This limited scope provides the WTO with much of its legitimacy, and ensures that member nations tolerate its very real interference with their national sovereignty. However, such narrow limits on the scope and basis of WTO judgments have led to policy decisions that run counter to local and even world public opinion. The individuals affected by WTO rulings are unlikely to prefer the narrow standards of trade expansion to broader standards addressing the environment, employment, human rights, and other areas affected by trade policy. The WTO has been criticized, with reason, for its undemocratic character; not only are its unelected panels of decisionmakers only very indirectly accountable, but the rules under which they operate forbid them to consider many of the most salient elements of the public will regardless of the formal structure of accountability.

The third and final argument that moral considerations ought not to be excluded *ex ante* from political deliberation lies in the empirical association between moral and political reasoning in practice. Though there is relatively little consensus on the moral content of political ideals, there is overwhelming consensus that political ideals have some moral content. This is of course an empirical statement, a fact which could be disproved at any moment. But while it is possible to imagine a politics devoid of moral content, the fact remains that the overwhelming majority of people, across differences of time and space, conceive of politics in moral terms such as "justice," "right," "respect," "human dignity," and so forth.[5] On the presence of moral reason in politics, Gutmann and Thompson write that "many theorists of democracy refuse to face

[5] For an insightful discussion of the implications of this Wittgensteinian point for the concept of justice, see Pitkin 1972.

up to this moral fact of political life. They assume that it is simply the result of misunderstanding, which could be overcome if citizens would only adopt the correct philosophical view or cultivate the proper moral character" (1996: 360). Deliberative theory does not seek to do justice to the moral element in political life by selecting the correct moral point of view and arguing for its universal adoption. Nor does deliberative theory throw up its hands with a culturally relativist stance, accepting either separatism or a severely limited scope for public action as the natural result of moral pluralism. By treating political settlements as provisional, the fact of moral disagreement as permanent, but individual moral disagreement on particular issues as at least provisionally amenable to democratic accommodation, deliberative theory improves on both its liberal neutralist and its participatory, not to mention its cultural relativist, predecessors.

While deliberative theory envisions a more expansive thematic scope for *democratic* decisionmaking than mainstream liberal theory, it restricts the sphere of *theoretical* prejudgment to the preconditions of substantive deliberation. By contrast, with a set of absolute normative standards such as mainstream liberal theory seeks to provide, a theorist might derive specific policy outcomes that follow from those standards. For example, since the United States Supreme Court handed down its decision in *Wisconsin* vs. *Yoder* (406 US 205 [1972]), theorists have been second-guessing the outcome, each by means of his or her preferred set of standards of political right. In an interesting example of what Cass R. Sunstein (1999) calls "incompletely theorized agreement," while most academic critics disagree with the court's ruling, which allowed an exception to the schooling requirement for Amish teenagers, these critical opinions are based on theoretical standpoints as divergent as utilitarianism and Kantian liberalism. Thus, over the past few decades, the reading public for political theory, such as it is, has enjoyed significant clarification of what sorts of policy outcomes may be justified by what sorts of philosophical premises. However, and, I would add, fortunately, political theorists have not reached even provisional consensus on which set of philosophical premises might be most authoritative, even for a particular time and place, much less universally. More problematically, the theoretical focus on correct policy outcomes has come at the expense of some very fundamental political problems, including the whole complex of questions around how legitimate liberal-democratic contemporary regimes in fact are. The theorist interested in reaching, say, the populist-utilitarian answer to the question addressed in *Wisconsin* vs. *Yoder* only complains about the outcome reached by the court, not the fact that the court itself as it currently operates constitutes an affront to most relatively strong conceptions of democratic value (but see Ginsburg 2001; and Shapiro 1999a).

Deliberative theory at its best focuses on institutions rather than outcomes. Gutmann and Thompson, for example, make a compelling argument for the progressive power of state-sanctioned norms of transparency and publicity, rather than direct regulation or privatization, with the case of environmental impact statements. The "practice of requiring agencies and companies to issue environmental impact statements before proceeding with major development projects is often more effective, according to one study, than other methods of enforcement. . . . Agencies and companies pay closer attention to environmental matters when they know they will have to defend their actions in detail in public" (1996: 98). Provisional theorists of deliberative democracy, with their focus on institutions and empowerment rather than on directing policy outcomes, are thus able to avoid the shortcomings of both the liberal and the participatory poles of alternative democratic theories. On the one hand, they avoid the neo-Lockean pitfall of overreliance on the private sector and the concomitant loss of democratic self-rule; on the other, they dodge the neo-Rousseauian specter of total government, with its twin dangers of tyranny and ineffectiveness.

Deliberative theory values democratic self-rule in a strong sense: access to and accountability from the policy structure for those affected by its policies. Such a principle raises interesting questions about membership and citizenship which I cannot address here. It does improve upon modernist democratic theory based on the national state by providing for substantive rather than merely formal empowerment. Classic social contract theory limits positive rights to an arbitrary group of empowered citizens, chosen by a combination of morally arbitrary qualities (geography, wealth, blood ties, and so forth) (Rawls 1996). Recent changes in the global political environment only make more urgent what was already a difficult democratic conundrum: why should control over policies affecting a large group be limited to some subset of that group?

Most deliberative democrats do not in fact draw this strong conclusion about the limited legitimacy of national citizenship from their premises about self-rule. There are certainly good reasons for cleaving to the national state in the absence of more promising institutional alternatives. Historically, the national state has at some times and places been the instrument of progressively more inclusive citizen empowerment (Marshall 1964). As the substantive scope of narrowly national policy contracts, new transnational actors are acquiring policy discretion, often without even the formal accountability that constrained national governments. Small wonder, then, that anti-globalization protests attract not only those opposed to expanded trade and its downstream effects, but also activists concerned with democracy's promise of self-rule. Not all developments associated with globalization have run up against deliberative principles, of course. As Margaret Keck and Kathryn Sikkink (1998) point out, for example, new transnational networks of advocacy groups in areas such as human rights and the environment have enjoyed occasionally surprising success in the global public sphere. However, the implications for global and local citizenship of deliberative theory have not yet been adequately explored by political theorists (for an exception outside deliberative theory, see Kymlicka 2000).

Political inclusion *within* the national state, however, is a topic that deliberative theorists have made a central part of their reconstruction of democratic theory. While standard liberal democratic theory has successfully championed formal political inclusion over a long series of progressively more inclusive policies, deliberative theorists point to a "reality gap" between merely formal and substantive political inclusion. According to Iris Marion Young, for example, both passive and active exclusions are unjust: "In principle, inclusion ensures that every potentially affected agent has the opportunity to influence deliberative processes and outcomes" (Young 1999: 155). The state has an obligation to ensure formal inclusiveness (via policies such as voting rights – in light of the experience of exclusion by large numbers of Black Floridians in the 2000 US presidential elections, this is clearly an ongoing project). But according to provisional deliberative theory, the state must also ensure "effective opportunity" to participate, a standard that reaches well beyond formal rights (Young 1999: 157).

Thus deliberative theory improves substantially on its neo-Lockean and neo-Rousseauian competitors. It avoids the charge of insufficient attention to liberty by means of provisional defense of the rights necessary to the deliberative process, and by restricting would-be paternalist prejudgment of policy outcomes. It avoids the opposite charge of being insufficiently critical by removing some liberal restrictions on the scope of democratic decisionmaking and by its attention to the institutional preconditions of substantive deliberation. For all its substantive advantages over previous democratic theories, however, deliberative theory remains saddled with some of the traditional shortcomings of standard liberal theory: an implausible model of political change that relies on unlikely accounts both of individual psychology and of the time-frame of shifts in social mentalities; a neo-paternalist commitment to the "right" policy outcomes based on noble but ultimately arbitrary absolute principles; and a too-thin conception of the reasoning subject that excludes the vast majority of potentially empowered democratic citizens, especially those with strong religious views. Gutmann and Thompson are certainly right to reject cultural relativism. Even as committed a communitarian as

Michael Sandel agrees that to "make justice the creature of convention is to deprive it of its critical character" (1998: xi). Gutmann and Thompson criticize Robert P. George's argument that one day permissive abortion laws will be viewed in the same light as permissive slavery laws are today with the Kantian point that privileging any particular historical standpoint cannot be justified: "Imagining the time in the future when one's argument might win the day does not add anything to the strength of one's argument in the present" (Gutmann and Thompson 1999: 270). This is certainly a valid point, and one that draws strength from the very reasonable assumption that changes in public morality do not always happen in a positive direction. However, Gutmann and Thompson entirely miss the force of historical and cultural arguments like George's. Instead, they expect that, given the right institutional background, public reason will trump social context and historical prejudice, and over the short run at that. A focus on an idealized individual reasoner produces such conclusions: why shouldn't a person, confronted with the shortcomings of a long-held view in the process of public debate, simply change her mind? But this is the wrong question. Rather than ask how individual reasoners of good will might form excellent political judgments, we should ask how the discursive environment changes over time such that some arguments become more or less possible to make.[6]

Deliberative theory's psychological implausibility and its historically informed solution: the historical public sphere

Deliberative theory's model of political change rests on an underlying psychological implausibility: under deliberative conditions, rational individuals are supposed to achieve immediate enlightenment. While reasoning in the public sphere can make profound differences in general public opinion, these changes are effective at the level of a public more than at the level of the individual, and over the medium and long term, rather than immediately.[7] Ian Shapiro illustrates the model's implausibility in the case of Gutmann's and Thompson's work with the point that, alluring "as this reasoning might be to you and me, I find it hard to imagine a fundamentalist being much impressed by it, particularly when she learns that any empirical claims she makes must be consistent with 'relatively reliable

[6] Such a move, by the way, is a shift from inquiry guided by Kantian absolutist ethics (what must a reasoner of good will conclude?) toward inquiry guided by Kantian provisional political theory (what are the institutional preconditions of a healthy public sphere, and how can it contribute to political progress?).

[7] A similar point is made by Kant in "An Answer to the Question: What is Enlightenment?" (VIII: 35–6).

methods of inquiry'. . . . The Gutmann/Thompson model works only for those fundamentalists who also count themselves fallibilist democrats. That, I fear, is an empty class . . ." (1999a: 30–1).

Deliberative theory's case for short-term individual enlightenment would be strengthened by examples of the phenomenon in action, but these are unfortunately in very short supply. Perhaps the strongest case in the Gutmann and Thompson volume, a study of deliberation around the Oregon health-care rationing debate, is also the most deeply flawed, in that very few of those actually affected by the policy (those dependent on the state for health care) actually participated in the public deliberation on the policies that affected them. Even James Fishkin, whose deliberative polling mechanism is the most promising of proposed deliberative institutions, admits that we "do not know to what degree deliberative opinion polls would contribute to thoughtful, self-reflective opinion formation" (1991: 83). Deliberative enlightenment seems unlikely to occur on the time-scale contemplated by most of its contemporary advocates. Theorists interested in the mechanism by which it might occur ought to turn their attention away from individual rational choice and toward the dynamics of argument in the public sphere.

There are a number of such literatures available, though I will restrict my discussion to a few examples here (Reinhart Koselleck and the *Geschichtliche Grundbegriffe* school; J. G. A. Pocock and those studying the "languages of political thought"; and Jane Mansbridge's "everyday talk" model). None of these schools of thought offer a model for deliberative theory to embrace and apply without significant revision. The historical schools are, of course, much more interested in describing changes in political mentality than in relating those changes to the degree of substantive political freedom enjoyed by a discourse's participants. However, these alternate models of a dynamic public sphere offer deliberative theory a way out of its psychological implausibility without sacrificing its advantages.

Reinhart Koselleck and the history of political concepts. Reinhart Koselleck and his fellow contributors to the enormous and very useful encyclopedia of modern political concepts, *Geschichtliche Grundbegriffe*, have traced the development of some important political ideas and their legitimating arguments, concentrating especially on the key period between 1750 and 1820 in Europe. By showing how various concepts shift their meanings over time, Koselleck illustrates the dynamics of differing styles of political argument. In fact, as certain terms acquire new political meaning, old strategies of legitimation become, as it were, linguistically impossible. For example, discussing the establishment of French revolutionary political language, Koselleck argues, "In France, the platform of revolutionary language, victorious after 1789, rapidly and effectively rendered the privileges of the estates incapable of legitimation" (1989: 659).

Koselleck and his colleagues include philological information from well before the modern period. However, a basic premise of their research is that the history of concepts has operated according to a different dynamic since the Enlightenment; now, concepts are not merely collected from reality, but expected to drive and respond to a constantly changing political environment. New political terms create expectations about change rather than describe current politics (2002: 128–9). In fact, "the lower their content in terms of experience, the greater the expectations they created – this would be a short formula for the new type of political and historical contexts" (2002: 129). Koselleck concludes: "Political and social concepts become the navigational instruments of the changing movement of history. They do not only indicate or record given facts. They themselves become factors in the formation of consciousness and the control of behavior" (2002: 129). Though the language Koselleck himself uses encourages the reader to think of linguistic phenomena as somehow autonomous historical forces, this is misleading. Koselleck is describing something very like the Kantian and neo-Kantian public sphere, in which concrete political actors participate in the creation of new modes of political thought. Koselleck illustrates this dynamic with the concept of emancipation.

> The following thesis may be ventured: with the introduction of the reflexive verb 'to emancipate oneself' (*sich emanzipieren*), a profound shift of mentality was, for the first time, foreshadowed and then brought about. While initially it was a word used by the cognoscenti, the poets and philosophers, who sought to liberate themselves from all pregivens and dependency, the new active word usage was expanded to increasingly refer to groups, institutions, and entire peoples. . . . Contained within the reflexive word usage was *eo ipso* a thrust against the estates system. (2002: 252)

These concepts do not change according to any internal, independent dynamic of their own. Instead, a received new concept is the result of a political battle fought by linguistic means in the public sphere, subject to standards of rationality to be sure, but also to forces of political mobilization, temporary strategic advantages, extrapolitical cultural shifts, and plain dumb luck.

J. G. A. Pocock and the languages of political thought. Like Koselleck and the *Geschichtliche Grundbegriffe* group, Pocock and his fellow investigators into the "languages of political thought" are concerned with the dynamics of political ideas. Rather than create histories of political thought, Pocock argues, we should consider ideas as events, and political history as the history of political discourse (Pocock 1987). Changes in political language shut out possibilities that used to exist, while opening up new ones heretofore unimaginable. While Pocock puts equal emphasis on investigating the effects of linguistic change on political possibility, on the one hand, and the "learning" of "political languages" to facilitate historical insight, on the other, deliberative democrats are of course relatively uninterested in the latter pursuit (except insofar as it allows us to understand independently interesting sources better). As Pocock and his colleagues establish their case for context-based readings of political ideas, then, they also provide deliberative theory with support for a model of a dynamic and potentially empowering public sphere. Pocock himself makes reference to this dual result of context-sensitive research in the history of political ideas; even as the historian is interested in old patterns and the ways they constrain change, "there is a process in the contrary direction; the new circumstances generate tensions in the old conventions, language finds itself being used in new ways, changes occur in the language being used, and it is possible to imagine this process leading to the creation and diffusion of new languages . . ." (Pocock 1987: 32).

The study of the languages of political thought provides an interesting model for a mechanism of the public sphere, but only a model. As Pocock notes, most material thus far studied by himself and his colleagues predates industrial society. Democratization, the information revolution, mass literacy, and mass media all contribute to new contexts for the same basic mechanism driving change. The question remains: what arguments are possible to make under what conditions? But the speed, medium, and scale of change should be fairly different in a postindustrial context.

Jane Mansbridge on "everyday talk". Mansbridge's model of the dynamics of public speech bears some resemblance to Koselleck's and Pocock's, but unlike the historical models, Mansbridge is focused on the contemporary world. The underlying mechanism of change in Mansbridge's account is composed of social activists and their innovations, but serious change occurs only as the activists and their ideas interact with the general public. She writes, "In social movements, new ideas – and new terms, such as 'male chauvinism' or 'homophobia' – enter everyday talk through an interaction between political activists and non-activists. Activists craft, from ideals or ideas solidly based in the existing culture, ideals or ideas that begin to stretch that base" (1999: 219).

Political possibilities change as vocabularies shift. For example, a world without a word for homophobia is a world in which the public may not have been encouraged to consider the moral status of prevailing anti-gay attitudes.

Like the other models, everyday talk is not perfect for deliberative purposes. Mansbridge is more interested in the political aspects of social life, such as the empowerment of women in oppressive family situations via the incorporation of new ideas into their moral vocabularies. She also does not address the drastically unidimensional motives of the big corporate media organizations which offer the most opportunities for idea transmission in the advanced industrial societies. But her model of the mechanism of change is compelling: "If parts of new ideals begin to win more general acceptance, many people, including non-activists, begin changing their lives in order to live up to those ideals in a better way." Mansbridge recognizes that powerful representatives of interested parties will try to promote or denigrate ideals that provide material advantages and disadvantages, but argues that "material loss and gain do not fully explain adherence to or rejection of an idea. People are governed in part by their ideals, and they often want to act consistently" (1999: 220).

Unlike most deliberative theorists, who focus on rational choice at the individual level, Jane Mansbridge offers a model for a dynamic public sphere based on interaction between social activists and non-activists. She does not restrict contributions to public discourse on the basis of their emotional content, but instead allows the full panoply of political language to count as contributions in the public sphere. As Marcus, Neuman, and MacKuen have also argued, the current scholarly attention directed exclusively toward instrumentally rational public opinion fails to comprehend the human political animal as it in fact operates in the world: emotional systems, they demonstrate, can promote ultimately rational political behavior (Marcus *et al.* 2000).

Deliberative theory's philosophical implausibility and its rationally informed solution: provisionalism

Some of the earliest calls for more deliberation in democratic practice arose out of critique of the state of really existing liberal democracy in the advanced industrial nations. For example, James S. Fishkin's long-running effort to incorporate citizen deliberation into the practice of public decisionmaking in Texas, the United States, and Great Britain arose at least partly from dissatisfaction with the institution of opinion polling. In place of public opinion surveys, with all their well-known flaws, Fishkin calls for opportunities for representative samples of the public to deliberate and inform themselves on a topic and come up with what the public "*would* think about the issue if it focused on it in a more substantial way"[8] (Fishkin 1995: 2, Fishkin's emphasis). Fishkin's deliberative democratic institutions do not seek consensus, but only to reveal the public will in all its diversity without the distortions of poor information, false attitudes, and other survey research shortcomings.

John S. Dryzek's early *Discursive Democracy: Politics, Policy, and Political Science* (1990) is also partly inspired by dissatisfaction with contemporary reliance on public opinion polls, and also offers an alternative aimed at reaching some more genuine public will ("Q methodology," an intersubjective, non-hierarchical survey technique) (Dryzek 1990: 173ff.). Dryzek's critique, however, goes beyond unsatisfactory levels of public debate in contemporary democracies to the dysfunctions of instrumental reason itself; unlike Fishkin, Dryzek works in the tradition of the Frankfurt School critics of modern society, including among others Max Horkheimer, Theodor Adorno, and Jürgen Habermas. Like Horkheimer and Adorno, Dryzek attributes many contemporary social problems to the nature of liberal democracy itself (rather than its incomplete realization), particularly insofar as liberal democracy conceives of the public good as the aggregate of many individual preferences (Dryzek 1990: 125). However, unlike Horkheimer and Adorno, and like Habermas, Dryzek sees the potential for democratic, maybe even liberal-democratic, redemption in the expansion of the public sphere. For Habermas and for Dryzek, both critical theorists of a deliberative democratic bent, the alternative to (unavailable) universal standards of liberal rights is not free-wheeling pluralism or cultural relativism, but recourse to the normative standards inherent in the deliberative procedures of the public sphere (Dryzek 1990; Habermas 1992).

Critical theory's commitment to absolute standards of political right, whether rooted in the ideal speech situation (Habermas 1973) or in a theory that posits the occlusion of genuine public right by an irrational or hyperrational system, does provide the basis for a grounded critique of contemporary society. However, societal criticism made on this kind of basis only has as much weight as its underlying principles. Thus far, no consensus has emerged on which set of basic principles democratic theorists ought to embrace (Rawlsian or Habermasian? neo-Marxist or neo-liberal? Post-Nietzschean or neo-Aristotelean? . . .). Ian Shapiro comments on the situation: "If there were an answer to [the question of which political institutions would be chosen by truly neutral reasoners], it might

[8] This is, of course, a classically Kantian formulation for approximating the general will. All existing states being imperfect (and right, therefore, being provisional rather than conclusive), the head of state must strive to make the decisions that *would be made* by the united will of the people (Kant VI: 328).

arguably provide a yardstick for political evaluation. . . . If, however, there is no compelling answer to the motivating question, as the existence of multiple pretenders to the neo-Kantian throne suggests is the case, then the enterprise is forlorn" (Shapiro 1996: 6).

The question is not whether standards are necessary (for critique, they are), but what *status* and *scope* apply to them. Provisional, as opposed to conclusivist, theorists of deliberative democracy recognize that the standards themselves are included among potential items for public deliberation (Gutmann and Thompson 1996: 351–4). Gutmann and Thompson, for example, explicitly identify deliberative theory as having "a provisional status" and say that it "expresses a dynamic conception of politics" (Gutmann and Thompson 1999: 276). They add: "this provisional status does not merely express some general sense of fallibility: it is specific to resolutions of deliberative disagreements in politics and carries implications for political practices and institutions" (Gutmann and Thompson 1999: 277).

Some deliberative theorists would also limit the *scope* of application of normative standards of public reason to procedural questions, rather than both deliberative procedures and policy outcomes.[9] I shall include among provisional theorists those who recognize at least the status if not the scope limitation on universal deliberative norms. Most in this group insist on the value of leaving people to determine their own policy outcomes. But even the most prominent[10] advocates of provisionalism in deliberative theory, Gutmann and Thompson, waver on this point. As I discuss below, Gutmann and Thompson make contradictory statements about scope limitations, insisting rightly on the wrongness of theoretically prejudged policy outcomes, but also insisting wrongly on the universal rightness of some very specific policies. I shall argue below that universal deliberative standards do apply to that very limited subset of policy outcomes which affect substantive self-rule on the part of individuals, and thus that Gutmann and Thompson are right to include slavery among legitimately forbiddable policy outcomes, right to exclude abortion rights from the narrow circle of universalist policies, but wrong to include medical rationing in that domain.

[9] See Knight 1999, which distinguishes among three types of deliberative theory: constraints on arguments themselves according to public reason (Rawls); substantive constraints on policy outcomes (such as Gutmann and Thompson's use of the ideals of liberty and opportunity); and Knight's favored version: derivations of substantive constraints from procedural norms (Knight, Knight and Johnson, Habermas, Bohman, *inter alia*). Note that Knight includes both provisional and conclusive theorists of deliberative democracy in his favored group.

[10] As measured by the Social Science Citation Index, as of February 2003.

Conclusivist (i.e., non-provisional) theorists of deliberative democracy do not, therefore, present interesting alternatives to liberal or participatory democratic theory. In other words, without provisionalism and the status/scope limitations such a perspective implies, deliberative democratic theory does not improve substantially on its predecessors. Cass R. Sunstein argues similarly for what he calls "incompletely theorized agreements" in the name of respect for historical persons: such agreements "are valuable when we seek moral evolution over time," since a rigid set of absolute principles "would disserve posterity" (Sunstein 1999: 132).[11] To respect the bedrock democratic principle of substantive self-rule, theories must remain provisional rather than conclusivist, leaving most issues to the active decisionmaking of citizens themselves.

Any democratic theory that prejudges the outcome of a particular policy area beyond those directly involved in guaranteeing self-rule violates the bedrock democratic principle of respect for what Frederick Schauer calls the "people's right to be wrong" (1999: 24). Anticipations of the correct outcomes for substantive policymaking respond to the understandable demand for justice as soon as possible. However, attempts to prejudge policy outcomes beyond those required for substantive self-rule end up foreclosing the most important elements of self-rule from citizens' purview. Provisional deliberative theory, then, ought to insist on the preconditions of citizen empowerment and maximum institutional democracy in the name of the bedrock democratic commitment to substantive self-rule.

This does not boil down to what Gutmann and Thompson call proceduralist democratic theory. Mere formal proceduralism has proved too weak to support the equally bedrock democratic commitment to justice (Shapiro 1999b). Again, from a provisional perspective, then, what matters most for democratic theory are the preconditions of substantive self-rule. For example, in a discussion of the Clinton health care plan, Ian Shapiro argues that its failure is due, "more than any other single factor, to the blank check the United States Supreme Court has given those who have large amounts of money, or the capacity to raise it, to shape the terms of public debate. It creates a reality in which, rather than compete in the realm of ideas, politicians actually must compete for campaign contributions" (1999a: 35). Similarly, Douglas Rae argues that attention to very specific local conditions of citizen empowerment is necessary to convert "intangible" into "tangible" freedoms (1999: 169). Norman Daniels's

[11] A rarely recognized Kantian point. Though in the sphere of ethics Kant is indeed something of an absolutist, in politics Kant denied any single generation the right to settle questions, including moral ones. See "An Answer to the Question: What is Enlightenment?" (VIII: 33–42).

work on deliberative theory and managed health care is also of interest here, as Daniels confesses himself a lapsed Rawlsian who no longer believes that absolute principles can guide institutions making moral decisions in the health-care area (1999: 199–200). Middle-level moral policy problems, such as health-care rationing, respond better to a provisional approach modeled on case law, with important decisions reached in a public forum. Rather than handing down correct moral policy answers from a position of ethical expertise, Daniels argues, health-care policymakers ought to be required to give reasons in public for their decisions. "Case law does not imply past infallibility. . . . It involves a form of institutional reflective equilibrium, a commitment to both transparency and coherence in the giving of reasons" (1999: 205). For these theorists, the problem illustrated by health-care policy in the United States is less its (very real) failure to build an acceptable system than that system's failure to make good on democratic principles of inclusiveness.

The issue of health care illustrates a contradictory tendency in mainstream deliberative theory. Despite a laudable commitment to provisionalism, most deliberative theorists are still too willing to prejudge the outcome of what should be democratically and dynamically determined policy choices. Gutmann and Thompson, for example, argue that medical rationing can contradict the deliberative principles of inclusiveness, reciprocity, and basic opportunity, since someone denied a chance at life will not be able to participate politically. Their main case study on this point is that of Dianna Brown, an Arizona resident denied an organ transplant because she could not afford one herself, and the state had decided to fund other medical needs instead (Gutmann and Thompson 1996). Compelling as Brown's need is, by prejudging the outcome of this case, Gutmann and Thompson narrow the scope of democratic deliberation intolerably. They do not recognize that a focus on substantive rather than merely formal freedoms entails the recognition that innumerable public decisions, and not only those explicitly and directly dealing with life and death, can result in the death of a potentially empowered individual. Substantively speaking, there is little difference between state policy on organ transplants, and state policy on automobile safety, environmental toxicity, junk food in the public schools, and thousands of other issues with demonstrable connections to individual life chances.

Between a formal proceduralism that privileges the status quo, and a neo-paternalism that privileges the views of the theorist above those of the polity, lies the sort of "thick proceduralism" recommended by a provisional rather than a conclusivist approach to political thought. Some conclusive theorists may plausibly argue that people acting through the institutions of deliberative democracy are more likely to reach the "right answers" to policy questions than current minimalist institutions allow. However, in the first place, this is an empirical question. One could come up with an arbitrary list of "correct" liberal democratic policy outcomes, measure their prevalence along with the level of institutional "deliberativeness" in a large-N cross-national study, and see whether the two variables are significantly related. Presuming consensus among theorists on at least some basic policy outcomes, large-scale investigation into the usefulness of deliberative institutions might proceed on the model of, say, Robert Putnam's ongoing study of the relation between social capital and general measures of societal well-being (2000). The results of such a study would certainly be interesting, but they would tell us nothing about the normative status of deliberative democratic institutions. In the second place, prejudging the outcomes of deliberation violates a bedrock precept of democratic theory in the deliberative mode, namely respect for persons as concrete historical individuals able to engage in substantial self-rule. As Iris Marion Young writes, "a theory of democracy in itself should have little to say about the substance of welfare policy, but should have a great deal to say about the institutions, practices, and procedures for deliberating about and deciding on a welfare policy" (1999: 156–7).

However, exceptions may be made for policies that relate to the preconditions of deliberation itself. On the issue of slavery, for example, Gutmann and Thompson plausibly object to Robert P. George's view that proslavery viewpoints have not always deserved moral disrespect (Gutmann and Thompson 1999: 270). The difference between the issue of slavery, for which there can be no moral respect for the pro side, and the issue of abortion, for which Gutmann and Thompson argue moral respect must be accorded all reciprocity-respecting viewpoints, is this: the institution of slavery contradicts the basic principles underlying deliberative democracy itself by excluding adults affected by public policies from deliberation over them (that is, from self-rule).

The line between provisional and conclusivist theories is admittedly a fine one: why exclude Habermasian conditions of ideal speech from the ranks of the provisional theorists while including Gutmann's and Thompson's conditions of deliberation? The answer is twofold. First, Habermas and other theorists of conclusive right make universalist claims for their principles, while Gutmann and Thompson and other theorists of their type make only contingent claims for theirs. "If you want the legitimacy accorded to the deliberative policymaking process, then you must respect these principles inherent in deliberativeness *per se*," a provisional deliberative theorist might argue.[12] Second, Gutmann and Thompson argue that even bedrock principles such as reciprocity are open to revision in the deliberative sphere: "Citizens learn more about the nature of reciprocity, as they reason according to its requirements. Because they regard the requirements of reciprocal reasoning, like its results, as provisional, their understanding of reciprocity has the capacity to correct itself over time" (Gutmann and Thompson 1996: 351).

Deliberative theory's institutional implausibility and its empirically informed solution: the global public sphere

Despite their continued interest, expressed in case studies such as that of Dianna Brown, in achieving moral justice in the very short term, Gutmann and Thompson do make some interesting longer-term institutional suggestions. I have already mentioned their call to broaden the policy tool of the environmental impact statement to require many more decisionmakers to give public reasons for their major development activities. They decry the emphasis in the United States on the courts as moral arbiters of last resort, and call for moral politics to break out of its judicial cage and into the general public sphere. But as many of their defenders, not to mention their detractors, have noted, Gutmann's and Thompson's book makes relatively few positive institutional suggestions; in fact, the whole book seems at times to be a demonstrative exercise in deliberative reasoning itself.

One reason Gutmann and Thompson may have shied away from making concrete institutional suggestions for improving the deliberative climate is the sheer implausibility of some of the proposals already on the table. The least implausible, most well-researched and well-documented innovation in deliberative institution-making is James Fishkin's project of deliberative polling, which has been tried in the state of Texas, and nationally in the US and UK. While deliberative polling has produced some very interesting findings, such as the finding that ordinary citizens of the state of Texas are quite willing to trade higher energy prices for more environmentally friendly and sustainable energy policies, it has also been criticized as ineffective. Studying the "Grenada 500" deliberative poll during the 1992 general election in Great Britain, David Denver and his colleagues (1995) conclude that two problems plagued the effort: a persistent problem with unrepresentative samples, and the failure of the deliberative experience to deliver on its promises. Fishkin and his

[12] This is why Mansbridge bases her version of provisionalism on the Kantian notion of a "regulative principle."

fellow investigators have paid a great deal of attention to the problem of representativeness, and may well be able to resolve that problem with adjustments to the incentive structure and the like. The second problem is more serious: the very short term (in deliberative polling, a matter of a few hours or at most a few weeks) may not be the right place to look for the kind of dynamic shift in public reason that deliberative theory leads us to expect. Denver *et al.* found no significant change in participants' views after the Grenada 500 deliberative poll; what's worse, they found a decrease in political knowledge and no increase in political sophistication (Denver *et al.* 1995: 152–3). I shall add that even a very well-established deliberative institution like the deliberative poll faces enormous obstacles in moving from a few trial runs to becoming part of the national political culture. Compare Fishkin's deliberative polling institution with some of the other suggestions offered by deliberative theorists (such as increasing access to the public sphere, campaign finance reform, or requiring public consultation on most public business); for the price of sequestering, compensating, and publicizing deliberative polling groups as a regular national political institution, might we not enact more direct political reforms aimed not at short-term, potentially paternalistic enlightenment but at the long-term empowerment of the citizenry to enlighten, in fact, to rule, itself?

The empirical record for deliberative institutions modeled on an individual level is not encouraging.[13] However, for institutions seeking deliberation at the level of a historical public sphere, there are some promising new findings. A recent and fine example of this sort of political science is the work by Margaret Keck and Kathryn Sikkink on transnational advocacy groups (1998). They argue that a new, important, and poorly understood entity has emerged in international politics: transnational advocacy networks. These networks, linking activists in different countries interested in moral causes such as human rights and environmental protection, can be shown to have concrete political effects under certain conditions. Keck and Sikkink identify, for example, a set of conditions leading to a "boomerang effect," in which domestic activists faced with an unresponsive government are able to work with international activists who put pressure on their more responsive home governments, which in turn are able effectively to pressure the originally unresponsive state (13). What sets transnational advocacy networks apart from more traditional political agents is primarily their commitment to "principled ideas," which they

[13] Not only are the empirical prospects grim, but even the theoretical prospects for deliberative institutions on the time scale contemplated are not at all encouraging. For a nicely reasoned effort in that direction, see Ferejohn (2000).

use to attempt to change perceptions of interest among more powerful actors (30). Keck and Sikkink use classic comparative methodology to examine their object of inquiry, setting up structured comparisons, for example, between failed and successful campaigns by advocacy groups (202 and *passim*). The logic of transnational advocacy groups' activity, however, was set out years ago by Kant in his theory of the concrete political effects of speech in the public sphere (Habermas 1989; Ellis forthcoming). Keck and Sikkink's empirical findings on the effectiveness of transnational advocacy groups complements recent work in neo-Kantian political philosophy on the norm of reciprocity and its political function (Rawls 1996; Gutmann and Thompson 1996). Keck and Sikkink's empirical findings about the transcultural value of the concept of human dignity (as opposed to human rights) has interesting implications for liberal theories of overlapping consensus. In fact, Keck and Sikkink themselves draw some broad and interesting theoretical conclusions from their empirical findings: "liberalism carries within it not the seeds of its destruction, but the seeds of its expansion. Liberalism, with all its historical shortcomings, contains a subversive element that plays into the hands of activists" (1998: 205).

Attention to the empirical literature on the dynamics of the public sphere should, therefore, provide encouragement for deliberative theory modeled on the longer term, while the paucity of evidence of short-term individual-level opinion change under deliberative conditions should lead traditional deliberative theorists to seek alternative mechanisms of political change. The long-run shifts in which arguments are politically possible to make in the public sphere are a far more promising avenue of inquiry for provisional deliberative theory than the hoped-for short-run shifts in public opinion.

Conclusion

As Hannah Arendt once wrote of an earlier debate in political science, "The trouble with modern theories of behaviorism is not that they are wrong but that they could become true" (1958: 322). Laitin's prediction of a theory-free political science is not impossible, but for the science to conform to its own standards, political life itself would have to change. It is certainly possible to imagine historical contexts that are more and less "theory-friendly," so perhaps a world in which conflicts over constructed political ideals do not occur is not unthinkable. Thus far, however, political science has not found such a world. In fact, as has been argued well by others (Green and Shapiro 1994; Kuhn 1970), particular and contestable theoretical perspectives inform every school of research, political science included. Provisional theory in all three modes – historical, rational, and empirical – is therefore needed to address the questions raised by the political world as it really exists.

REFERENCES

Arendt, Hannah. 1958. *The Human Condition*. Chicago: University of Chicago Press.

Benhabib, Seyla (ed.). 1996. *Democracy and Difference: Contesting the Boundaries of the Political*. Princeton: Princeton University Press.

Bohman, James. 1997. "Deliberative Democracy and Effective Social Freedom: Capabilities, Resources, and Opportunities," in Bohman and Rehg (eds.), pp. 321–48.

Bohman, James and William Rehg (eds.). 1997. *Deliberative Democracy: Essays on Reason and Politics*. Cambridge, MA and London: MIT Press.

Brunner, Otto, Werner Conze, and Reinhart Koselleck (eds.). 1972–. *Geschichtliche Grundbegriffe: historisches Lexikon zur politisch-sozialen Sprache in Deutschland*. 8 vols. Stuttgart: Klett-Cotta.

Burns, Nancy, Kay Lehman Schlozman, and Sidney Verba. 2001. *The Private Roots of Public Action: Gender, Equality, and Political Participation*. Cambridge, MA: Harvard University Press.

Cohen, Joshua. 1997. "Deliberation and Democratic Legitimacy," in Bohman and Rehg (eds.), pp. 67–92.

Daniels, Norman. 1999. "Enabling Democratic Deliberation: How Managed Care Organizations Ought to Make Decisions about Coverage for New Technologies," in Macedo (ed.), pp. 198–210.

Denver, David, Gordon Hands, and Bill Jones. 1995. "Fishkin and the Deliberative Opinion Poll: Lessons from a Study of the Granada 500 Television Program." *Political Communication* 12: 147–56.

Dryzek, John S. 1990. *Discursive Democracy: Politics, Policy, and Political Science*. Cambridge: Cambridge University Press.

2000. *Deliberative Democracy and Beyond: Liberals, Critics, Contestations*. Oxford and New York: Oxford University Press.

Ellis, Elisabeth. Forthcoming. *Kant's Politics: Provisional Theory for an Uncertain World*. New Haven: Yale University Press.

Elster, Jon. 1997. "The Market and the Forum: Three Varieties of Political Theory," in Bohman and Rehg (eds.), pp. 3–34.

Ferejohn, John. 2000. "Instituting Deliberative Democracy," in *Designing Democratic Institutions: Nomos XLII*. New York: New York University Press. 75–104.

Fiorina, Morris. 1999. "Extreme Voices: The Dark Side of Civic Engagement," in *Civic Engagement and American Democracy*. Skocpol and Fiorina (eds.). Washington, DC: Brookings.

Fishkin, James S. 1991. *Democracy and Deliberation: New Directions for Democratic Reform*. New Haven and London: Yale University Press.

1995. *The Voice of the People: Public Opinion and Democracy*. New Haven and London: Yale University Press.

Fishkin, James S., Robert C. Luskin, and Roger Jowell. 2000. "Deliberative Polling and Public Consultation." *Parliamentary Affairs* 53: 657–66.

Foucault, Michel. 1977. *Language, Counter-Memory, Practice*. Ithaca, New York: Cornell University Press.

Galston, William A. 1999. "Diversity, Toleration, and Deliberative Democracy: Religious Minorities and Public Schooling," in Macedo (ed.), pp. 39–48.

George, Robert P. 1999. "Law, Democracy, and Moral Disagreement: Reciprocity, Slavery, and Abortion," in Macedo (ed.), pp. 184–97.

Ginsburg, Ruth Bader. 2001. "In Pursuit of the Public Good: Access to Justice in the United States." *Washington University Journal of Law and Policy* 7: 1–15.

Green, Donald P. and Ian Shapiro. 1994. *Pathologies of Rational Choice Theory: A Critique of Application in Political Science*. New Haven and London: Yale University Press.

Gutmann, Amy, and Dennis Thompson. 1996. *Democracy and Disagreement: Why Moral Conflict Cannot be Avoided in Politics, and What Should be Done About It*. Cambridge, MA: The Belknap Press of Harvard University Press.

1999. "Democratic Disagreement," in Macedo (ed.), pp. 243–80.

2000. "Why Deliberative Democracy is Different." *Social Philosophy and Policy* 17(1) (Winter): 161–80.

Habermas, Jürgen. 1973. "Wahrheitstheorien." In *Wirklichkeit and Reflexion: Walter Schultz zum 60. Geburtstag*. Helmut Farhenbach (ed.). Pfüllingen: Neske.

1989. *The Structural Transformation of the Public Sphere: An Inquiry into a Category of Bourgeois Society*. Cambridge, MA: MIT Press.

1992. *Faktizität und Geltung: Beiträge zur Diskurstheorie des Rechts und des demokratischen Rechtsstaats*. Frankfurt am Main: Suhrkamp.

1996. "Three Normative Models of Democracy," in Benhabib (ed.), pp. 21–30.

1997. "Popular Sovereignty as Procedure," in Bohman and Rehg (eds.), pp. 35–66.

Isaac, Jeffrey C. 1995. "The Strange Silence of Political Theory." *Political Theory* 23(4): 636–52.

Kant, I. 1902–. *Kants gesammelte Schriften. Preussischen Akademie der Wissenschaften*. Berlin and Leipzig: Walter de Gruyter & Co.

Keck, Margaret E. and Kathryn Sikkink. 1998. *Activists Beyond Borders: Advocacy Networks in International Politics*. Ithaca and London: Cornell University Press.

Knight, Jack. 1999. "Constitutionalism and Deliberative Democracy," in Macedo (ed.), pp. 159–69.

Knight, Jack, and James Johnson. 1994. "Aggregation and Deliberation: On the Possibility of Democratic Legitimacy." *Political Theory* 22(2) (May): 277–96.

1997. "What Sort of Equality Does Deliberative Democracy Require?" in Bohman and Rehg (eds.), pp. 279–320.

Koselleck, Reinhart. 1972–89. Geschichtliche Grundbegriffe and Introduction, in *Geschichtliche Grundbegriffe: Historisches Lexikon zur politisch-sozialen*

Sprache in Deutschland, O. Brunner, W. Conze, and R. Koselleck (eds.). Stuttgart: Ernst Klett Verlag.

1985. *Futures Past*. Keith Tribe (trans.). Cambridge, MA: MIT Press.

1988. *Critique and Crisis: Enlightenment and the Pathogenesis of Modern Society*. Oxford: Berg.

1989. "Linguistic change and the History of Events." *Journal of Modern History* 61: 649–66.

2002. *The Practice of Conceptual History: Timing History, Spacing Concepts*. Todd Samuel Presner *et al.* (trans). Stanford: Stanford University Press.

Kuhn, Thomas S. 1970. *The Structure of Scientific Revolutions* (2nd edn.) Chicago: University of Chicago Press.

Kymlicka, Will. 1996. "Three Forms of Differentiated Citizenship in Canada," in Benhabib (ed.), pp. 153–70.

2000. *Politics in the Vernacular: Nationalism, Multiculturalism and Citizenship*. Oxford: Oxford University Press.

Laitin, David D. 1998. "Toward a Political Science Discipline: Authority Patterns Revisited." *Comparative Political Studies* 31(4): 423–43.

2001a. "The Political Science Discipline." Remarks accompanying a paper (Laitin 2001b) delivered at the annual meeting of the American Political Science Association, San Francisco, CA, September.

2001b. "The Political Science Discipline." Paper delivered at the annual meeting of the American Political Science Association, San Francisco, CA, September.

Macedo, Stephen (ed.). 1999. *Deliberative Politics: Essays on Democracy and Disagreement*. Oxford and New York: Oxford University Press.

Mansbridge, Jane. 1996. "Using Power/Fighting Power: the Polity," in Benhabib (ed.), pp. 46–66.

1999. "Everyday Talk in the Deliberative System," in Macedo (ed.), pp. 211–42.

Marcus, George E., W. Russell Neuman, and Michael MacKuen. 2000. *Affective Intelligence and Political Judgment*. Chicago: University of Chicago Press.

Marshall, T. H. 1964. *Class, Citizenship, and Social Development: Essays*. Garden City, NY: Doubleday & Co.

Pagden, Anthony. 1987. "Introduction," in *The Languages of Political Theory in Early-Modern Europe*. A. Pagden (ed.). Cambridge: Cambridge University Press.

Phillipson, Nicholas and Quentin Skinner (eds.). 1993. *Political Discourse in Early Modern Britain*. Cambridge: Cambridge University Press.

Pitkin, Hanna Fenichel. 1972. *Wittgenstein and Justice: On the Significance of Ludwig Wittgenstein for Social and Political Thought*. Berkeley: University of California Press.

Pocock, J. G. A. 1987. "The Concept of a Language and the *metier d'historien*: Some Considerations on Practice," in Pagden (ed.).

1989. *Politics, Language, and Time*. Chicago: University of Chicago Press.

Putnam, Robert D. 1993. "The Prosperous Community." *The American Prospect* 4(13), March 21.

1996. "Unsolved Mystery, the Tocqueville Files." *The American Prospect* 7(25), March 1 and April 1.

2000. *Bowling Alone: The Collapse and Revival of American Community*. New York: Simon & Schuster.

Putnam, Robert D., with Robert Leonardi, and Rafaella Y. Nanetti. 1993. *Making Democracy Work: Civic Traditions in Modern Italy*. Princeton: Princeton University Press.

Rae, Douglas. 1999. "Democratic Liberty and the Tyrannies of Place." in Shapiro and Hacker-Cordón (eds.), pp. 165–92.

Rawls, John. 1971. *A Theory of Justice*. Cambridge, MA: Harvard University Press.

1980. "Kantian Constructivism in Moral Theory." *The Journal of Philosophy* 77(9): 515–73.

1985. "Justice as Fairness: Political not Metaphysical." *Philosophy and Public Affairs* 14(3): 223–51.

1996. *Political Liberalism* (2nd edn). New York: Columbia University Press.

1997. "The Idea of Public Reason" and "Postscript," in Bohman and Rehg (eds.), pp. 93–141.

Richter, Melvin. 1989. "Understanding Begriffsgeschichte." *Political Theory* 17(2): 296–301.

1995. *The History of Political and Social Concepts: A Critical Introduction*. New York: Oxford University Press.

Romer, Paul M. 1993. "Two Strategies for Economic Development: Using Ideas and Producing Ideas." Proceedings of the World Bank Annual Conference on Development Economics 1992. 63–91.

Sandel, Michael J. 1998. *Liberalism and the Limits of Justice* (2nd edn.) Cambridge: Cambridge University Press.

Schauer, Frederick. 1999. "Talking as a Decision Procedure," in Macedo (ed.), pp. 17–27.

Shapiro, Ian. 1999a. "Enough of Deliberation: Politics is about Interests and Power," in Macedo (ed.), pp. 28–38.

1999b. *Democratic Justice*. New Haven: Yale University Press.

1996. *Democracy's Place*. Ithaca and London: Cornell University Press.

Shapiro, Ian and Casiano Hacker-Cordón (eds.). 1999. *Democracy's Edges*. Cambridge: Cambridge University Press.

Sunstein, Cass R. 1999. "Agreement without Theory," in Macedo (ed.), pp. 123–50.

Young, Iris Marion. 1999. "Justice, Inclusion, and Deliberative Democracy," in Macedo (ed.), pp. 151–58.

17 What have we learned?

Responses by *Robert A. Dahl, Truman F. Bewley, Susanne Hoeber Rudolph, and John Mearsheimer*

Robert A. Dahl: Complexity, change, and contingency

I am simply going to set out, perhaps somewhat more dogmatically than is appropriate, a set of observations. I think some of you will disagree with some of them. Some of you may disagree with all of them. If it turns out that most or all of you agree with all of them, then I am wasting your time and my own by being here.

I begin with a question and observation: What is our subject? What is political science? For those of you who have read some of my work over the years it won't come as a surprise that my answer to the question is this: political science is the systematic study of relations of power and influence among human beings. My observation is: if you define political science in this way, then the study of politics is perhaps a subject of greater complexity than any other area of scholarly research, writing, and teaching.

There are, I believe, at least three reasons why the study of politics is a subject of exceptional complexity.

First, consider the number of possible relations of power and influence among relevant units. Perform a simple-minded thought experiment: identify some of the possible relations of power and influence among two people. You'd come up pretty quickly with a list of five possible relations, maybe ten, maybe more; but let's be modest and settle for ten. Now imagine ten people and once again assume, say, ten possible relations of power and influence among them. Then go up to 100 people, and the possible number of relationships. Perhaps you might now move to 400 or 500 persons, which perhaps approaches the limit with which we can interact with individual people. Now let's add relations between individual persons and associations. Then proceed to relations among groups and associations; now proceed to relations among individuals, associations, and the special kind of association we call a state. Now add international organizations. If you aren't already totally exhausted, consider all the possible relationships among individuals, associations, states, and international organizations.

No doubt we could go further, but my point is clear enough: the world we study is extraordinarily complicated.

The second factor that makes the study of politics so difficult is that neither the units themselves nor their relationships are static. They *change*. They sometimes change in very brief periods of time and they certainly change over long periods of time.

Finally, *contingency* plays an extraordinary role in human affairs. I believe that developments of enormous consequence often turn on contingencies that can't be confidently predicted before they occur nor confidently shown to be inevitable after they occur. I'm not referring to simple-minded "what-if?" scenarios that anyone can create. I'm thinking about serious historical possibilities. For example, it was very far from certain that Lenin would ever arrive in St. Petersburg. One can imagine all sorts of plausible scenarios in which he never reached Russia in 1917. If he had not arrived in Russia, would the Bolsheviks have gained power? I seriously doubt it. Yet if the Bolsheviks had not won power, the history of the world would have been profoundly different. So, too, with Adolf Hitler. From Henry Turner's *Hitler's Thirty Days to Power* (1996) we can readily construct a highly a plausible scenario in which Hitler never achieved power. As Turner's careful and fully documented research reveals, from November, 1932 through January, 1933 a very small number of persons made decisions that resulted in Hitler's accession to power, and that outcome was in no sense inevitable. That he gained power was highly contingent, Turner shows, on a series of decisions made by a few human beings who might readily have made a different decision, one that would have deprived Hitler of office. Yet the unintended consequence of their decisions was Adolf Hitler's seizure of power, his dictatatorship, World War II, the Holocaust, and the death of many millions of human beings.

My point, then, is that highly consequential historical contingencies add immeasurably to the complexity of the world with which we must deal – a complexity on which we must not only base our descriptions but, so far as possible, our explanations, generalizations, and predictions.

You may challenge my assumption about the complexity of the world with which we, as political scientists, must deal; but if you are prepared to assume that I am not wildly off the mark, the question arises: how ought we to deal with a subject of such daunting complexity?

Alas, I not only don't have a good answer, I'm afraid I don't even have a plausible answer. However, I do want to offer some general observations.

I believe there are some things we ought *not* to do. To begin with, we shouldn't pursue political science as if it were physics. Instead, we should cure ourselves of what some people have called "physics envy." If, after

25 centuries, Aristotle's physics is irrelevant to modern physics, while political scientists are still influenced by and can find something useful in Aristotle's *Politics*, is this – as perhaps many physicists believe – simply because political scientists are less competent scientists than physicists?. I don't believe so. I believe the explanation is that the subjects of political science are infinitely more complex than the subjects of physics. To be sure, physics is a complex *subject*. But the *subjects* of physics are essentially quite simple. Yet even physics has not been able to overcome some fundamental problems that arise out of the complexity of the world that they study. After nearly a century, no one, as best I can tell, has been able to reconcile standard physics with the weird aspects of quantum mechanics. Although I can't claim to much depth of knowledge about physics, as best I can judge if you begin to reflect about some of the possibilities in quantum mechanics – simultaneous action at a distance, for example – the world it portrays can't be reconciled to the ordinary world of human experience, nor, what is more, to the world of standard physics. I note that some very eminent physicists and cosmologists have recently envisioned the existence of multiple universes, each one operating on somewhat different laws, even greatly different laws. Move over, political science!

To bring the issue closer to home, let's try to imagine a physicist trying to cope with information about, let's say, 200 universes, each different in some degree from the rest. And what's more, each of these universes contains systems that continue to change. Does that remind you a bit of a world with 200 different countries that undergo constant change and variations in their political systems and relationships?

Second, we shouldn't try to emulate economics, or at least economics as it was predominantly conceived until quite recently. Like physics, economics is a complex subject, and throughout much of its history standard economics made the subjects of economic study very simple, perhaps excessively simple. That period may now be coming to an end as behavioral economics and other approaches introduce much of the complexity and contingency that characterize political science. As our Yale colleague Robert Shiller has shown in his recent book *Irrational Exhuberance* (2000), the assumption of rational actors in standard economic theory doesn't help much to explain economic bubbles.

Third, I'm inclined to think that trying to create a general theory of politics may be a waste of scholarly time. At best, a general theory of politics might be entertaining, it might be amusing, it might even be insightful, but my guess is that it would also prove to be a superficial gloss on the complexities of political life. With luck, it might create a passing intellectual fad. At worst, it would probably turn out to be a great bore.

Fourth, we shouldn't assume that the complexity of our field can be competently dealt with by any single theoretical or methodological approach. Let's continue, then, to accept the need for theoretical and methodological diversity.

Finally, we should avoid excessive reductionism. By reductionism I mean explaining the behavior of complex systems by narrowing causation down to a single factor, an often distant factor in a lengthy causal chain. To illustrate what I would regard as a rather extreme form of reductionism, let me quote the words of a natural scientist in a recent book review: "The wolf shall lie down with the lamb, and the leopard with the kid, but the true forces of Darwinian competition are as powerful as ever *and they are responsible for the ultimate biological achievement in cooperation – the body politic*" (my emphasis). Can we really explain, let us say, the institutions that protect human rights today by taking a giant leap back across the long causal chain to genes and Darwinian competition?.

So: to return to my question. If these paths lead nowhere, and we're confronted by actions and relations of extraordinary complexity, how are we to proceed? Although as I said I don't have a satisfactory answer, I'd like to conclude with a half dozen observations, each, I fear, with implications vast enough to nurture an entire conference. First, I think we might occasionally ask ourselves, as some of you have asked here, what is political science for? What good is it? Other than the opportunities for jobs, material benefits, and a pleasing professional life, why should anyone pursue the study of politics? My answer, which may be yours as well, is that in the end the only justification for political science is that we believe it could help to achieve good ends.

Second, and following from the first, we shouldn't be misled by the term *science* in "political science." I believe there will always be, and there should always be, an important role in the study of politics for those among us who pursue ethical questions. What good ends can be served by politics? In what does human good consist? Moving closer to empirical issues, what means will best serve these ends? On what grounds do we justify tradeoffs among various good ends? And so on.

Third, and reflecting something I said earlier, ours will always be a world of high uncertainty, high risk, and danger. I mean the world out there, the world we live in. We're a little more aware of the dangers in our world than we may have been not long ago, before the era that ended with 9/11. In a world of danger and uncertainty, a very small increase in reasonably reliable knowledge may sometimes result in a very large gain for human well being. In short, we should not disparage small improvements in the reliable knowledge we help to create. They just might make a significant difference.

Fourth, it seems obvious to me, as it does to I think most of you, that we need a division of labor. Not everyone will or should pursue our questions in the same systematic way. Yet because of the enormous complexity of political life the divisions among so-called fields and subfields of political science are to some extent arbitrary. By redefining its major subfields not long ago the Yale Department of Political Science recognized that in the face of increasing knowledge the older distinctions no longer served the ends of teaching and learning that might be better served by a new alignment of subjects. No doubt, changes may be desirable in the future in the way that we think about our subfields.

Fifth, at the risk of excessive repetition, we don't need and can't afford just one approach to the study of politics. We need many approaches. The study of politics has been an intellectually untidy field. It will remain untidy: much more untidy than physics; more untidy than psychology; probably more untidy than economics; and maybe even more untidy than the new behavioral economics.

Finally, a question and a possible answer. Are we who are engaged in the study of politics accumulating knowledge – or are we just going around in circles? Given the complexity of our field, the speed with which important changes take place in the political world, the occurrence of contingent events of extraordinary consequence, does political science, the study of politics, move ahead? Or are we all running as fast as we can just to keep up with this complex and ever-changing world?

When I reflect on the work of the scholars represented in this volume and others in the larger field of political science, and then compare the knowledge and understanding that is available today with the field of political science to which, over half a century ago, I began to devote my teaching, research, and writing, my answer is unambiguous. Political science is incomparably better now than it was then. It's richer, it's broader, it's more systematic, it's more solidly grounded in human experience.

Yes, our field is dauntingly complex, yet we're far more aware of the complexities than we were 50 years ago and we understand them better. Difficult as it is to grapple with these complexities, political scientists have responded, and I feel pretty confident that the coming generation of political scientists will continue to respond, to a field of activity that will remain inordinately complex and yet undeniably crucial for achieving good human ends.

Truman F. Bewley: The limits of rationality

I am going to interpret these essays in terms of my own experience as an economist. I have heard a great deal of perplexity about the uses of mathematics, the rationality hypothesis, whether people know their own motives, and about the power of statistical analyses. These are questions that arise in economics as well. I've been surprised not to read more about the things that I think would arise naturally in political science: reciprocity and fairness and justice as motivators, as well as a sense of a group. These are things that economists are finding are strong motivators. One problem with the rationality hypothesis is that it implies a model that mimics reality, when you reduce everything to assumptions about individual interactions. It's like Newtonian physics, if you want to use that analogy, where everything is in terms of interaction among particles. That reductionism may apply in some circumstances, but it is probably false to think it applies in general.

I think that a lot of the issues and problems can be resolved by getting closer to the subject, and that involves collecting one's own data. That's been my experience, anyway. The contexts in which people behave rationally and understand their motives and the areas where statistical data are useful become clear when you get close to the subject you are studying. I think it's a natural thing to do. You imagine chemists – they work with the things they're dealing with in their laboratories. In economics there has been a tendency to suppose that you shouldn't ask people what they're doing or what their motives are because they'll deceive you, or they won't know what their own motives are, they don't understand them, they'll exaggerate their own roles, and glorify themselves, and so on. So what you do is put a distance between yourself and the subject matter. This can be healthy in some circumstances, but it can also be unhealthy. It's like treating the economy as if it were like some distant galaxy and then you are really stuck with using only very indirect theoretical and statistical methods to understand it. I think those are very useful but I think you need to have more immediate experience.

One thing I wanted to note is that I think the distinction that came up earlier in this volume between mathematical and verbal reasoning is a false distinction. If you go very far in mathematics, you're using definitions and proofs, and that's all verbal. There's a lot of care in defining the concepts and making the arguments, but people make mistakes all the time, they say "clearly" when something is not clear or is wrong, and the mistakes are brought out by debate, by interaction among mathematicians, just as they are in any verbal discussion. Of course you want to be careful in your reasoning but I don't think that mathematics has any priority in that. It elaborates the language, if you like. Mathematics can be thought of as a language.

Now, I turn to my own recent work. I have sought to explain the phenomenon of downward wage rigidity and also to discover whether this is really a valid phenomenon. The available statistical data are really not much help. There are a lot of time series data on pay, on hours of work, and they specify roughly the industry in which somebody works, whether they stay with the same employer, the kind of job they have, but it actually gives you very little information or no information about the explanation of wage rigidity or even whether it is a valid phenomenon. After all, individual workers have changes of job assignments, they change shifts, they change from full-time to part-time work, and their pay may go down, so you may observe individual pay reductions which are in fact not pay reductions in the sense of pay for a certain kind of work. And it's also hard to reconcile wage rigidity with economic theory, because you imagine that if an unemployed worker is willing to work for a firm for less than the going wage for that particular kind of work in the firm, then they could reach a deal and the worker could be hired at lower than the going wage. This paradox has perplexed economists for a long time.

What I did was to resist the refusal to ask decisionmakers what they're doing, what their experiences have been, and why they make decisions. And I simply interviewed well over 300 business people and labor leaders and people who deal with the unemployed as counsellors. Of course, I don't think these respondents lie, but you have to treat what they say critically. You are better able to do so if you talk to a great variety of people in different circumstances, and what you try to do is make sense of what they say in terms of some kind of overall, logically coherent story. I like ancient history, and I have often wondered how wonderful it would be just to be able to talk to one Sumerian – you could learn a great deal. And it is the same thing if you just go talk to one businessman. So that was my initial idea – to talk to three or four, but I kept going.

Let me briefly try to explain what I learned. Most jobs involve an important amount of training and not much supervision. For such jobs employers are very concerned that their workers have good morale. Morale is a mood conducive to good work – not necessarily happiness – it is also identification with the objectives of the employer. So this is some sort of group attribute, it is not reducible really to individual motives. A worker with good morale is like a soldier, willing to make sacrifices for the employer, because the worker believes in the employer's objectives, believes in the outfit, in the firm that he's working for. Good morale is sustained by a spirit of reciprocation and fairness between the worker and the organization and among the workers themselves. It is something that is difficult to maintain.

One part of the reciprocation is that good work is rewarded with pay raises. Cutting pay hurts people's feelings, even if everybody's pay is cut people associate pay increases with a reward for doing good work, so when pay is there is some kind of offense, a lack of reciprocation, I call it the "insult effect" – a slap in the face is what people call it. And another factor which I call the "standard of living effect," is that people are hurt financially by pay reductions and then they seek somebody to blame and they blame the employer. Employers worry about this because their work effort is impaired, they might get higher labor turnover, they get a bad reputation as an employer, it's harder for them to hire people. I did all of these interviews during a recession in the early nineties and the worry was that as soon as the recession was over people would quit and it would be hard to hire new workers. They didn't really have any place to go then. Another important thing is that pay cuts do occur and they are accepted if they're viewed as fair. So fairness again enters. They're fair if they're going to save a lot of jobs, if the firm is an unusual firm that can lower its product prices as a result of lowering pay and then sell a lot more, or if it's about to go out of business and close it's doors. So such pay reductions occur, but they're unusual except when you have a real disaster as you did during the Great Depression. I tried to find firms that cut pay and it was difficult to do so.

Something I noticed in my interviews is that employers were very rational in their reasons for resisting pay cuts. It's actually the employers who resist pay cuts, because of the constraints imposed by workers' attitudes and their probable reactions to pay cuts. Workers are probably not rational in the economists' sense, in the sense of pursuing their own self-interest, for they allow morale to effect their work effort, they indulge feelings of being offended when their pay is cut, or when their pay is lower than co-workers who are equally qualified or less qualified than themselves, and in taking revenge for a reduced standard of living. All these things are irrational, but they are very natural. But in a relationship of close cooperation and reciprocation these feelings of fairness and identification arise automatically. So if managers' pay is cut they react as their workers would and so they understand their workers' reactions. But they can detach themselves from the situation and behave rationally when they're thinking of making the decision of whether to cut pay, or numerous other similar decisions. The context affects the degree of rationality of the decisionmaker. It's also very important in a firm, which is like a little polity, to realize that the whole is held together by ideas of justice and fairness and identification with a group, and when that comes unstuck, the firm can fall apart, and there's a real fear on the part of the managers that things will come unstuck.

Statistical information can be very useful, but I think the data for this problem you have to collect yourself. For instance, there are many kinds of jobs where you don't require much training and you can easily supervise the work. In that setting, employers are much less interested in morale, and they're much more likely to resort to coercion or firing people who don't do their work properly. In that kind of setting I think I saw that pay was more flexible downward, that pay cuts were more likely. But there is no data to verify that. A good test of the theory would be to check this observation statistically by collecting data on time series for pay for different kinds of jobs – those that require lots of training and those that require less training and so on.

We have read much about contingency in this volume. Well, anything can happen in a firm. What I've been talking about are generalizations, but they don't always apply. Employers are idiosyncratic and they'll make strange decisions, from which the business community learns. People do reduce pay or do very odd things in firms, even important firms, and that has to do with the personality of the leader or the internal political circumstances of the firm.

Susanne Hoeber Rudolph: Toward convergence

When asked to sum up what we have learned from the essays in this volume, we were offered the possibility that we might conclude we had learned nothing. My expectations in these matters are moderate. I expected to rearrange my mental furniture and add a few pieces.

I acquired some new words, no trivial gain. To acquire new words means loading on board – cognitively and/or normatively – the baggage attached to those words. We've read a lot about ordinary language, but language is of course never ordinary. That is Wittgenstein's point. Words carry with them an inherited set of usages and conventions. It took me several pages through the contribution by Donald Green and company before I recognized that the noun "observation" wasn't the old friend that I have known so long. It was a stranger encrusted with the conventions of a well-developed literature that I have not previously accessed. I needed Gary Cox as a tourist guide to Heckman-land to familiarize myself with the ethnography of that particular culture.

The essays forced me to think about the question of convergence: convergence of the diverse modes of inquiry, literatures, and epistemologies that are at the center of this conference. Convergence raises several issues. There is a broad spectrum of possible attitudes toward convergence. Margaret Levi enacts one end of the spectrum – she is optimistic; convergence was well launched. The synergy of diverse methods would improve knowledge and insight. Others are incredulous about friendship pacts across methodological positions. They did not exist, and why should they? Convergence raises questions of both an empirical and a normative kind: Is there convergence? Should there be convergence? Convergence of what?

By convergence we might be talking about a variety of pairings among modes of inquiry, some of which might work better than others. Shall we imagine a convergence of formalism and political ethnography? Shall we imagine a convergence between Bruce Bueno de Mesquita's rationalism and Frances Fox Piven's policy studies? Shall we imagine a convergence between Gary Cox's formalism and John Mearsheimer's case approach?

Before we talk about convergence we presumably have to specify the elements that might converge. This is no straightforward task. On one side we see what is inadequately summed up as the qualitative or interpretive approach. We talk about case method, ethnography, historical institutionalism, theory in its non-formalized incarnation. On the other side we see modes of inquiry permeated by the scientific paradigm, and talk about formal approaches and quantitative reasoning. But these categories are not homogenous wholes. A hypothetical process of convergence would presumably engage sub-packages of traits and practitioners. To operationalize convergence requires disaggregation of the contesting epistemic communities.

I've made myself two lists, each of which represents an ideal type of the contending perspectives. Deploying quotation marks to disown full responsibility for the words, I have placed "scientific" on one side and "qualitative/interpretive" on the other side.

"Scientific" Mode of Inquiry	"Interpretive" Mode of Inquiry
certainty	skepticism; contingency
parsimony	thick description
cumulative knowledge	non-linear succession of paradigms
causality	meaning
singularity of truth	multiplicity of truth
universal/homogeneous	contextual/heterogeneous
objective knowledge	subjective knowledge

I won't have time to elaborate these distinctions here, but have tried to do so elsewhere.

It is well known that such a construction is a heuristic meant to start a conversation more than an image of "reality." Attributes do not obediently stay on one side or another of a heuristic boundary. Overlaps appear. For example, it would be absurd to suppose that causality, which I've put on the side of the scientific paradigm, is not a matter of great interest and concern to people in the interpretive and qualitative tradition.

Still, causality figures more in the scholarly imagination of authors who embrace the scientific trope, and meaning in the imagination of qualitative interpretive scholars. The two approach causality in different ways. The "scientist" starts with the language of hypothesis, continues by employing protocols modeled on those prevailing in the natural sciences – e.g. not selecting on the dependent variable, increasing variables that might falsify the hypothesis – and finishes by eliminating competing hypotheses. The interpretivist starts with a question (or with a "descriptive" project in which causality is embedded) and accumulates answers according to the different protocols that guide proof in historical or ethnographic modes. She ranks answers according to their probability, but may not resolve the question which of the plausible answers is unequivocally THE solution.

But back to the question: Should there be convergence? We see a variety of bridge-building projects in progress that imply convergence is desirable. David Collier's efforts to form an organized section of the APSA for what he calls qualitative methods, with a manifesto that suggests that statistical methods and formal reasoning might well inform qualitative methods, is one example. The Report of the APSA Working Group on Graduate Education, a work in progress, is another. Mixing modes of inquiry is essential for many political projects. I am myself something of a utility hitter, having used survey research and statistical approaches in the work that Lloyd Rudolph and I have produced on public spheres and on the Indian economy. We have used historical approaches in work on sovereignty.

I see several possible outcomes for convergence. "Bridge building" can designate a collaborative and synthetic process. On the other hand, it can signify a hegemonic and exclusivist project. Bridge-building can evoke the scholar utilizing a variety of modes of inquiry as the problem calls for them. For example, good survey research in a new environment – Yemen – would move toward a meaningful questionnaire by calling for participant observer fieldwork in order to discover the categories of thought, the plausible imaginaries, the normative environment of the relevant population. It would call for experience in sampling procedure and statistical probabilities to design the sample. It would take statistical and discursive interpretation to elicit the meaning and implications of the project. Here cultural and quantitative reasoning are at work.

But bridge-building initiatives can sound like assimilation projects. I have heard it suggested that "qualitative methods" can profit from the "rigor" of quantitative studies. The suggestion implies either that there are no protocols or inadequate protocols of methodological procedure for interpretive projects. Assimilation projects fall into bad odor when they are situated within environments of asymmetric power and function as hegemonic initiatives. Convergence can be an invitation to fruitful cross fertilization or to paradigmatic invasion, occupation, and conversion. Therefore it's not clear that everyone would cheer at the news of convergence. There is the danger that attempts to systematize interpretive methods would shear them of their thick description, their empathy, their normativity, in the interests of parsimony and impersonal objectivity.

Finally, there is the empirical question: Is it happening? What evidence is there of convergence? Let me return to the question of language, and to the fact that when I learn something conceptual at a conference, it often boils down to language acquisition. Is there any evidence of consequential language exchanges across paradigm boundaries? Terms typical of an approach are the patient mules that bear the burden of whole methodological procedures or perspectives. There has been significant migration of vocabularies across epistemes which we can read as measures of exchange. Game theory has exported the terms "prisoner's dilemma," "free rider," "payoffs," "public goods," "models," for regular use in other epistemic communities. Ethnographic and interpretive methods on their part have exported "narrative," "discourse," "rhetoric," "stories," "hermeneutics." Sometimes a vocabulary item becomes a vehicle of disparagement, as when formalists throw the term "ordinary language" at interpretivists, or exemplars become "anecdotes." Sometimes ownership of a term becomes a *casus belli*, as in the tug of war over who owns the prestige-bearing noun, "theory."

Now where does this leave me? The status of language exchanges helps us see where we are but doesn't tell us where to go. By this measure there is clearly exchange among paradigms. It signals that there is some exchange of valued currency, but no common monetary system, some mutual epistemic legibility, but no community of modes of inquiry. I end on the unsatisfactory note that there is enough communication to provide choices for intentional graduate students to imagine alternatives, but not enough to signal active collaboration. Without fruitful collaboration and convergence, we need at least legibility and civility.

John Mearsheimer: A self-enclosed world?

Let me begin with a word about self-identification, which highlights how complicated the world of political science really is. Susanne divided that world in half, or into two parts. But I find myself on both sides of the divide. Basically I view myself as a rational choice theorist who does not use math. I start with simple assumptions about the nature of the international system and from them I make deductions about state behavior. As both Lloyd and Susanne Rudolph know, much to their chagrin I might add, I love simple models that explain how the world works. I like to say to students that I know the world is complicated, but please give me a simple theory that tells me how it works. And even if the theory is wrong, if it is simple and elegant, I am likely to be impressed. I don't know why; God just hardwired me that way.

However, because I have used case studies to test and help refine my theories, and because I know a lot of history, I usually am described as a qualitative person or a case study person who is located on the opposite side of the divide from the rational choice people. But again, I am basically a rational choice person who happens to use cases and who thinks that knowing a lot of history is important for developing smart theories. I say all of this to highlight how complicated the world of political science can be.

One other point of self-identification is relevant here: I am something of a methods maven. I have taught a graduate-level methods course at Chicago for about ten years. It is really a philosophy of science/methods course. Moreover, I frequently tell graduate students to take all the methods courses they can, because I think methods are very important, and students should be spun up on a wide variety of them. One can go overboard, but sound methodological training is indispensable for doing first-rate social science.

Having said enough about myself, I now want to assess whether we have dealt with the central issue that Ian Shapiro and Rogers M. Smith asked us to focus on in this volume. In their initial letter to me – and I assume that all the other participants received essentially the same letter – they said that they wanted us to talk about the relationship between methods and problems. By problems it was clear that they were talking about providing answers to important real-world questions, or offering theoretical insights about how the world works. That's what problem-driven research is all about. We all agree, I think, that almost every scholar does both. All of us are concerned about both methods and solving problems. But the really important issue – which Ian and Rogers identified in their letter – has to do with the balance between the two. It was implicit in their letter that they believe that our discipline privileges methods over problems, and that this bias is unhealthy for our discipline. The question is: How well have the contributors to this volume dealt with that important issue?

I have two responses to that question. First, we have barely addressed the issue that Ian and Rogers laid out in their letter. Second, many of the essays have borne out their point that our discipline privileges methods over problem-solving.

Except for Rogers Smith, I do not think that a single person explicitly addressed the question of the relationship between methods and problem-solving in the way that was set out in the invitation letter. What we have done instead, and I think this is reflected in Susanne's comments, is focus on methodological – or epistemological if you like – differences among us. We have had the so-called formalists on one side of the debate and the qualitative scholars or critical theorists on the other side of the divide. Inside the formalist camp there has been a debate between rational choice advocates and large-N scholars like Donald Green that has been every bit as adversarial as the debates between formalists and qualitativists.

In all of these debates among the panelists, however, the focus has been almost exclusively on methodological issues. There has been little sense in the discussions that there is a fascinating world out there, that fundamental changes are taking place as we enter the twenty-first century, and that we should be committed to trying to understand that emerging world. It seems to me that there are all sorts of important problems in the world that we should be excited about studying, and that we ought to think about methods in terms of how useful they are for helping us address those concrete problems. Scholars should ask: Does my theory or my methodological approach provide important insights about important issues? But I did not hear participants talking in those terms. Instead, we seemed to be obsessed with methods.

Let me come at this matter from a different angle to drive my point home. Rudra Sil asked: "Who is our audience?" This is a great question. When we write a book or an article, who are we appealing to? Who is going to read it? Who is going to care about it? How are we going to spread our ideas to others? Who are those others? I do not think anybody answered Professor Sil's important question, and I think the reason nobody answered his question is that our discipline operates on the assumption that we only talk to each other. We operate in a self-enclosed world. The fact is that political scientists in recent years have tended to marginalize themselves from the wider world. We do not engage those outside our discipline in important ways. To use the word "policy" – which is actually a synonym for politics – to describe the work of a prospective hire, is to doom that person. You never want to say that a job candidate does policy-oriented work, even if it is first-rate, because it deals mainly with real-world issues. Instead, we prefer to hire individuals on the basis of their methodological proclivities and skills. In short, discussions about hiring always have a heavy focus on methodology; usually little attention is paid to what scholars have to say about the real world.

Let me take this point a step further by considering the subject of writing op-eds for newspapers, or articles for popular journals like the *American Prospect* and *Foreign Affairs*, or appearing on the *Lehrer NewsHour*. If you do such things, they are likely to be held against you in a hiring meeting or a promotion meeting. It is just not the kind of work we academics are supposed to do. That kind of thinking, I might add, applies to distinguished senior scholars as well as young scholars.

I remarked to Alan Ryan that over the course of my twenty years at Chicago, the group of political science colleagues who I have found to be the most interesting to talk with about politics is the political theorists – and here I am talking about the political philosophers. The reason is that they have tended to be much more interested in the real world than most of my other colleagues. For example, Stephen Holmes, Bernard Manin, and Nathan Tarcov, all political theorists in my department at one time, were remarkably knowledgeable about nitty-gritty political issues. If one wanted to talk about NATO expansion or the deployment of SS-20s to Europe, talking with them made for great conversation. They were all deeply interested in the real world.

In my conversation with Alan, I asked him why he writes for the *New York Review of Books*. He said that he is trying to communicate to a wider audience than just his academic colleagues. I know few other academics who write for the *New York Review of Books*, or would consider doing so. Why? Because we live in our own little world where we talk mainly to each other. Of course, not every political scientist fits that description, so I am obviously overstating the case somewhat. But I do believe that there is a lot of truth in my claim, and that it has been reflected in what we read in this volume, where, again, the emphasis has been almost exclusively on methods. And this happened despite the marching orders that Rogers and Ian gave us to write about the relationship between methods and problems.

I would like to say a few words about my own views on the relationship between methods and problem-solving. First, I think we should privilege problem-solving over methods. I do not believe that methods are unimportant, which is why I said earlier that I teach a methods course and I encourage students to take methods courses. Nevertheless, methods are merely tools for answering important questions. Second, I think we should reach out to a wider audience than our fellow political scientists. Of course, we should speak to each other, but there is a wider world out there which should care about what we have to say and we should care about communicating with it. We should do this because we have a social responsibility to our fellow citizens to help them understand how the world works.

I do not think I have ever met a political scientist who has wrestled with the question of what he or she is doing in this business. In other words, what is the purpose of being a political scientist? Is it simply to talk with other political scientists? I often say to myself that I do not know how most political scientists get out of bed in the morning with a bounce in their step, since most of what they do all day is talk to each other. When most of them die and are put in the grave, the only people likely to care will be their families and their former students. Shouldn't they instead feel like they have big ideas that they want to communicate to the wider world? By the way, I am not talking about going down to Washington and sitting on George Bush's knee and telling him how to make policy. I am talking about writing articles and books that reach a larger audience than just one's academic colleagues. I am talking about writing pieces that all sorts of educated people can read.

Third, I think that the best social science theories, whether it is in economics, sociology, or political science, are those that are linked closely with important real-world problems. The idea that there is an intellectual world where theory dominates and a practical world where theory is useless is one that I reject. These are not separate realms. Indeed, the essence of our enterprise should be to come up with interesting and innovative theories that people find compelling because they help them understand how the world operates. Some critics of this perspective argue that focusing on real-world problems leads to journalism. That criticism is actually mentioned in the letter that Ian and Rogers sent to us laying out the purpose of this volume. I think the charge is wrongheaded; there is no reason why privileging problem-solving over methods should turn scholarship into journalism.

Let me highlight my claim with two examples. The first involves Albert Wohlstetter, who was a university professor at Chicago, which is a very prestigious position. He was appointed to that exalted position when he was at the Rand Corporation. He did not have a Ph.D. and he was made a university professor mainly on the basis of a single article that he wrote in *Foreign Affairs* in 1959. Nevertheless, that appointment was a smart one, because of the great importance of the ideas he put forth in that one article. Specifically, Wohlstetter developed one of the key strands of deterrence theory, a simple, elegant strand of deterrence theory that has informed our debates about the subject since the 1950s. Now, the relevant question for us is: How did Wohlstetter come to develop his seminal theory?

Wohlstetter was a mathematical logician by training whose career at Rand was rather undistinguished until the early 1950s when he became involved in a project designed to study an important policy issue. The US Air Force was debating whether it should station its growing fleet of B-47 bombers in Europe or leave them in the United States. Rand assigned Wohlstetter and a handful of other analysts to study the issue and come up with recommendations. In the process of doing this basing study, Wohlstetter and his colleagues invented a framework or theory for thinking about the problem. Specifically, they came up with the distinction between first-strike and second-strike, as well as concepts like crisis stability, vulnerability, and survivability. All of these notions, as well as the theory of deterrence put forth in the article, were path-breaking and truly important for helping us think about nuclear deterrence. But they were not developed by isolated academics locked away in an ivy tower looking to invent a theory of deterrence. Instead, they were developed by first-rate minds engaged with a truly important real-world problem.

Another important scholar whose work fits the same mold is Thomas Schelling, who is something of a god for most rational choice scholars, including me. I think one of the great crimes of academia is that Schelling has not gotten a Nobel Prize for his work on deterrence theory. I say that as someone who disagrees with some of Schelling's key ideas. The point I want to make here, however, is that his seminal writings on deterrence during the 1950s and 1960s came out of policy-oriented research that he did at Rand, where he was examining many of the same issues that concerned Wohlstetter, who, of course, was his Rand colleague. All of this goes to show that the claim that focusing on real-world problems will lead to journalism and not serious scholarship is mistaken.

I want to conclude with a final point about math. I have no problem with scholars who use math in their work. I think it sometimes facilitates the production of elegant theories, and as I said before, I like elegant theories. But I agree with Truman Bewley that the use of math is not a necessary condition for coming up with good theories. John Roemer and Bruce Bueno de Mesquita have said, in effect, that if you do not use math you are not doing real social science and that the resulting theories will not be worth much. I would note that both Schelling and Wohlstetter used little math in their work, yet they both produced seminal theoretical works.

We actually tried to hire Thomas Schelling in the political science department at Chicago many years ago, but the appointment was resisted by the economic department, where there was a strong feeling that his work was not rigorous enough, which means that he was not an applied mathematician. But who cares whether he uses math? The key point is that he invented a body of important theoretical ideas that have profoundly influenced how huge numbers of scholars and practitioners think about deterrence, and how we think about international relations more broadly. For a certain body of economists, however, and I think for Bueno de Mesquita and Roemer as well, if you are not using sophisticated math to develop and articulate your theories, there is something wrong with your work. Again, I am not saying that there is anything wrong with using math, but I do think it is wrong to argue that math is necessary for developing good theory.

My bottom line is that I thank Ian and Rogers, who to their great credit, gave us a very important issue to deal with in this volume. The discipline of political science needs to think about the audiences it seeks to reach and the importance of solving problems versus focusing on methods. We failed, however, to respond to their directive and address these weighty issues.

Index

Printed in the United States
By Bookmasters